Lake Hydrology

Lake Hydrology

An Introduction to Lake Mass Balance

WILLIAM LEROY EVANS III

 JOHNS HOPKINS UNIVERSITY PRESS *Baltimore*

Printed in Canada on acid-free paper
9 8 7 6 5 4 3 2 1

Johns Hopkins University Press
2715 North Charles Street
Baltimore, Maryland 21218-4363
www.press.jhu.edu

Library of Congress Cataloging-in-Publication Data

Names: Evans, William L., III, author.
Title: Lake hydrology : an introduction to lake mass balance / William LeRoy Evans III.
Description: Baltimore, Maryland : Johns Hopkins University Press, 2020. | Includes bibliographical references and index.
Identifiers: LCCN 2020013108 | ISBN 9781421439938 (hardcover) | ISBN 9781421439945 (ebook)
Subjects: LCSH: Lake hydrology.
Classification: LCC GB1605 .E936 2020 | DDC 551.48/2—dc2 3
LC record available at https://lccn.loc.gov/2020013108

A catalog record for this book is available from the British Library.

Johns Hopkins University Press uses environmentally friendly book materials, including recycled text paper that is composed of at least 30 percent post-consumer waste, whenever possible.

CONTENTS

PREFACE

This text is written for the graduate student or advanced undergraduate whose goal is to pursue a career in the natural sciences as a working scientist in the "real world" but not too far from the realm of academics. As such, it is designed to benefit working professionals as well as academics and researchers who are involved in teaching limnology, lake management, lake remediation, or investigation of lake systems.

Limnology is the study of lakes, rivers, and wetlands as ecological systems focusing on the biota and biochemistry of limnologic systems. In contrast, yet in conjunction, this text introduces the hydrology and physics of flow into and out of lake systems as it pertains to lake stage, groundwater and lake bottom interaction, hypsometry and evaporation, and transpiration (evapotranspiration) and surface flow. The physical models described in the text are designed to be incorporated into applied mass balance or water budget investigations to facilitate better estimates of lake behavior over time. Therefore, this treatment of physical lake behavior can be utilized as is or incorporated within an advanced coursework in limnology, lake-basin hydrology, lake management, or another related discipline.

The first chapter is an introduction to the basic scientific principles of dimensional analysis and units of measurements, followed by a description on the properties of water and the hydrologic cycle in chapter 2. When describing the process of overall lake mass balance, the drainage basin affecting the lake system is the basic unit of analysis defining flow into and out of a lake system. As such, chapter 3 is a discussion on the physical structure and nature of drainage basins followed by the origin and classification of lakes that include the physical characteristics that define their behavior in terms of mass balance, such as thermal stratification and evaporation or lake morphometry for storage determinations and hypsometry.

Chapter 4 on evapotranspiration begins the description of the physical processes in the hydrologic cycle that directly influence flow into and out of lakes. This is followed by topics on surface and groundwater flow.

Chapters 5 through 8 investigate surface flow to lakes and are an attempt to consolidate an extremely broad range of topics on key flow processes that contribute to the lake mass balance. Precipitation is the major factor controlling the hydrologic cycle of a region and subsequent drainage basins. Rainfall amounts, temporal and areal distribution, flow paths, and residence time of water in the catchment from precipitation delineate the mass balance of surface and groundwater discharge. This knowledge is essential for water resource assessment and management and forecasting the probability of storm events requiring the need for flood control.

Chapter 5 begins with describing storm events and how rainfall is measured over a drainage basin. The relationships between rainfall (and snowfall) and water conveyance in a watershed is at the very core of wetland hydrology and mass balance. The type and distribution over time of a precipitation event—where rainfall is deposited on a variety of surfaces—includes its surface flow paths, infiltration, and residence time within the entire drainage basin as it flows to the lake, which is discussed in Chapter 6.

Chapter 7 then proceeds to examine the fundamentals of estimating stormwater runoff rates and volumes to a lake. These methods and models are designed to simulate rainfall-runoff processes along with knowledge of the hydrological and geological character of the catchment to accurately define mass balance exchanges to wetlands and lakes within the watershed. Chapter 8 concludes the investigation of surface flow with a succinct overview of basic stream characteristics and streamflow as it pertains to discharge into or out of a lake system within a drainage basin.

Chapter 9 covers the basics of groundwater flow to lakes into and out these systems via the underlying subsurface in relation to physical hydrodynamic forces that drives these processes. In general, lake and wetland systems are surface expressions of the water table and are in direct hydraulic connection with the water—

table aquifer. Thus, the equations of fluid mechanics that provide the basics for the quantitative description of groundwater flow to and from lake systems will mainly involve saturated laminar flow conditions where flow occurs along a gradient from high to low energy.

The section on lake seepage and hypsometry (chapters 10 and 11) are primarily a result of my own graduate research in lake hydrogeology and contain data collected on Lake Jackson, north of Tallahassee, Florida. Lake Jackson is a karstic lake system with wide-ranging fluctuations of lake stage and variable lake bed hydraulic conductivity. Lake Jackson and other karstic lakes in the area experience intermittent periods of lake drainage due to the periodic activation of sinkholes (ponors) within the lake basin. Even though this type of lake is restricted to karstic regions of a specific type of lake basin, the theory and methods of investigation developed as a result of this research are applicable to most lake systems.

Chapter 11 provides a basic introduction to lake hydrological modeling with discussions on deterministic, stochastic, and geostatistical methodology. It is in this chapter where the mathematical models of mass balance, seepage, and hypsometry are developed. To take the student through the modeling process, a step-by-step hypothetical situation of population growth and urbanization with potential impacts to a lake system is included as an example of a typical scenario encountered in the fields of limnology, environmental science, or engineering. For more complex modeling requirements, this textbook provides numerous references as well as web links for downloading various water budget models, such as the US Environmental Protection Agency's BASIN model and US Geological Survey's MODFLOW.

In keeping with the stated objective of this book, emphasis is placed on the basic physical principles of fluid flow as it pertains to lake systems. Many of the topics presented in the text are discussed in much greater detail in engineering fluid mechanics, hydrogeology, and hydrology coursework as well as courses in fluid physics referenced throughout the manuscript. This text is designed to be beneficial to students with various scientific and engineering backgrounds. Hence it contains basic explanations on the physical parameters that influence lake behavior as well as the mathematics that describes these systems. The appendix includes an introduction to calculus integration, partial differential equations, and statistical methodology such as linear regression for the student that is not familiar with these subjects.

ACKNOWLEDGMENTS

First and foremost, I owe a debt of gratitude to my major professor, Dr. Jon David Furbish, who inspired me in the art of science and taught me a different approach to hydrogeology through the power of mathematics, statistics, and physics. I am also very appreciative for Dr. Sam Upchurch's review of this book's manuscript and early comments. Several individuals helped me improve my early drafts and guided me with their knowledge and experience on science publications: Dr. Tom Messimer, Dr. Ralph Dougherty, Dr. Donald Robson, Dr. Sean McGlynn, Jack Sams, Hope Setac, Richard Green, Karl Bertelsen, and the staff at the Florida Geological Survey. I would also like to thank the staff of John Hopkins University Press—Esther Rodriguez, Andre Barnett, Tiffany Gasbarrini, Joanne Haines, and Hilary Jacqmin—for all their guidance and help through this process. I am especially grateful to Joanne Haines as the manuscript editor for her attention to detail and great suggestions, resulting in a textbook far better than originally submitted. I am extremely appreciative of Judy Jericho for her patience and support from the early stages of this endeavor and throughout its completion.

Finally, I am grateful for the love and support of my parents, my mother June Timmons Evans and father Colonel William L. Evans Jr., especially during my period as a "gradual" student (as my Dad would say I gradually got out of school).

Lake Hydrology

CHAPTER 1

Introduction

1.1. Limnology

Limnology, from the Greek *limme* meaning lake and *logos* meaning knowledge, is the multidisciplinary science of terrestrial water systems such as rivers, lakes, and wetlands. The majority of work done in limnology has concentrated on biology and ecology (Wetzel 1975). Of practical importance have been investigations designed to understand limnologic systems in order to manage, regulate, and remediate these valuable aquatic resources (Kalff 2001). To effectively do this work, chemistry, geology, hydrology, physics, and other scientific disciplines must be involved to adequately describe lake behavior.

Hydrology is the scientific study of water that involves the occurrence, distribution, movement, and chemical and physical properties of the waters of the earth and their relationship with the environment within each phase of the hydrologic cycle. The water cycle, or *hydrologic cycle*, is a continuous process by which water is purified by evaporation and transported from the earth's surface to the atmosphere and back to the land and oceans (see section 2.2). Lakes interact with all components of the hydrologic cycle, including precipitation, evaporation, plant transpiration, infiltration, groundwater flow, seepage, runoff, and streamflow.

Unlike most limnologic texts that emphasize lake ecology and biology, this text is designed to rigorously describe the physics of water flow into and out of lake systems as it pertains to lake stage as well as storage as it relates to components of the hydrologic cycle. The water budget—or fluxes of water in and out of a lake system—in concept appears deceptively straightforward as a simple mass balance: where inflow equals outflow, plus or minus lake storage. However, due to the complexity and interaction of the various components of the hydrologic cycle, and utilization of flow to a lake on the scale of a drainage basin (or sub-basin), the ability to accurately quantify the various hydrological components is limited.

Physical models have been developed and are discussed in detail using storm events and subsequent surface and groundwater flow with lake bottom interactions, hypsometry, lake hydraulics, and evapotranspiration. These techniques are designed to be incorporated into applied lake mass balance models to better ascertain estimates of inflow and outflow over time.

The lake mass balance discussed in chapter 11 is based on endorheic, or closed lake, basins with large variations in lake stage usually associated with karstic processes and interaction with groundwater. However, the theory and explanation of lake behavior, mathematical and statistical analyses and modeling, and field and laboratory methodology described in this text are applicable to the majority of lake and wetland systems (chapter 3).

When determining the cause and variation in stage for a certain lake, it is necessary to understand the physical and hydraulic controls on the lake system. Fluctuations of lake levels for closed lake basins as well as exorheic lakes are controlled primarily by the amount of precipitation within the lake basin over time. Generally, a lack of rainfall results in reduction of lake levels and heavy rainfall results in rising lake levels. Lake stages are also influenced by surface and groundwater runoff, evapotranspiration, seepage, and exchange of water between one or more aquifer systems that interact with the lake basin. Thus, a lake level is the net result of the mass balance between inflow and outflow over time. The balance between input and output is continually changing; hence the level or stage of a lake is either rising or falling.

Analysis of lake behavior and stage fluctuation is usually accomplished by estimating the quantities of water that move into the lake from various sources and those that move out of the lake to various sinks. Records of lake-level fluctuations, local rainfall, estimates of evaporation, and general knowledge of the hydrology and geology of the area are required when estimating the water balance for a lake. For most lakes, the data necessary to make such quantitative computations is usually not available (Hughes 1974b). When such estimates are used in a water-balance equation, the difference between the computed and observed changes in lake level may be considerable and vary erratically or systematically (Hughes 1974a).

1.2. Dimensions, Units, Measurements, and Mathematical Conventions

The dimension of a physical quantity can be expressed as a product of the basic physical dimensions mass, length, time, electric charge, and absolute temperature. There are two systems of dimensions commonly used when describing physical processes that involve certain quantities as being fundamental to the system. These two systems are the mass, length, and time [*MLT*] system and the force, length, and time [*FLT*] system. Both systems are related by Newton's second law of motion, $F = ma$, where each side is dimensionally homogeneous with dimensions of MLT^{-2}. Force is a derived quantity (see below) in *MLT* systems while mass is a derived quantity in the *FLT* system. The measurement of any quantity is made relative to a specific standard or unit. For example, we can measure length in units such as inches, feet, miles, centimeters, meters or kilometers.

The International System of Units (abbreviated SI from the French Systeme Internationale d' Unites) is the designated system of scientific measurements worldwide. In addition, the Metric Conversion Act of 1975 (later amended by the Omnibus Trade and Competitiveness Act of 1988, the Savings in Construction Act of 1996, and the Department of Energy High-End Computing Revitalization Act of 2004) designated the metric system as the preferred system of weights and measures for United States trade and commerce and directed federal agencies to convert to the metric system to the extent feasible. Therefore, SI units are used throughout this text and are given in Table 1.1 for quick reference.

Fundamental quantities such as mass, length, time, and temperature can be divided into two categories: base and derived. The corresponding units are called base units and derived units. In the interest of simplicity, scientists want the smallest number of base quantities possible consistent with a full description of the physical world. This number happens to be seven, and all other quantities used to describe the physical world can be defined in terms of these basic seven quantities (Table 1.1).

Table 1.1. International System of Units

Quantity	Dimension	Base unit	Symbol
length	[L]	meter	m
mass	[M]	kilogram	kg
time	[t]	second	s
electric current	[A]	ampere	A
thermodynamic temperature	[K]	Kelvin, degrees Celsius	K, °C
luminous intensity	[I_v]	candela	cd
amount of substance	[mol]	mole	mol
Quantity	**Dimension**	**Derived unit / Formula**	**Symbol**
area	[L^2]	square meter / m^2	
volume	[L^3]	cubic meter / m^3	
speed, velocity	[Lt^{-1}]	meter per second / $m(s^{-1})$	
acceleration	[Lt^{-2}]	meter per second squared / $m(s^{-2})$	
mass density	[ML^{-3}]	kilogram per cubic meter / $kg(m^{-3})$	
specific volume	[L^3M^{-1}]	cubic meter per kilogram / $m^3(kg^{-1})$	
luminance	[I_vL^{-2}]	candela per square meter / $cd(m^{-2})$	
force	[MLt^{-2}]	Newton / $kg(ms^{-2})$	N
pressure	[$ML^{-1}t^{-2}$]	Pascal / $N(m^{-2})$	Pa
stress	[$ML^{-1}t^{-2}$]	Pascal / $N(m^{-2})$	Pa
energy	[$ML^{-2}t^{-2}$]	Joule / N(m)	J
power	[ML^2t^{-3}]	Watt / $J(s^{-1})$	W
specific heat	[$L^2K^{-1}t^{-2}$]	Joule per kilogram-kelvin / J(kg-1)(K^{-1})	
viscosity dynamic	[$ML^{-1}t^{-1}$]	Pascal-second / Pa·s	μ
viscosity kinematic	[L^2t^{-1}]	square meter per second / $m^2(s^{-1})$	
amount of substance concentration	[$molL^{-3}$]	mole per cubic meter / $mol(m^{-3})$	

When describing the dimension of a quantity, a reference is made to the types of units or base quantities that make it up. The dimensions of area, for example, are always length squared and by convention are abbreviated using square brackets: [L^2]. Velocity can be measured in units of meters/second, miles/hour, or others but are always a function of length divided by time [L/t] or [Lt^{-1}]. Table 1.2 denotes symbols and their descriptions and dimensions used specifically in this text as they apply to hydrology as well as statistical notation.

1.3. Dimensional Analysis

Dimensional analysis is the scrutiny of relationships between different physical quantities or processes by examining their fundamental dimensions (e.g., mass, length, or time, see Table 1.1). Dimensional analysis offers a method for reducing complex physical problems to the simplest or most efficient form prior to obtaining a quantitative solution. Physical quantities that are measurable and have the same dimensions, such as length (L) or (l), even if

Table 1.2. Definition of symbols

General	
L	Length
A	Surface area [L^2]
V	Volume [L^3]
t	Time
v	Velocity [Lt^{-1}]
ρ	Density
p	Pressure
ps	Gauge pressure ($ps = 0$)
g	Acceleration of gravity, approximately 9.81 meters per second per second.
J	Concentration gradient of molecules in Fick's Law of Diffusion
$\dot{m}_v$	The change in mass with respect to time [L^3t^{-1}]
R	Reynolds number for laminar and turbulent flow
u	Couette flow velocity [Lt^{-1}], the flow of a viscous fluid in the space between two surfaces.
τ	Shear velocity
NGVD	National Geodetic Vertical Datum of 1988 is the vertical datum established for vertical control surveying in the United States of America
Lake Systems	
Q	Flow = Rate of discharge [L^3t^{-1}]
P	Precipitation [L]
E	Evaporation [Lt^{-1}]
T	Transpiration [Lt^{-1}]
Y	*liquid-air surface tension* (force/unit length)
Ø	contact angle
ET	Rate of evapotranspiration [Lt^{-1}] (combination of E and T)
PET	Daily potential evapotranspiration in mm/day [Lt^{-1}]
q_s	Lake seepage [Lt^{-1}]
α	Lake seepage constant K_A/L
S	Storage (lake) [L^3]
▼	Piezometric or potentiometric surface
h_b	Height of lake bed overlying aquifer
I_0	Light intensity over a specified wavelength just below the lake surface
R_n	Net radiation
I_z	Light intensity at depth z
η_d	Vertical attention (or extinction) coefficient
z_{eu}	Thickness of the euphotic zone
Y	Psychrometric constant (0.67 mb/°C for elevation 99.0 m NGVD)
λ	Latent heat of vaporization of water (58.4 calories/cm/mm)
Aquifer Systems	
h	Hydraulic head [L]
h_a	Height of potentiometric surface of the aquifer [L]
K	Hydraulic conductivity [Lt^{-1}]
K_A	Effective hydraulic conductivity [Lt^{-1}]
K_b	Effective hydraulic conductivity of ponor [Lt^{-1}]
q	Average linear velocity [Lt^{-1}]
S_s	Specific storage [L^{-1}]$^-$
S_a	Aquifer storativity
S_y	Specific yield
T	Transmissivity the ability of an aquifer to transmit water [L^2t^{-1}]
α	Aquifer compressibility
Φ	Fluid potential
σ_e	Effective stress
σ_T	Total stress

(*continued*)

Table 1.2. *Continued*

Aquifer Systems	
ψ	Capillary pressure head
θ	Soil moisture content of the soil matrix
n_p	Porosity
n_e	Effective porosity
H_0	Hydraulic head at time zero for falling-head permeameter
H_i	Hydraulic head at initial time i for falling-head permeameter
T_c	Viscosity correction temperature of water (Celsius, C) for falling-head permeameter
Mathematical Symbols (see appendix for basic discussion on derivatives and integration)	
Σ	Summation
π	Pi: the ratio of a circle's circumference to its diameter, which is a mathematical constant commonly approximated as 3.14159.
e	Mathematical constant, where *e* is an irrational and transcendental number approximately equal to 2.718281828459 (2.303)
ln	Natural log base of the natural logarithm value of e
∇	Gradient (locally steepest slope) of a vector or scalar field
Δ	Delta: change of any variable quantity in mathematics and the sciences (more specifically, the difference operator)
df	Total derivative of function *f*
∂f	Partial derivative of function *f*
$\int$	Integration of function *f*
x,y,z	Three primary directions in the (*x,y,z*) Cartesian coordinate system with *x,y* in the horizontal plane and *z* the vertical plane (e.g., lake depth, or topographic elevation).
Statistical Symbols (see appendix for basic discussion on statistics and linear regression)	
R	Coefficient of determination for least squares regression
ε	Random component or error term
Θ_0	Y-intercept
Θ_1	Fixed coefficient of regression
Θ_2	Fixed coefficient of regression
Σ	Summation: sum of all values in a range of series
r	Residuals

the units of measure are different (e.g., meters versus feet), can be directly compared to each other.

An equation that precisely describes a physical relationship must be dimensionally homogeneous. This means that the dimensions on both sides of the equation are the same. Checking for dimensional homogeneity is a common application of dimensional analysis. Dimensional analysis is routinely used as well to check the plausibility of derived equations and computations. This analysis can also be used to develop a reasonable hypothesis for complex physical situations that can be tested by experiment or by more related theories on how a system operates.

Additionally, dimensional analysis can be used to describe physical relationships and can be beneficial when modeling lake systems. For example, when examining mass balance of a lake system, it is important to know the amount, or volume, of water entering a lake basin area after a precipitation event. The amount of rainfall in a catchment area has dimensions of volume $[L^3]$ and can be described as

$$\text{Volume} = (\text{height of water collected in a rain gauge}) \times (\text{basin area}).$$

The first check in the calculation is to determine whether the equation is dimensionally homogeneous:

$$[L^3] = [L] \times [L^2] = [L^3].$$

However, dimensional homogeneity does not guarantee that the equation will give an accurate result. Errors in data collection or descriptive parameters of the variables in question can all affect the outcome. If the dimensions are not homogeneous, then there is a basic flaw in the derivation of the original equation. For example, what if the area of the rain gauge was included in the equation above? This would result in a unit volume representing precipitation, and when multiplied by the basin area the outcome would be a heterogeneous dimension of $[L^5]$.

1.4. Spatial Coordinates

The Cartesian coordinate system is a process of graphing positions of two points on a two-dimensional surface or plane or three points in a three-dimensional space. The Cartesian coordinates x, y, and z are used to denote conventional spatial direction where z is in the vertical axis in relation to the earth. The Cartesian plane consists of two perpendicular axes with a central point of origin (0,0) in an east-west direction (x-axis) and north-south direction (y-axis) and where the values are negative west and south of the origin in the x and y direction, respectively.

The x and y axes represent number lines with equal increments along each axis (different axes can have different increments and represent different variables). Adding a third axis perpendicular to the xy plane, and by convention passing through the point of origin, creates a three-dimensional space usually denoted as the z-axis. For example, each increment on the x-axis might represent two units while each increment on the y-axis represents three

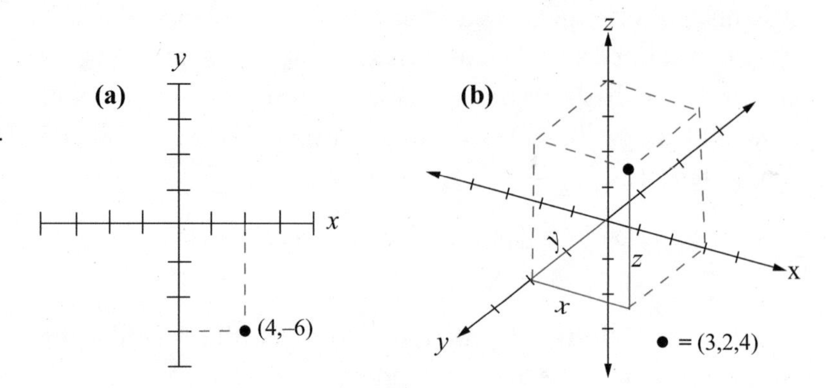

Figure 1.1. Cartesian coordinate system with two-dimensional *x* and *y* axes (a) and three-dimensional *xyz*-space (b).

units (Figure 1.1a). Points or coordinates are indicated by writing an opening parenthesis, the x-value, a comma, the y-value, and a closing parenthesis, in that order. Thus in Figure 1.1a, the point is $(x,y) = (4,-9)$. The origin is usually, but not always, assigned the value (0,0).

In addition to the east-west (x) and north-south (y) positions, coordinates can be further determined by *up-down* (z) displacements from the origin with each x, y, and z plane perpendicular to each other (Figure 1.1b). As is the case with the x and y axes, the z-axis is a linear number line. Points or coordinates are indicated by writing an opening parenthesis, the x-value, a comma, the y-value, another comma, the z-value, and a closing parenthesis, in that order. If each increment on the x, y, and z axes represent one unit, for example, the point (x,y,z) in Figure 1.1b is (3,2,4). The origin is usually, but not always, assigned the value (0,0,0).

The Cartesian coordinate system can be used to demonstrate various physical relationships, mathematical equations, and functions. By convention $y = f(x)$ denotes that the dependent quantity y is a function f of the independent variable, or coordinate, x. Where the variable component is a function of more than one independent variable, for example velocity v_i (see Table 1.2) as a function of space (the three coordinate positions) and time (t), it is conventional in physics to denote this as $v = v_i(x,y,z,t)$. For lake systems, the Cartesian coordinate system can be used to demonstrate relationships between lake levels (or stage) with lake area and lake volume in an (x,y,z) format, such as in Figure 1.2.

In xyz-space, the volume of a bowl-shaped lake can be estimated by an applicable mathematical function, such as the graph

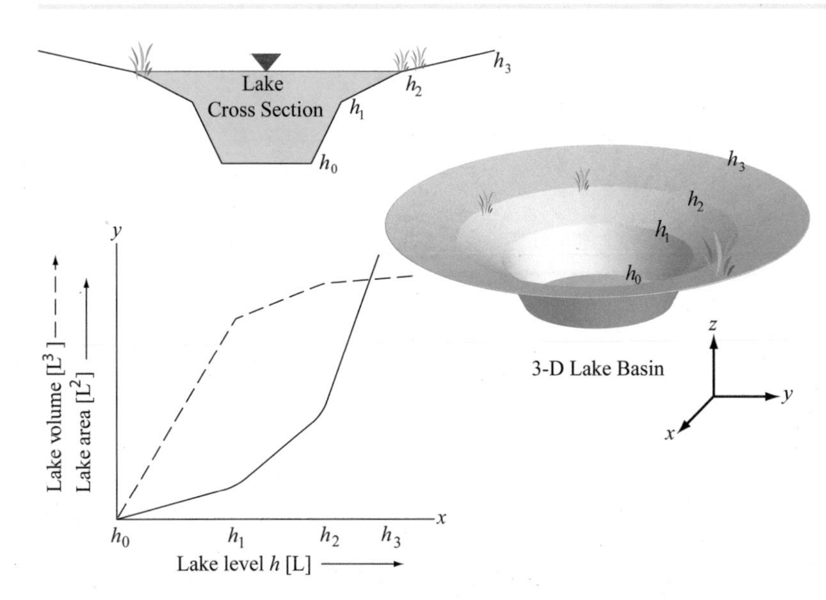

Figure 1.2. Hypothetical relationship between lake volume, lake area, and lake level using the Cartesian coordinate system.

of the function $f(x,y) = x^2 + \frac{1}{4}y^2$. This function represents an elliptic paraboloid and can be demonstrated using Cartesian coordinates where z by convention is the vertical or near vertical coordinate axis in reference to the earth's surface and is a function $f(x,y)$ (Figure 1.3) (see also appendix, section 1.4, Figure A1.5). Hypsometry is the three-dimensional representation of the lake bottom as represented by dimensional contour lines or curves of constant elevation as it relates to a datum such as sea level (see section 9.2). Thus, the physical parameters of a lake's hypsometry—such as volume, lake level, and area—can be described and modeled utilizing partial derivatives (Anton 1980, 887–89) (see sections 3.6 and 3.7).

1.5. Mathematics and Statistics

I recommend readers have knowledge of basic limnologic principles and such mathematics as elementary differential equations. Ideally, for students and scientists alike, this includes completion of two to three semesters of calculus as well as completed coursework in statistics and probability with some knowledge of regression analyses.

The use of such mathematics as partial differential equations and regression statistics for describing various scientific and engineering processes are usually included in the respective curriculum for the specific disciplines involved. Therefore, specific

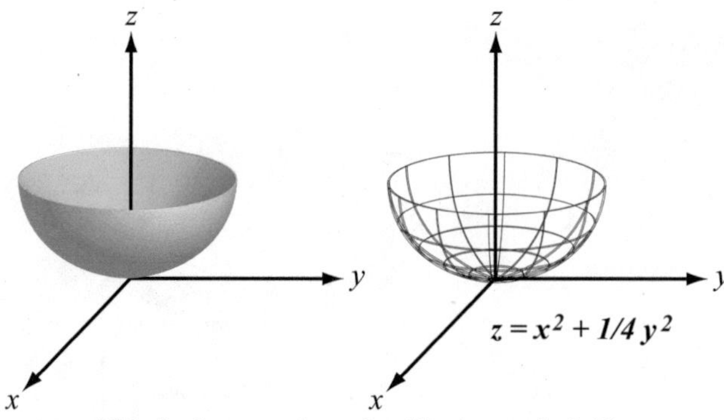

Figure 1.3. Using the partial derivative to determine the volume of an elliptic paraboloid.

subjects and texts such as physics, hydrogeology, thermodynamics, or fluid dynamics, as fundamental examples, are without a discussion of the basic mathematics involved. Thus, the "good" student finds it necessary to learn or revisit the mathematics topic in question, often expending much time and effort muddling through vast amounts of material prior to finding the information relevant to the specific lesson. As this text was written with a variety of disciplines and academic levels in mind, which may or may not stress the mathematics and statistics needed to understand the hydrology of lake systems, or if the reader is like how I was as a student, a young and not-so-disciplined undergraduate or lacking the background coursework, he or she may simply "prevaricate" this part of the text and move on without fully understanding the material. Thus, the "basic" mathematics relevant to describing the hydrology of lakes or similar natural systems discussed throughout this text is provided in the appendix.

CHAPTER 2

Water and the Hydrologic Cycle

2.1. Water and Its Properties

No water, no rain, no creeks, no streams, no rivers, no lakes, no oceans, no limnology, no limnologist, no limnology text, no life as we know it. Water is the basic component of all biotic metabolism, including photosynthesis. The majority of basic physical chemical interactions that drive the geological processes of the earth at or near its surface use water as a major component. The online edition of the *American Heritage Dictionary of the English Language* defines *water* as:

> Water (wo'tar, wot'ar) n. 1. A clear, colorless, odorless, and tasteless liquid, H_2O, essential for most plant and animal life and the most widely used of all solvents. Freezing point 0°C (32°F), boiling point 100°C (212°F), specific gravity (4°C) 1.0000; weight per gallon (15°C) 8.338 pounds (3.782 kilograms). 2a. Any of various forms of water: *waste water*. . . . 3a. A body of water such as a sea, lake, river, or stream.

Water is a covalently bonded compound of two hydrogen atoms and one oxygen atom where the hydrogen bonds are located on the same side of the oxygen at 104.5 degrees apart, resulting in a strongly dipolar molecule (Figure 2.1). Because of the complex structure of the hydrogen bonding and the weak bonding between water molecules, liquid water is more a liquid crystal than a true fluid, resulting in its unusual properties. Water is a *dihydride* of oxygen. Comparing it with dihydrides of the elements in the same family on the periodic table as oxygen—that is, hydrogen sulfide (H_2S), hydrogen selenide (H_2Se) and hydrogen telluride (H_2Te)—we find that many of the physical properties are anomalous. At atmospheric pressure and room temperature (25°C) the dihydrides of Se, Te, and S are all gaseous, but water remains a liquid unless the temperature is increased to 100°C. In addition,

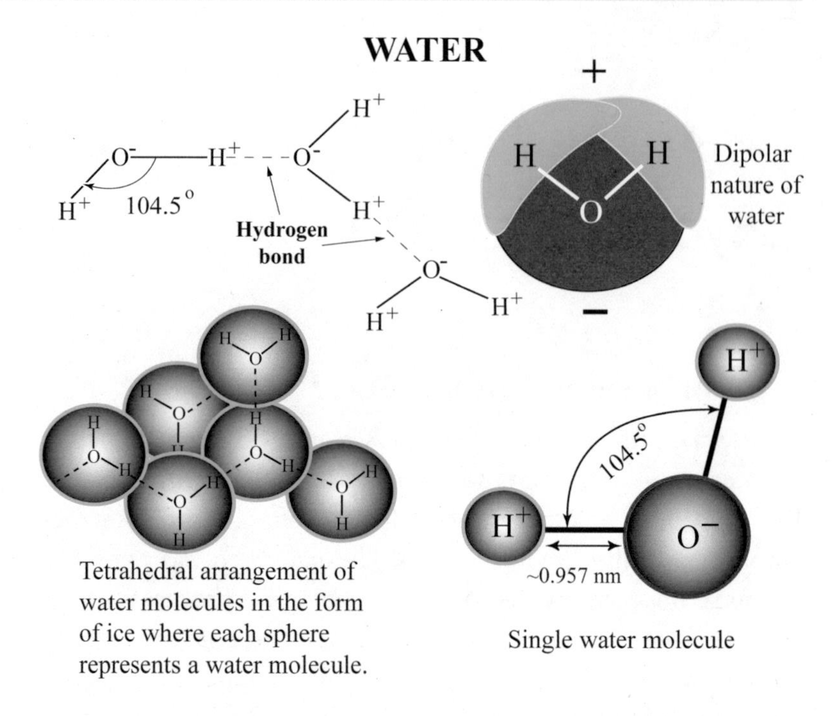

Figure 2.1. The physical makeup of water.

its surface tension and dielectric constant are much higher than would be predicted from properties of the other dihydrides.

It is far denser than its related elements with a maximum density at 4°C (Figure 2.2). Its freezing point is lower than expected, and it freezes to form ice, a substance less dense than the liquid water from which it forms. Ice has a well-defined structure at 1 atmosphere pressure (ice-I) with the following characteristics: (1) every hydrogen atom is in an oriented H-bond between two oxygen atoms, which are spaced about 2.76 Å apart; (2) every oxygen is bonded to four hydrogens in an undistorted tetrahedron (Figure 2.1); (3) the oxygen atoms lie in a network of puckered hexagonal rings with Hs cementing the O network together; and (4) the network is full of holes, thus it is less dense than water (Pauling 1959). At 4°C the density of ice is 0.915 grams per cubic centimeter (g/cm^3) while the density of water at that temperature equals 0.999 g/cm^3 (Figure 2.2). There are several different phases of ice that are stable at different temperature and pressure conditions. This last property has far reaching ramifications because if solid water were denser than liquid water, it would freeze from the bottom up and, consequently, life in its present form would not exist, as all bodies of natural water would freeze solid whenever

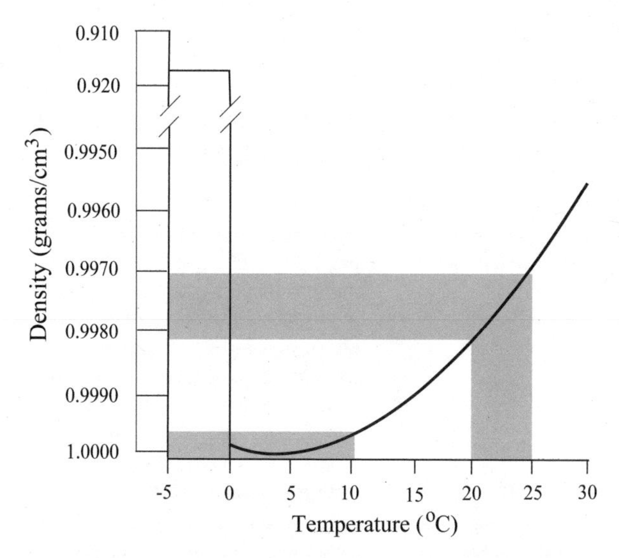

Figure 2.2. Density-temperature relationship for distilled water. Gray-shaded areas illustrate relative difference in density for 5°C temperature changes. Modified with permission from WOW (2004).

temperatures fell below the freezing point of water. Hence, water's physical characteristics are unique (Table 2.1) and responsible for the aquatic systems as well as all life on earth.

The minerals that make up earth's crust are largely inorganic solids with positively and negatively charged ions existing in a crystal lattice structure held together by electrostatic bonds. Because of its dipolar character and high dielectric constant, water can surround positively charged ions with its negative charge and negatively charged ions with its positive charge thus neutralizing the electrostatic bond that maintains the integrity of crystal structures. This process of disassociation and hydration (surrounding ions with other water molecules) breaks down the various mineral structures, removing their ions into solution. The solubility of these essential salts and nutrients is necessary to sustain life. Water dissolves, to some extent, a various amount of virtually every solid or gas it encounters and places it back into the mix of the hydrologic cycle.

Water molecules are the only substance on earth that exist in all three physical states of matter: solid, liquid, and gas. During changes of state, massive amounts of heat exchange can occur, which plays an important role in the redistribution of heat energy in the earth's atmosphere. In terms of heat being transferred

Table 2.1. Physical and chemical properties of water

Property	Comparison with other matter	Importance
Density	Under atmospheric pressure the maximum density of water is ~ 4°C; expands upon freezing (0°C) with a 9% increase in volume.	Allows lake stratification and surface freezing versus bottom freezing.
Melting and boiling points	Atypical: Both properties are very high when compared to other hydrides.	Allows water to exist in liquid phase in a range of 0–100°C. Water exists in three phases within the critical temperature range that accommodates life.
Viscosity	Moderate	Influences water mixing and sedimentation rate of particles and is resistant to organism movement
Surface tension	Very high Highest of all substances	Increases the difficulty of surface waves breaking, thereby slowing the rate of heating and cooling in lake systems.
Radiation absorption	Large in the infrared region, but moderate in the photosynthetic/visible region.	Allows greater heat absorption in surface water but reduced surface absorption at shorter wavelengths, which allows greater penetration of photosynthetic available radiation.
Solvent properties	Dipolar nature makes it an excellent solvent for salts and polar organic molecules.	Important in dissolution and transport of dissolved substances from catchments and the atmosphere to aquatic systems.
Specific heat or heat capacity	Highest of any liquid other than ammonia.	Moderates or buffers temperature extremes. Water's high specific heat allows for the moderation of the Earth's climate and helps organisms regulate their body temperature more effectively.
Molecular viscosity (a measure of resistance to distortion, or flow) = 10^{-3} Newton(s)(m^{-2})	Less than most other liquids at the same temperature.	Water flows readily to equalize pressure differences.

Source: Modified from Sverdrup, Johnson, and Fleming (1942); Berner and Berner (1987); and Kalff (2002).

into the atmosphere, approximately three-quarters of this process is accomplished by the evaporation and condensation of water (Kalff 2001).

2.2. The Hydrologic Cycle

The hydrologic cycle is the fundamental concept in limnology that evaluates the water budget, or mass balance, of water movement and storage on earth and in the atmosphere. The cycle consists of three principle phases: (a) precipitation, (b) evaporation and transpiration (commonly referred to as evapotranspiration, or ET), and (c) surface and groundwater flow. Each phase involves transport, temporary storage, and a change in state or phase (gas, liquid, or solid) of the water. Solar energy drives the cycle with gravity and other forces playing an important role in transport. The formation of water vapor and transport of vapor and liquid in the atmosphere is primarily influenced by solar energy. Precipitation and flow of surface- and groundwater are mostly driven by gravitational forces. Figure 2.3 illustrates the phase and flow-system concept of the hydrologic cycle while the schematic in Figure 2.4 is more representative of the approach used when modeling mass balance. The global water budget (Table 2.2) adds further insight into the water resources of our planet.

The accurate treatment of earth's hydrologic cycle is central to the scientific investigation of climate dynamics as the global water cycle is an integral part of the earth's energy cycle and thus plays an essential role in determining large-scale circulation and precipitation patterns. The intricately linked hydrologic cycle and global energy processes consist of a complex web of feedback systems operating over a vast continuum of time and space scales.

All three phases of water in the climate system are strongly radiatively active atmospheric constituents in all forms. Clouds, composed of water in the liquid and solid phases, play a dominant role in regulating the energy budget of the planet and behaviorally remain a major source of uncertainty in our ability to project the effects of climate change (e.g., Stephens and Webster 1981; Cess

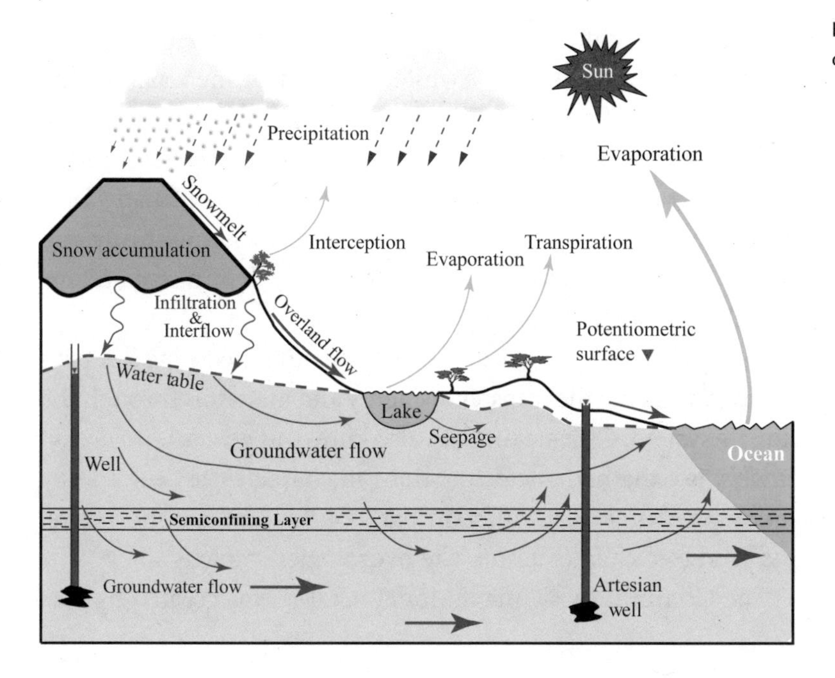

Figure 2.3. Hydrologic cycle.

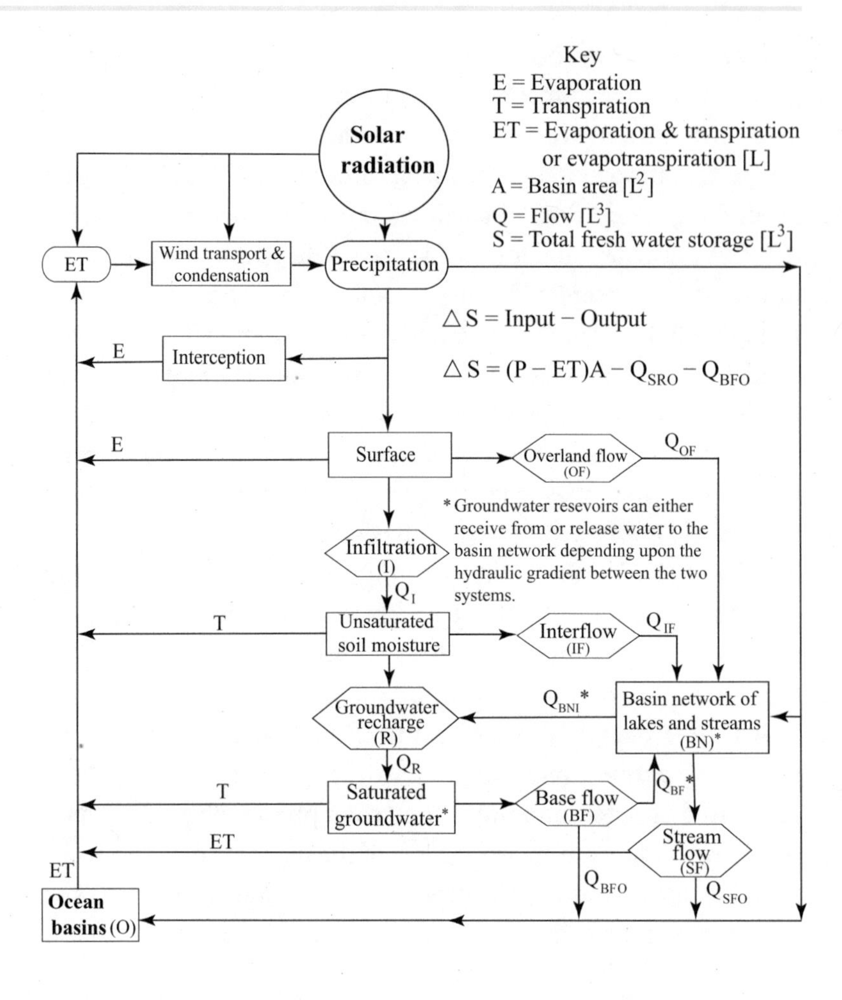

Figure 2.4. Schematic representation of the hydrologic cycle.

et al. 1990; Houghton et al. 2001). They cool the earth by reflecting solar radiation back to space while at the same time warming the planet by absorbing thermal radiation emitted from the surface and lower regions of the atmosphere. The processes responsible for phase transitions of water also contribute to the diabatic forcing of the earth's dynamical circulations and are key to the overall energy budget, particularly for the thermally driven circulations in the Tropics (Chahine 1992).

Climate models are based on well-documented physical processes to simulate the transfer of energy and materials through the climate system. Climate models, also known as *general circulation models*, use mathematical equations to characterize how energy and matter interact in different parts of the ocean, atmosphere, and land (see chapter 11 on lake hydrological models).

The Community Climate Model (CCM) was created by Na-

Table 2.2. Global water budget

Saltwater	Percentage global water (%)	Freshwater	Percentage global water (%)
Oceans	97.542	Glaciers/ice	1.806
Saline lakes/seas	0.004	Ground/soil: root	0.001
		Zone 0–763 meters	0.303
		762–3810 meters	0.331
		Fresh Lakes	0.010
		Reservoirs	<0.001
		Atmosphere	0.001
		Rivers	<0.001
Total saltwater	97.546	Total freshwater	2.454

Source: Modified with permission from US Environmental Protection Agency (EPA 2019).

tional Center for Atmospheric Research in 1983 as a freely available global atmosphere model for use by the wider climate research community. The formulation of the CCM has steadily improved over the past two decades, computers powerful enough to run the model have become relatively inexpensive and widely available, and usage of the model has become widespread in the university community and at some national laboratories.

The Community Climate System Model (CCSM) is a coupled climate model for simulating the earth's climate system. Composed of four separate models simultaneously simulating the earth's atmosphere, ocean, land surface, and sea ice, and one central coupler component, the CCSM, allows researchers to conduct fundamental research into the earth's past, present, and future climate states. The most recent version, CCSM4, was made available to the community on April 1, 2010, from the CCSM website (www.cesm.ucar.edu/models/ccsm4.0).

For example, the hydrological budget over land is a balance between precipitation, evapotranspiration, runoff, and storage changes in soils or snow. Simulated evapotranspiration in CCSM version 3 is 63% of precipitation where ground evaporation is the largest component of the simulated evapotranspiration (59%) followed by canopy evaporation (28%) and transpiration (13%) (Hack et al. 2006). Observational estimates of the partitioning of global evapotranspiration, however, suggest that transpiration should be the dominant component followed by ground evaporation and canopy evaporation (Hack et al.). In contrast, Choudhury et al. (1998), using a process-based biophysical model of evaporation validated against field observations, found that the partitioning was 52% transpiration, 28% ground evaporation, and 20% canopy evaporation.

The dominant form of runoff in CCSM3 is surface runoff (52% of total runoff); followed by drainage from the soil column (41%); and runoff from glaciers, lakes, and wetlands (7%). This latter runoff term is calculated from the residual of the water balance for these surfaces. This term may also be nonzero for other surfaces as well because the snowpack is limited to a maximum snow water equivalent of 1000 kg m^2 (Hack et al. 2006).

Zonal annual average values of the land surface hydrologic cycle are shown in Figure 2.5. Hack et al. (2006) found that CCSM3 simulation overestimates precipitation north of 45°N, generally underestimates it in the northern Tropics, and overestimates it in southern portions of South America. The meridional distribution of evaporation over land areas generally tracks the precipitation

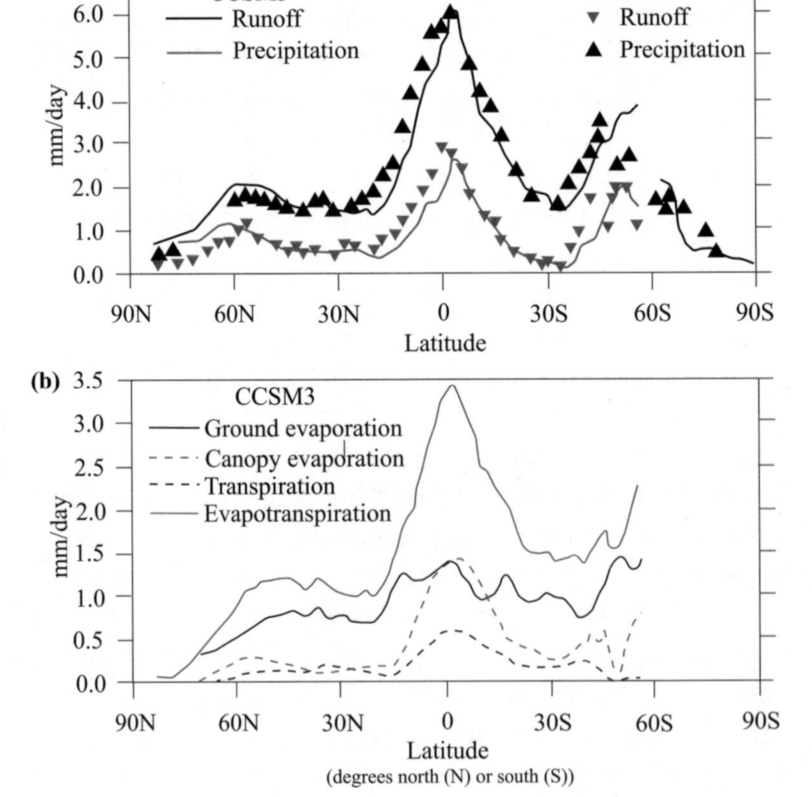

Figure 2.5. Latitude zoned averaged annual mean precipitation and evaporation for Community Climate System Model (CCSM3) (a) and observations and latitude zoned averaged annual mean land hydrology values from CCSM3 (b). Modified from Hack, J.J., J.M. Caron, S.G. Yeager, K.W. Oleson, M.M. Holland, J.E. Truesdale, and P.J. Rasch. 2006. "Simulation of Global Hydrological Cycle in the CCSM Community Atmosphere Model Version 3 (CAM3): Mean Features." *Journal of Climate* 19 (11): 2199-221. © American Meteorological Society. Used with permission.

with a maximum in the Tropics. Runoff biases also track precipitation biases, suggesting that improvements in the simulated precipitation would lead to improvements in the simulation of runoff. At high latitudes, the primary active hydrological component is runoff. At other latitudes, ground evaporation generally dominates. An exception to this is in the deep Tropics (10°S–10°N) where canopy evaporation is equally important. Transpiration is the smallest component of evaporation at all latitudes (Hack et al.).

The processes of the hydrologic cycle operate on a wide range of time and space scales and are exceedingly difficult to quantify observationally. The most reliable observations of the hydrologic cycle are limited to relatively long time and large spatial scales. As such, current observational data provide relatively weak constraints on the formulation of hydrological processes in global models (Hack et al. 2006). This is significantly true as it pertains to lake basin studies. Hence, the focus of this text will be the hydrologic cycle as it relates to more restricted spatial and temporal scales, such as watershed areas during a storm event and how it impacts a lake system.

2.3 Mass Balance of Water

The *law of conservation of mass*, also known as the *principle of mass conservation*, is that the mass of a closed system (i.e., completely isolated) will remain constant over time; that is, it is said to be *conserved* over time. The mass of an isolated system cannot be changed as a result of processes acting inside the system. This implies that for any chemical process in a closed system, the mass of the reactants must equal the mass of the products. This concept is attributed to the experiments of Antoine Lavoisier (1789), the father of modern chemistry, where he explained the departure of material during a combustion process.

The hydrologic cycle can be expressed quantitatively by utilizing the principle of mass conservation (also see section 9.9 as it applies to groundwater flow). In its most recognized form, the law states that "mass can neither be created nor destroyed," thus the mass flow in and out of a system or control volume must be in balance. For the conservation of a fluid mass, such as applied to a simple system (Figure 2.6), may be given as:

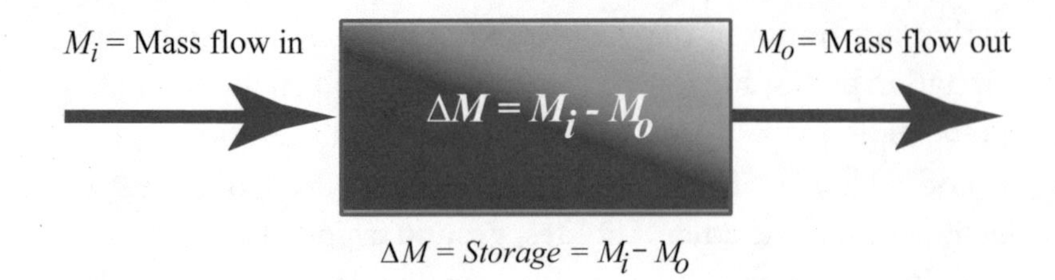

Figure 2.6. Simple two-dimensional representation of conservation of mass.

Change in mass storage with time =
mass inflow rate – mass outflow rate.

The mass flow in and out of a control volume (through a physical barrier) can be expressed as

$$\Delta M = \rho_i v_i A_i \Delta t - \rho_o v_o A_o \Delta t \tag{2.1}$$

where

ΔM = change in storage mass [M] in the system,
ρ = density [M/L^{-3}],
v = velocity [Lt^{-1}],
A = area [L^2], and
Δt = change in time (t).

In a complex three-dimensional lake system, the statement of mass conservation can be applied to a domain of any size, such as a cube of porous material representing seepage into or out of lake through lake bottom sediments (or flow via a ponor or stream) as well as a cube representing volume into a lake system directly by precipitation, overland flow, and loss due to transpiration and evaporation. This cube represented by a Cartesian coordinate system is an *elemental control volume* where the changes in x (Δx), y (Δy), and z (Δz) equals the unit volume (Figure 2.7).

Using Figure 2.7 as mass moving into and out of a lake system via a porous medium, mass inflow rate through the cube face

$$ABCD = \rho_w q_x \Delta y \Delta z. \tag{2.2}$$

The density has units of [ML^{-3}], and where $q = Q/A$, the *specific discharge* has a velocity of [Lt^{-1}] (see section 9.6 for details) such that $\rho_w q_x \Delta y \Delta z$ has the units of mass per time.

In complex natural systems, mass outflow rates are often different than the input rates and can be expressed through

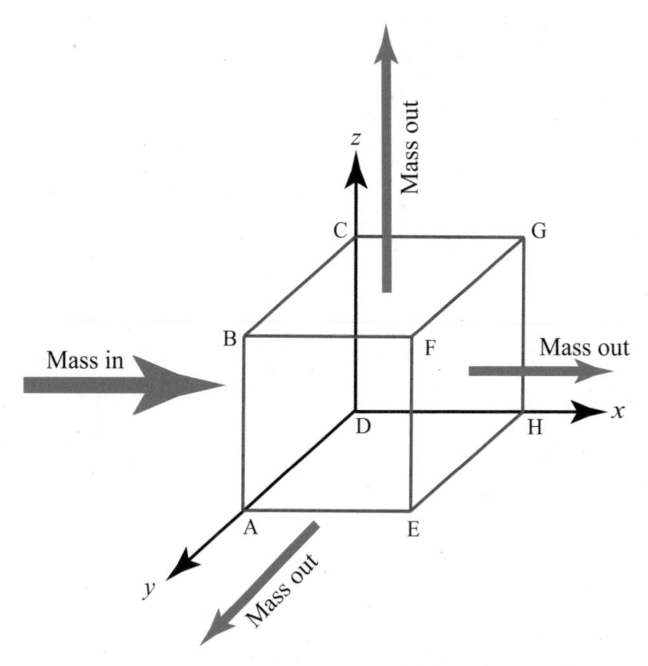

Figure 2.7. Flow in and out of a representative volume.

$$\text{EFGH} = \left[p_w q_x + \frac{\partial (p_w q_x) \Delta x}{\partial x} \right] \Delta y \Delta z. \tag{2.3}$$

The net outflow rate is the difference between the inflow and the outflow, or subtracting equation (2.2) from equation (2.3), which results in the net outflow rate through

$$\text{EFGH} = -\frac{\partial (p_w q_x) \Delta x \Delta y \Delta z}{\partial x} \tag{2.4}$$

And correspondingly the outflow rates through cube face CDHG and BCGF would be

$$\text{CDHG} = -\frac{\partial (p_w q_y) \Delta x \Delta y \Delta z}{\partial y} \tag{2.5}$$

and

$$\text{BCGF} = -\frac{\partial (p_w q_z) \Delta x \Delta y \Delta z}{\partial z}. \tag{2.6}$$

To obtain the net outflow rate through all the faces, add the results of each face, or

$$\text{CDHG} = -\left[\frac{\partial(p_w v q_x)}{\partial x} + \frac{\partial(p_w q_y)}{\partial y} + \frac{\partial(p_w q_z)}{\partial z}\right]\Delta x \Delta y \Delta z \cdot \quad (2.7)$$

The net outflow rate per unit volume is obtained by dividing by $\Delta x \Delta y \Delta z$, or

$$\text{BCGF} = -\left[\frac{\partial(p_w q_x)}{\partial x} + \frac{\partial(p_w q_y)}{\partial y} + \frac{\partial(p_w q_z)}{\partial z}\right]. \quad (2.8)$$

The right-hand side of the conservation of mass statement is a change in mass storage with respect to time. The mass occupies the lake in the unit volume where density ρ_w is mass per unit volume of fluid (V_w) $[ML^{-3}]$ and porosity n for a fully saturated medium is the fluid volume per unit total volume (V_w/V_{Total}) such that $p_w n$ is the mass per unit total volume. Therefore, the equation of mass conservation becomes

$$-\left[\frac{\partial(p_w q_x)}{\partial x} + \frac{\partial(p_w q_y)}{\partial y} + \frac{\partial(p_w q_z)}{\partial z}\right] = \frac{\partial(p_w n)}{\partial t}. \quad (2.9)$$

Stating that the net outflow per unit volume equals the time rate of change in fluid mass per unit volume. Assuming that the fluid density does not vary spatially, the density term on the left-hand side can be taken out as a constant, so the equation (2.9) becomes

$$-\left[\frac{\partial(q_x)}{\partial x} + \frac{\partial(q_y)}{\partial y} + \frac{\partial(q_z)}{\partial z}\right] = \frac{1}{p_w}\frac{\partial(p_w n)}{\partial t}. \quad (2.10)$$

Similarly, when determining the conservation of fluid mass directly into and out of a lake system, the terms q and n used for flow through a porous medium are simplified to units of velocity v and volume V_w.

By assuming a constant fluid density, equation (2.10) now expresses volumes of fluid (V_w) per unit versus mass per unit volume, where the two are related by $\rho_w = [ML^{-3}]$. For the left-hand side of the equation, $q_x = Q_x/A$, which is a volumetric flow of fluid per unit time per unit area. Thus, the left-hand side describes the net fluid outflow rate per unit volume and the right-hand side represents the rate of change fluid volume that is

$$\frac{\partial(V_w/V_T)}{\partial t}. \quad (2.11)$$

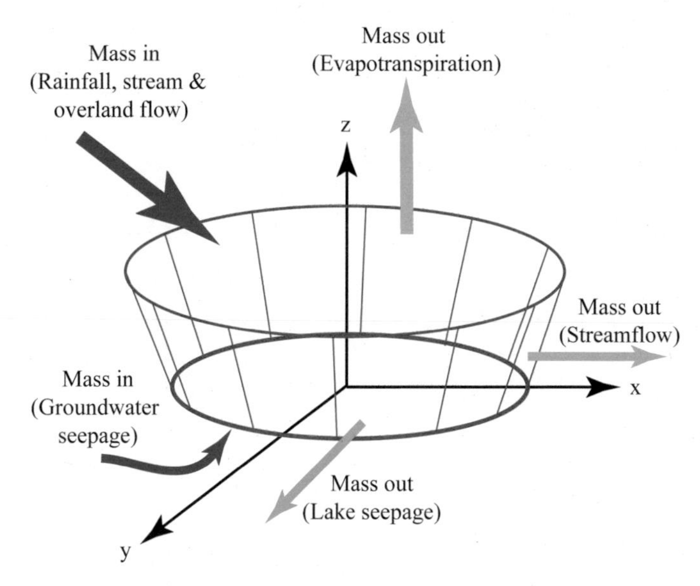

Figure 2.8. Idealized lake basin representative volume.

In summary, equation (2.11) states that the net fluid outflow rate for the unit volume equals the time rate of change of fluid within the unit volume. Equations (2.9), (2.10), and (2.11) can be used to estimate lake mass balance as a homologous comparison of an *elementary control volume* cube with a *lake basin representative volume* by substituting inflow and outflow lake variables such as precipitation and evapotranspiration for the geometric cube faces (Figure 2.8).

CHAPTER 3

Drainage Basins, Lentic Systems, Lake Morphometry, and Lake Volume

3.1. Drainage Basins

3.1.1. Introduction

A *drainage basin*, also known interchangeably as a *watershed*, *catchment basin*, or *river basin*, if discharging into a river, is the area of land draining into a stream, lake, reservoir, wetland, or mouth of a bay at a specified geographic location. A watershed is an area of land that drains all the streams and rainfall to a common outlet such as the outflow of a reservoir, lake, mouth of a bay, or any point along a stream channel. Ridges, hills, or topographic highs that separate two watersheds are called *drainage divides*. The watershed consists of surface water—lakes, streams, reservoirs, wetlands, and all the underlying groundwater. Large catchment areas such as a major river basin usually consist of smaller watersheds all converging to the main river.

In water resource studies, basins have long been acknowledged as the appropriate unit of analysis for water resource management because the basin concept of water passing through a stream cross section at the basin outlet originates as upstream precipitation on the basin. As catchment characteristics such as geology, topography, ecological systems, land cover, and land use all control magnitude, water quality, and timing of flows, studying a drainage basin as a system and obtaining hydrological information on a basin scale results in a more thorough understanding of the whole system as well as the interaction between different processes (McKinney et al. 1990).

Functionality of lakes and wetlands is intricately linked to the topography, hydrologic character, and *ecosystems* within the drainage basin. For example, the geomorphology (i.e., topography or slope) and size of the basin influences stream and lake size and shape, concentration of wetlands, and underlying hydrogeology. In general, the average slope or relief of catchments decreases as catchment area increases whereas smaller drainage basins tend to be relatively steep. In large catchments with low relief, the volume of stream and overland flow to wetlands and lakes normally declines with increased area (Kalff 2002). Geology, soil, vegetation, and land use all influence the hydrology of a catchment especially in terms of precipitation and runoff.

Precipitation is the major factor controlling the hydrologic cycle of a region and subsequent drainage basins. The ecology, geography, and land use of a region are all dependent upon the function of the hydrologic cycle. Rainfall amounts, temporal and areal distribution, flow paths, and residence time of water in the catchment from precipitation delineates the mass balance of surface- and groundwater discharge, as well as water quality, erosion, and sedimentation effects to rivers, streams, lakes, and other wetland systems.

When investigating lake mass balance, topographic data is essential. Detailed analysis of basin topography and land use is necessary for a cost effective and accurate portrayal of surface water behavior for a given area. The elements for catchment surface water analysis includes boundary or extent of the catchment, slope data throughout the basin, underlying geology and hydrogeology, soil types, vegetation, drainage topology (i.e., drainage flow paths and network of channels), channel cross sections, overland or runoff flow paths, and land use characterization such as native, agricultural, urban, roadways, and drainage structures (e.g., levees, culverts, retention areas).

3.1.2. The Physical Template of Watershed Structure and Composition

The basic physical processes that form watersheds consist of *climate*, *geomorphology*, and *hydrology* of a specific area. These processes are in a constant cyclic flux, with each other as well as with other environmental factors. Thus, the physical template or structure of a watershed is a result of the varying combinations of climatic, geomorphic, and hydrologic processes (Figure 3.1).

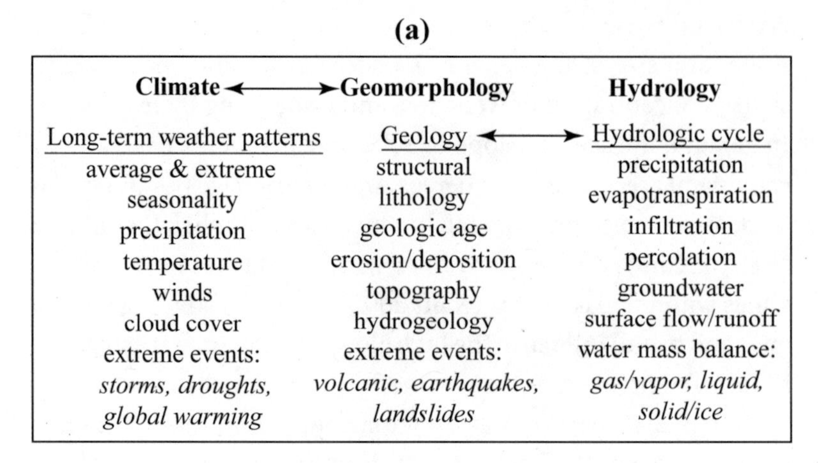

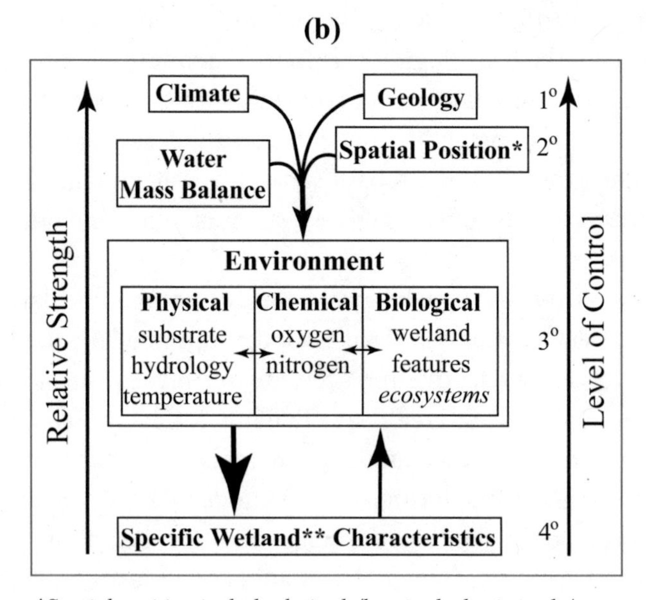

Spatial position includes latitude/longitude, basin scale/areas, and regions such as arctics, tropics, and deserts

***Wetlands include lotic/flowing systems (streams & rivers) and lentic/still systems (lakes & swamps)*

Figure 3.1. Watershed structure and formative process determined by physical template. The physical template includes various components of climate, geomorphology, and hydrology and the relative strength or influence of each on a watershed (a). Strength of influence is represented as the relative size of the arrows and the level of control is indicated as primary (1°), secondary (2°), and so on (b). Modified from O'Keefe, Elliott, and Naiman (n.d., figure 8); Naiman (1992).

Climate, although sometimes used synonymously with *weather*, is distinct from weather as it refers to long-term combinations of both average and extreme conditions of temperature, rainfall (including type and amount), humidity, wind, and cloud coverage measured over an extended period of time (O'Keefe, Elliott, and Naiman, n.d.). Weather, although more conducive to lake mass balance determination, refers to present-day environmental conditions such as current temperatures and meteorological events over a narrow window of time. Thus, weather trends observed over an extended period of time will eventually establish climatic regimes. Climatic conditions influence ecological communities, vegetation, streamflow magnitudes and seasonality, and other watershed characteristics.

Geology can be broadly defined as the science that deals with Earth's physical structure and substance, its history, and the processes that act on it. *Geomorphology*, a subset of geology, is the study of landforms on Earth and the processes that change them over time. *Fluvial geomorphology*, a subdiscipline of geomorphology that investigates how flowing water shapes and modifies Earth's surface through erosional and depositional processes, is essential when studying watershed formation, composition, and changes over time. Fluvial geomorphology as it pertains to stream and river surface flow; alteration of stream channels, flood plains, and associated upland transitional zones; and overall changes in watershed structure over time is especially important for investigating and effective long-term management of wetland systems in a watershed.

Hydrology in reference to watersheds is the study of water as it relates to the hydrologic cycle (see section 2.2) and its impacts to the water basin. This includes the distribution, circulation, and behavior of all forms (liquid, gas, and solid) of water at or near the Earth's surface, the chemical and physical properties at play, and all water-related interactions with the environment.

These three processes are not separate systems but instead interact and influence each other. For example, basins with high precipitation will have increased stream and river flow and subsequent erosion and deposition, the extent of which depends on the geomorphological topography and slope of the area. Volcanic eruptions can influence climate by increased atmospheric ash, thus lowering temperature on a global scale. On a watershed scale, lakes can be formed within the depressions of volcanic cones and calderas, or lava flows can block streams (see section 3.3.1c).

The *Rosgen stream classification system* (Rosgen 1996) is a widely used method for classifying streams and rivers based on common patterns of fluvial morphology As previously discussed, stream morphology is a direct reflection of (and influences in turn) the drainage basin morphology it is associated with. As part of Rosgen's classification, Level I designates ten types of watersheds (*valley types*) that can be incorporated into mass balance investigation for lentic systems.

The Rosgen system includes four levels in the hierarchical assessment of channel morphology. Rosgen (1996, 2011) has summarized these levels as follows:

> *Level I* . . . is the geomorphic characterization and identifies *valley types*, which integrate structural controls, fluvial process, depositional history, climate, and broad life zones. Level I also rapidly categorizes streams at a broad level on the basis of valley landforms and observable channel dimensions defining one of eight stream type letter designations (A, B, C, D, DA, E, F, and G), including channel pattern (multiple-thread versus single-thread channels), entrenchment ratio, width-to-depth ratio, sinuosity, and slope and channel pattern as delineative criteria in the initial eight broad groupings of stream types. . . . *Level II* is the morphological description that classifies stream types within certain valley types using field measurements [from specific channel reaches and fluvial features]. . . . *Level III* assesses stream condition to predict river stability (e.g., aggradation, degradation, sediment supply, streambank erosion, and channel enlargement).

Level IV involves the validation of all components of Levels I–III, including classification and process measurements.

For *lentic system* (lake and wetland) mass balance investigations, lotic flow into or out of a lake basin is best measured nearest the entry or exit point at the lake. Thus, Level I evaluation of the watershed characteristics is the most expedient characterization for lake and wetland mass balance research. Figure 3.2 illustrates the principles of Rosgen's classification system as it pertains to lentic systems. The lotic flow in the drainage network box represents surface flow from both stream and rainfall events into and out of a wetland system. Table 3.1 is a summary of Rosgen's valley types, and Figure 3.3 is an example of valley types VIII and II, lakes associated with these types of basins, and the effect of climate on valley type II.

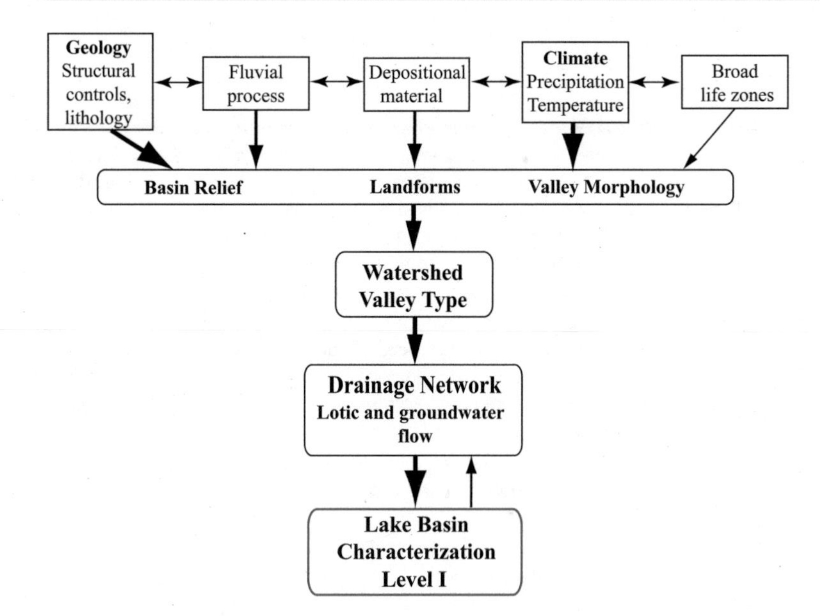

Figure 3.2. Physical characterization of watersheds and flow to lakes based on Rosgen's classification of valley types. Modified with permission from Rosgen (1996) and US Environmental Protection Agency (n.d.).

3.2. Lentic Systems

Lentic systems are defined by Kalff (2002) as "standing water systems (ponds, lakes) in which flow is primarily imposed by wind and heat and is not primarily unidirectional. In contrast, a *lotic system* is primarily a unidirectional flowing system (streams, rivers) imposed by gravity." (See section 4.6.8 for more on lotic systems).

As defined, lentic systems generally include lakes and ponds. A lake's *morphometric* structure (section 3.5) has a significant impact on its biological, chemical, and physical features. Lentic systems may be freshwater bodies, while others have varying levels of salinity (e.g., Great Salt Lake and the Dead Sea). Most basin-type wetlands are also generally grouped within lentic systems; these are areas of constant soil saturation or inundation with distinct vegetative and faunal communities.

A lake is "any relatively large body of slowly moving or standing water that occupies an inland basin of appreciable size" (Lane 1998). Definitions that precisely distinguish lakes, ponds, swamps, and even rivers and other bodies of nonoceanic water are not well established. It may be said, however, that rivers and streams are relatively fast moving; marshes and swamps contain relatively large quantities of grasses, trees, or shrubs; and ponds are relatively small in comparison with lakes.

Table 3.1. Rosgen valley types

Valley Type	Geomorphology	Description
I	Notched canyons, rejuvenated side slopes	Valley is V-shaped, confined, and often structurally controlled and/or associated with faults. Elevational relief is high, valley floor slopes are greater than 2%, and landforms may be steep, glacial scoured lands and/or highly dissected fluvial slopes. Valley materials vary from bedrock to residual soils occurring as colluvium, landslide debris, glacial tills, and other similar depositional materials.
II	Moderately steep, gentle side slopes often in colluvial valleys	Moderate relief, relatively stable, moderate side slope gradients, and valley floor slopes that are often less than 4% with soils developed from parent material (residual soils), alluvium, and colluvium. Cryoplanated uplands dominated by colluvial slopes are typical of the land types that generally comprise valley type II in the northern Rocky Mountains.
III	Alluvial fans and debris cones	Primarily depositional in nature with characteristic debris (colluvial or alluvial fan landforms) and valley floor slopes that are moderately steep or greater than 2%.
IV	Gentle gradient canyons, gorges, and confined alluvial valleys	Consists of the classic meandering, entrenched or deeply incised, and confined landforms directly observed as canyons and gorges with gentle elevation relief and valley floor gradients often less than 2%. Generally structurally controlled and incised in highly weathered materials often associated with tectonically "uplifted" valleys.
V	Moderately steep valley slopes, U-shaped glacial trough valleys	The product of a glacial scouring process where the resultant trough is now a wide, U-shaped valley with floor slopes generally less than 4%. Soils are derived from materials deposited as moraines or more recent alluvium from the Holocene period to the present. Landforms locally include lateral and terminal moraines, alluvial terraces, and floodplains. Deep, coarse deposition of glacial till is common, as are glaciofluvial deposits, with the finer size mixture of glaciolacustrine deposition above structurally controlled reaches.
VI	Moderately steep, fault-controlled valleys	Termed a fault-line valley; structurally controlled and dominated by colluvial slope building processes. The valley floor gradients are moderate, often less than 4%. Some alluvium occurs amid the extensive colluvial deposits, and stream patterns are controlled by the confined, laterally controlled valley. Sediment supply is low.
VII	Steep, highly dissected fluvial slopes	Consists of a steep to moderately steep landform, with highly dissected fluvial slopes, high drainage density, and a very high sediment supply. Streams are characteristically deeply incised in either colluvium and alluvium or residual soils. The residual soils are often derived from sedimentary rocks such as marine shales. Depositional soils associated with these highly dissected slopes can often be eolian deposits of sand and/or marine sediments. This valley type can be observed over a variety of locations, from the provinces of the Palouse Prairie of Idaho and the Great Basin or high deserts of Nevada and Wyoming to the Sand Hills of Nebraska and the Badlands of the Dakotas.
VIII	Wide, gentle valley slope with well-developed floodplain adjacent to river terraces	Most readily identified by the presence of multiple river terraces positioned laterally along broad valleys with gentle, down-valley elevation relief. Alluvial terraces and floodplains are the predominant depositional landforms, which produce a high sediment supply. Glacial terraces can also occur in these valleys but stand much higher above the present river than the alluvial (Holocene) terraces. Soils are developed predominantly over alluvium originating from combined riverine and lacustrine depositional processes.

Table 3.1 continued

Valley Type	Geomorphology	Description
IX	Broad, moderate to gentle slopes, associated with glacial outwash	Observed as glacial outwash plains and/or dunes, where soils are derived from glacial, alluvial, and/or eolian deposits. Due to the depositional nature of the developed landforms, sediment supply is high.
X	Very broad and gentle slopes, associated with extensive floodplains Deltas	A unique series of landforms consisting of large river deltas and tidal flats constructed of fine alluvial materials originating from riverine and estuarine depositional processes. Valleys or delta areas are often seen as freshwater and saltwater marshes, natural levees, and crevasse splays. There are four morphologically distinct delta areas that produce different stream types or patterns.

Source: Description from Rosgen (1996) with permission.

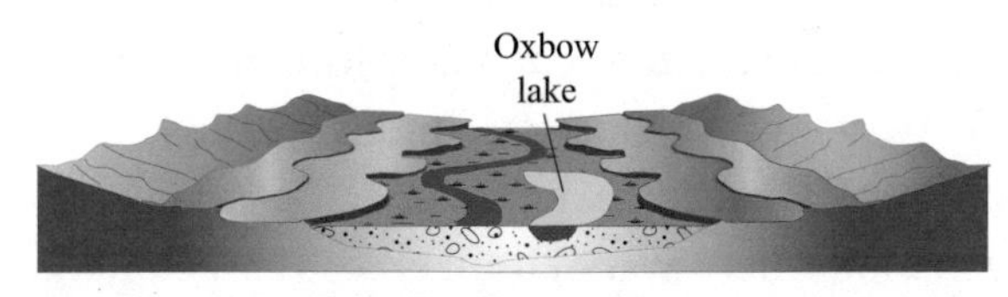

Valley Type VIII - Wide gentle-sloping valley with well-developed floodplain adjacent to river terraces

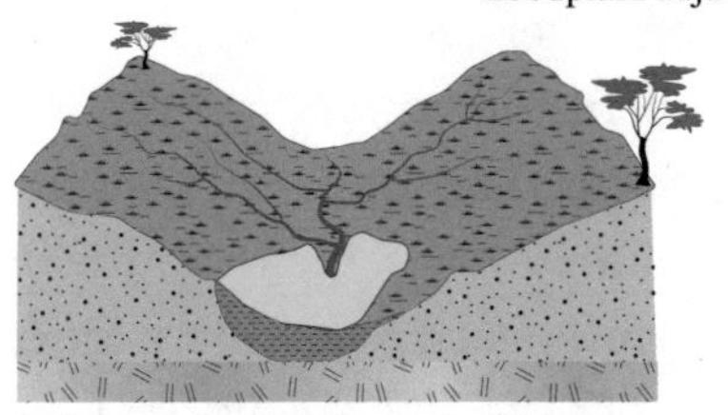

Valley Type II - Moderate rainfall climate

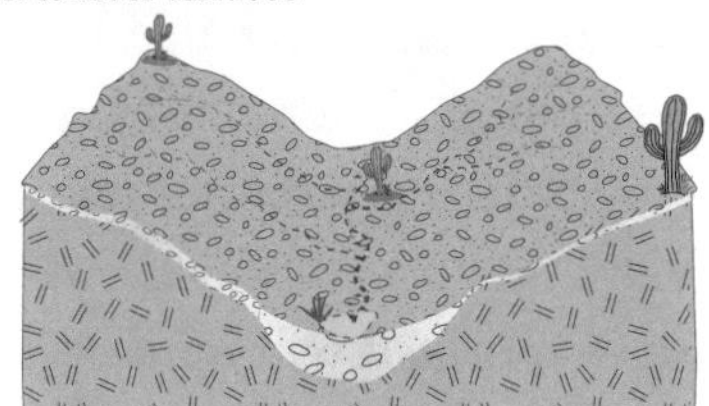

Valley Type II - Low rainfall arid climate

Valley Type II - Moderately steep, gentle slopes often in colluvial valleys. This valley exhibits moderate relief, relatively stable, moderate slope gradients with valley floor slopes often less than 4% and typically found in the northern Rocky Mountains.

Figure 3.3. Examples of valley types VIII and II from Rosgen's Level I classification system. Modified with permission from Rosgen (1996) and US Environmental Protection Agency (n.d.).

The term "lake" or "pond" is not established on a specific naming resolution or scientific criteria and can be considered arbitrary. Overall, lakes tend to be larger and/or deeper than ponds, but numerous examples exist of ponds that are larger and deeper than lakes. One classic distinction is that sunlight penetrates to the bottom of all areas of a pond in contrast to lakes, which have deep waters that receive no sunlight at all. Another is that ponds generally have small surface areas and lakes have large surfaces. So, a combination of surface area and depth are considered from a technical perspective for separating ponds from lakes (Biggs et al. 2005).

In short, no definitive line exists between lakes and ponds and the basic quantitative principles and mathematics discussed for describing mass balance in this text apply to lakes and ponds as well as wetlands of various types. For our purposes here, lakes are defined simply as an area filled with water usually in a depressed topographic area or basin, surrounded by land, and distinct from rivers or streams that feed or drain the lake.

Lakes and ponds are almost always connected with streams in the same watershed, but the reverse is not nearly as often true. Lakes with no stream or subsurface *seepage* outflow are known as *endorheic* (closed basin) and are usually saline (e.g., Utah's Great Salt Lake and the Dead Sea bordered by Jordan to the east and Israel and the West Bank to the west). Endorheic regions are usually in the interior of a landmass, far from an ocean in areas of relatively low rainfall where evaporation exceeds precipitation. Their watersheds are often confined by natural geologic land formations such as a mountain range, cutting off water egress to the ocean. In these regions rivers arise but lose themselves in dry watercourse or by entering *terminal* (closed) *basins* (Martonne and Aufrère 1928). As no water flows out of endorheic lakes, the water that is not evaporated will remain in the closed lake indefinitely. Since all inflows contain concentrations of salts and the evaporating water contains none, salts accumulate in these lakes. Endorheic water bodies include some of the largest lakes in the world, such as the Caspian Sea, the world's largest saline inland sea.

Exorheic or open lakes are in higher watersheds or those characterized by lower surface evaporation and have stream or groundwater (seepage) outflow that drain into a river, or other body of water that ultimately drains into the ocean (Kalff 2002). Exorheic lakes form in areas where precipitation is greater than evaporation. Because most of the world's water is found in areas of exceedingly effective rainfall, most lakes are open lakes whose water eventually reaches the sea. *Transitional lakes* and wetlands oscillate between open and closed systems as a result of climatic changes.

3.2.1. Origin and Classification of Lake Systems

Knowledge on lake origins and morphology provide information on geology, lithologic makeup, and structural control of the lake basin as well as sedimentation patterns and other physical factors useful in determining lake behavior in terms of drainage, seepage, rainfall inputs, evaporation, and fluxes in volume. The geology of a drainage basin expressed as structural control and lithologic composition exerts a major dynamic effect on the basin's physical characteristics such as relief, slope, shape, drainage density, stream and wetland patterns, erosion, and sedimentation. Variation in rock types and soils influence stream morphology due to differences in lithologic porosity, permeability, and erodibility (sedimentary vs. crystalline rocks). Porous and permeable lithology increases infiltration, facilitating storage and baseflow, while impermeable lithology increases runoff discharge. The type of rock affects the rate of weathering, sediments yields, and chemical solute supplied to streams and lakes (Schumm 1977).

In his classic monograph on limnology, Hutchinson (1957) provided a comprehensive analysis and classification of lake systems based on origin, formation, morphology, and worldwide distribution. His widely accepted classification recognizes eleven major lake types that are divided into seventy-six subtypes. The eleven major lake types are *glacial lakes, tectonic lakes, volcanic lakes, solution lakes, landslide lakes, fluvial lakes, aeolian lakes, shoreline lakes, organic lakes, anthropomorphic lakes*, and *meteorite (extraterrestrial impact) lakes*. The following is a brief synopsis on the major lake types described in detail by Hutchinson (1957).

3.2.1a. Major Lake Types

GLACIAL LAKES Activity from glaciers and continental ice sheets have produced a wide variety of lake subtypes (numbers 23–42 described by Hutchinson), including different types of lakes within many of the subtypes. Kalff (2002) narrows down four types of glacial activity that can form lakes: (1) ice barriers, (2) glacial erosion, (3) glacial deposition, and (4) a mixture of glacial and nonglacial processes (Figure 3.4). In areas of massive glaciation, the underlying bedrock can subside and partially rebound upon glacial retreat causing depressions and lakes. Thus, it is often difficult to define clear-cut distinctions between different types of glacial lakes and lakes influenced by other activities.

Glacial lakes contribute about three-quarters of the lentic bodies of water with areas greater than 0.01 km^2 worldwide with uncounted millions of smaller lakes created by glaciation (Kalff

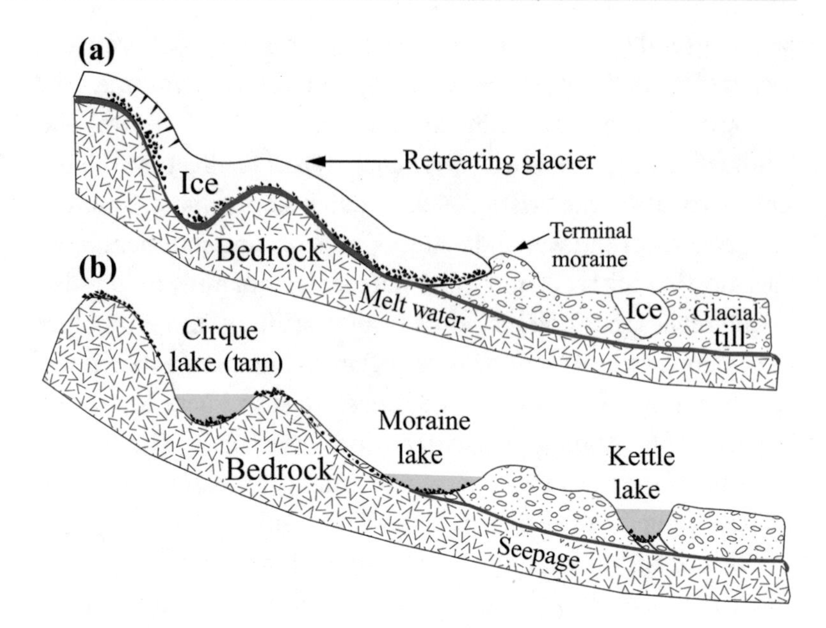

Figure 3.4. Three examples of glacial lakes (b) formed because of glacial activity (a). This would be classified as a type V valley under Rosgen's classification system (Table 3.1). Modified with permission from Rosgen (1996) and US Environmental Protection Agency (n.d.).

2002). Of the processes that form these lakes, Pleistocene glacial activity has been the most important mechanism for lake formation in North America and northern Europe (Hutchinson 1957). These areas, referred to as *lake districts*, contain lakes created by similar processes. While the individual lakes in a lake district often share similar geologic features, they are often quite unique in their chemistry, ecology, and parameters determining mass balance. In Northern Wisconsin and Minnesota, for example, many of the lakes were formed by the same glacial processes, but the individual biological, chemical, and physical characteristics of lakes even just a few miles apart can be dramatically different. In these lakes, landscape position of the basin, characteristics of the watershed, and morphometry of the basin are usually more important than method of basin formation for describing the ecological features of a lake as well as determining the lake mass balance.

TECTONIC LAKES Tectonic lakes are formed from depressions created by the lateral and vertical movements and interaction of the Earth's crust or lithospheric plates deep underground. These movements include faulting, tilting, folding, and warping. Crustal extension has created an alternating series of parallel grabens and horsts that form elongated basins alternating with mountain ranges often resulting in endorheic regions in interior landmasses with low precipitation (Figure 3.5).

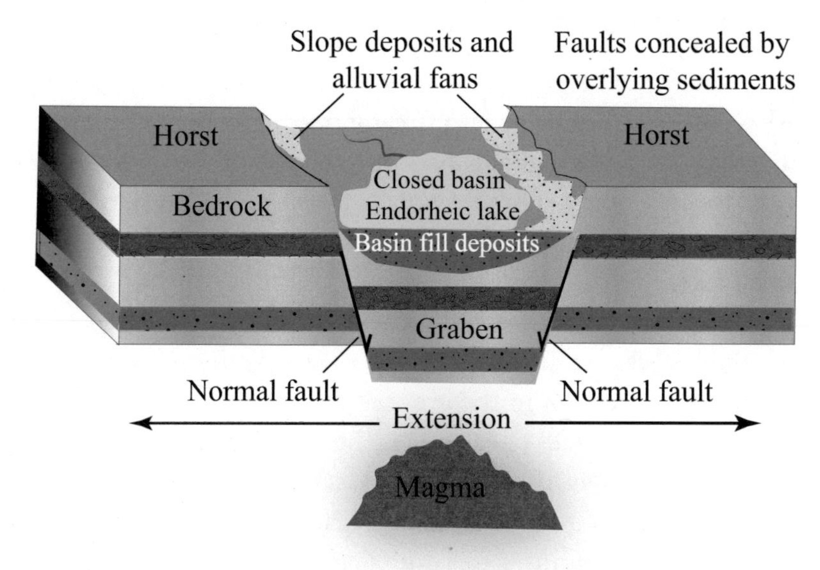

Figure 3.5. Tectonic lake formed as a result of graben and horst faulting from lithospheric extension and formation of a catchment valley, type VI (Table 3.1). Modified with permission from Rosgen (1996) and US Environmental Protection Agency (n.d.).

Tectonic lakes are among the largest and most voluminous lakes in the world. Lake Baikal and Lake Tanganyika hold 32% (18% and 14%, respectively) of the all the freshwater contained in the world's lakes (Kalff 2002). The largest saline lakes are also tectonic lakes; the Caspian Sea contains about 75% of the world's saline lake water. Lake Baikal and the East African Great Lakes are example of lakes formed in grabens.

VOLCANIC LAKES The formation of volcanic lakes can occur through a variety of processes that result first in the formation of a crater. Kalff (2002) lists craters as formed by three types of volcanic activity: (1) gas and magma being expelled directly during an eruption through the volcanic cone of underlying material, (2) hot magma encountering groundwater and causing an explosion or the degassing of magma, and (3) the collapse of the lithology overlying a partially empty magmatic chamber. The first two processes typically result in cone-shaped craters less than 2 km in diameter but often deep. These *cone lakes* are formed after the craters eventually fill with water from precipitation and runoff. Oregon's Crater Lake, the seventh deepest lake in the world with a maximum depth of 608 m and circular area of 54 km^2, is an example of the third process. These cover-collapse basins are generally much larger than the first two types and are called *calderas*, or *caldera lakes* if filled with water.

Lava flows, or *lahars*, can form lakes by moving into existing river valleys and forming dams as the lava cools and solidifies. Also, lakes can form from lava streams or lahars when the surface magma cools and eventually becomes solid while the hotter interior lava continues to flow. Eventually the surface of the hardened lava collapses, forming a depression that fills with water, creating small lakes.

SOLUTION LAKES Lakes can form when underground deposits of soluble rocks are dissolved by water running through the area, making a depression in the ground. Rock formations made of sodium chloride (salt) or calcium carbonate (limestone) are most likely to be dissolved by acidic waters. Once the groundwater has dissolved the rocks below the surface, the top of the land caves in, usually forming a round-shaped lake, called a *solution lake* (Figure 3.6). Typically, the depressions are deep enough to extend below the groundwater table and are permanently filled with water. Solution lakes are common in the Dinaric Alps in southern Europe, in the Balkan Peninsula, and in the United States (Michigan, Indiana, Kentucky, and particularly Florida).

The techniques and lake mass balance theory discussed in this text have been primarily developed utilizing research on solution lake systems in Florida, which contains approximately 7,800 lakes with a surface area greater than 4 km^2 (Meyers and Ewel 1990). Lakes originating by the process of dissolution of limestone

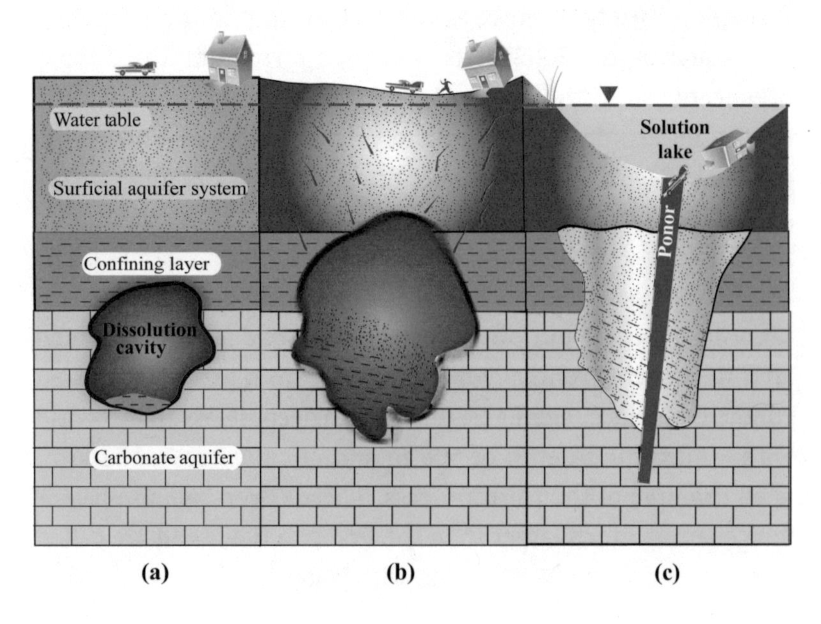

Figure 3.6. Solution lake formation from karstic processes: (a) creation of cavity due to dissolution of underlying carbonates, (b) cavity increasing in size and creating subsidence, and (c) collapse of overburden forming a lake.

are known as karst lakes (Sinclair and Stewart 1985). The German word *karst* originated from the limestone region northeast of Trieste Bay, in the extreme northern part of the Adriatic Sea. Karst is from a pre-Indo-European word *karra* meaning *stony* (Gams 1993). The Winter Park sinkhole (Figure 3.7) is a recent and well-documented example of this formative process (Bryan, Scott, and Means 2008).

Karstic lake basins are the predominant form of lake system in Florida (Meyers and Ewel 1990). Karst lakes characteristically are shallow with no surface outlet. These basins often contain remnant or active sinkhole features (*ponors*) within the lake. *Ponor* is an eastern European term meaning "swallow hole." This expression was originally used in karstic terrains of eastern Europe where these features drain water from lakes and streams into the subsurface. When active, these ponors are assumed to be a saturated continuous hydrologic connection to the underlying units resulting in a wide range of water-level fluctuations in association with wet and dry periods.

Figure 3.7. Winter Park sinkhole. Photo by Richard Deuerling (Lane 1981, 26).

LANDSLIDE LAKES Landslide lakes result from gravity flow of large quantities of material that fall from the sides of steep valleys or mountainous areas in the form of rockfall, mudflow, or avalanche into the beds of rivers or streams blocking the flow as a natural dam creating a lake. Landslides usually occur as a result of an excessive meteorological event, such as excessive rainfall, that reduces the stability of a slope or undercuts the lithology due to erosion. Lakes that are formed by landslides are usually temporary because they generally are susceptible to erosion by the flow of the river or stream.

Earthquakes and volcanoes can cause large landslides and more permanent lakes due to the greater quantity of landslide material. Earthquake Lake in the Gallatin National Forest of southwestern Montana was formed after an earthquake measured at 7.5 on the Richter magnitude scale caused an 80-million-ton landslide on August 17, 1959 killing 28 people and creating a landslide dam on the Madison River (Hebgen Dam). The lake is 58 m deep and 9.7 km long. The earthquake was the most powerful to hit the state of Montana in historic times (US Geological Survey 2009).

FLUVIAL LAKES *Fluvial lakes* are produced by running water from rivers or streams due to the high energy and erosive capabilities that can carve lake basins. These lakes include plunge pool lakes, fluviatile dams, and meander lakes (oxbow). The most common type of fluvial lake is the crescent-shaped oxbow lake (see Figure 3.3). The curved shape is a result of stream meandering in a river valley. The flowing river forms a sinuous shape as the outer sides of bends are eroded away more rapidly than the inner sides. Eventually a horseshoe bend is formed and the river cuts through the narrow neck. This new passage then forms the main passage for the river and the ends of the bend become silted up, thus forming a bow-shaped lake (Hutchinson 1957).

AEOLIAN LAKES When the low-lying land among dunes in a desert is lower than the groundwater level, the water gathers there and forms an aeolian lake. Aeolian lakes are lakes produced by wind action and consist of lake basins dammed by windblown sand. These interdunal lakes lie between well-oriented sand dunes; erosion creates shallow depressions that typically contain water seasonally or during periods of rainfall.

SHORELINE LAKES Shoreline lakes are formed along the coastline or between islands and mainland mainly due to the deposition of sediments by rivers, wave action, or ocean currents that result in the creation of a smaller water body separated from a larger water body by such deposits. For example, when estuaries are blocked or beach ridges grow by the action of sea currents, shoreline lakes are created. Similarly, the meeting of two spits dividing a larger lake results in a shoreline lake.

ORGANIC LAKES Organic lakes are created by the actions of plants and animals. Overall, they are relatively rare in occurrence and quite small in area. The basins in which organic lakes occur are associated with beaver dams, coral lakes, or dams formed by vegetation.

ANTHROPOMORPHIC LAKES Such lakes are "artificial" lakes created as a direct or indirect result of *hominid* activities. The most common origin of anthropogenic lakes is the creation of reservoirs by damming a river or stream. Such reservoir lakes serve several purposes, such as the generation of hydroelectricity, storage of water for future needs, and agriculture.

METEORITE (EXTRATERRESTRIAL IMPACT) LAKES Meteorite lakes, which are also known as crater lakes, are lakes created by catastrophic extraterrestrial impacts by either meteorites or asteroids. Over time, precipitation accumulates in the natural depression, creating a lake. Lonar Crater Lake, a saline soda lake located in the Indian state of Maharashtra, is an example of a meteorite lake. Studies of the sediments at the bottom of such lakes often yield valuable information about extraterrestrial objects.

On a human time scale, the lakes discussed above may be thought as permanent. Geologically defined, however, lakes are temporary bodies of water or ephemeral features on the landscape. They are found in depressions in the Earth's surface in regions where water is available to fill the basin. Over time, lakes fill with sediments and organic material while outlets tend to erode the lake rim away.

Additional methods for classifying and categorizing a lake system other than *lake origin* are variable and include *thermal stratification*; *photosynthesis*, *solar radiation*, and *light attenuation*; *oxygen saturation*; *salinity*; *nutrient concentration*; and other morphological and physical characteristics. These areas of classification are discussed throughout the limnologic literature.

The following categories of lake classification have implications for mass balance as it applies to plant growth, transpiration, and evaporation, the last of which is a major source of water loss from lentic systems (chapter 4). Photosynthesis, evaporation, and thermal stratification are driven by solar radiation, and transpiration is related to photosynthesis and plant growth by such nutrient concentrations available for plant growth as nitrogen, phosphorous, and calcium.

One of the most essential sets of properties influencing lake behavior is the interaction of light, temperature, and wind-mixing. The absorption and attenuation of light by the water column are major factors controlling temperature and potential photosynthesis. Photosynthesis provides the food that supports much of the food web in the lake ecosystem. It also provides much of the dissolved oxygen in the water. Solar radiation is the major source of heat to the water column and is a major factor determining wind patterns in the lake basin and water movements. Winds create the waves and currents necessary to produce the turbulent flow that is important for the distribution and mixing of chemicals, oxygen, and organisms throughout the water column.

Light intensity at the lake surface varies seasonally and with cloud cover and decreases with depth down the water column. The deeper into the water column that light can penetrate, the deeper photosynthesis can occur. Photosynthetic organisms include those suspended in the water (*phytoplankton*), algae attached to surfaces (*periphyton*), and vascular aquatic plants (*macrophytes*).

3.3. Solar Radiation

The sun transmits energy to the Earth in the form of *electromagnetic radiation* (EMR) wavelengths between 100 and 3,000 nanometers (nm). Approximately 97% EMR is split between visible light (380–740 nm) and many other wavelengths that the human eye is not sensitive to, such as *ultraviolet radiation* (100–400 nm) and *infrared radiation* (700–3,000 nm), the latter of which is responsible for heating the Earth. Solar radiation is responsible for photosynthesis and produces the heating gradient at different latitudes, which in turn drives climate, winds, ocean currents, and other such meteorological factors as rainfall and evaporation.

The amount of direct solar radiation that reaches the water surface varies with the angular height of the sun and, therefore, with time of day, season, and latitude (Wetzel 2001). The quantity and quality of light also vary with the transparency of the atmosphere and the distance the light must travel through it; therefore, it varies with altitude and meteorological conditions (e.g., cloud coverage) (Wetzel and Likens 2000). Much of the light is reflected from the water surface and is, therefore, unavailable to the aquatic system, although some can be backscattered to the water surface indirectly.

Photosynthetically available radiation (PAR) is between 400 and 700 nm. Limnologists are particularly interested in the depth of the water column where *phytoplankton photosynthesis* is greater than *respiration*. This zone is known as the *euphotic*, *photic*, or *trophogenic* zone and is defined as the zone where PAR is ≥ to 1% (Kalff 2002). The zone directly below the euphotic zone is where phytoplankton photosynthesis equals respiration and is called the *compensation depth*. Where the PAR is < 1% is called the *profundal* zone.

3.3.1. Light Attenuation and Lake Depth

The diminution or reduction of radiant energy with depth by both scattering and absorption mechanisms is termed *attenuation*, whereas absorption is defined as diminution of light energy with depth by transformation to heat (Westlake 1965).

When PAR strikes the surface of a lake it begins to attenuate, or decrease vertically, as it passes through the water column. Vertical attenuation of PAR is expressed as a percent reduction through a water layer of specified depth (Kalff 2002):

$$\frac{100(I_0 - I_z)}{I_0} \tag{3.1}$$

where I_0 = the light intensity over a specified wavelength just below the surface and I_z = the light intensity at depth z. Vertical attenuation in very clear lakes is exponential, but in highly colored or turbid lakes attenuation is much more rapid (Figure 3.8c). In general, approximately 53% of the total light energy is transformed into heat in the first meter (Wetzel 1975).

Limnologists find it more useful to express PAR within the lake water column as a vertical attenuation (or extinction) coefficient (η_d) rather than a percentage (Kalff 2002). Coefficient η_d is the slope of the line formed when the natural logarithm of energy flux is plotted against depth, expressed as

$$I_z = I_0 e^{-\eta d} \tag{3.2}$$

where I_z = the energy flux or photon flux at depth z (meters), I_0 = intensity a few centimeters below the lake surface, e = the base of the natural logarithm value of 2.303, and η_d = the vertical extinction coefficient.

Then converting equation (3.2) to a natural logarithmic (i.e., $\log_{10}$) form for convenience

$$\eta_d = \frac{\ln I_0 - \ln I_z}{z} \tag{3.3a}$$

or

$$\eta_d = \frac{2.303(\log_{10} I_0 - \log_{10} I_z)}{z}. \tag{3.3b}$$

The thickness of the euphotic zone z_{eu} can be calculated by setting I_0 at 100% and I_c (*compensation depth*) equal to 1% such that

$$z_{eu} = \frac{\ln 100}{\eta_d} = \frac{4.6}{\eta_d}. \tag{3.4}$$

Light attenuation is commonly determined by using one of two methods: (1) direct measurement of surface and underwater solar radiation, or (2) measuring water clarity or transparency to estimate the light extinction coefficient, which is a measure of how quickly light is attenuated (Wetzel and Likens 2000).

Measurements of surface and underwater irradiance (light availability) include shortwave radiation (100–400 nm), reflected longwave radiation (infrared, 700–300 nm), and photosynthetic active radiation (PAR, 400–700 nm) and are measured using a *pyrheliometer*, *pyranometer*, *net radiometer*, or *quantum sensor* (refer to Green, Robertson, and Wilde [2015] for information about measurements of solar radiation). Light attenuation can be measured using the surface (terrestrial) radiation sensor along with the underwater radiation sensor. The depth where the underwater sensor reaches 1% of the surface intensity identifies the bottom or thickness of the photic zone (i.e., equation 3.4). The extinction coefficient is the absolute value of the slope of the natural log of solar radiation with depth.

Water clarity, on the other hand, is commonly measured using a *Secchi disk*. The Secchi disk is a weighted disk with alternating black and white quadrants, 20 cm (8 in.) in diameter and attached to a line with distance marked in meters or feet. The disk is slowly lowered into the water column until it is no longer visible. The depth at which the disk is no longer visible is referred to as the Secchi depth. Secchi depths should be measured on the shady side of the boat. The thickness of the photic zone is measured by multiplying Secchi disk depth by a factor of approximately 2.5 (Welch 1948; Horne and Goldman 1994; Wetzel and Likens 2000). Light penetration, as measured with a Secchi disk, can vary considerably when the sun is at extreme angles. Therefore, if only one sample is to be collected in a day, the Secchi measurement generally is made between 1000 and 1500 hours (Green, Robertson, and Wilde 2015).

Water clarity is influenced by such factors as water color, algal and zooplankton concentrations, and suspended sediments. These factors can vary seasonally or between locations within an individual lake where the data are collected. For example, inorganic *turbidity* composed of suspended solids, such as silts and clays, may impact water clarity in the area where a stream or river flows into the lake while phytoplankton is the dominant factor affecting clarity at the lower end or center of the lake. Phytoplankton concentration typically increases during summer months and drops drastically during the winter.

3.3.1a. Biological Communities

Based on light attenuation and PAR analysis, lakes can be categorized into three distinct biological communities or zones (Figure 3.8a). Within the euphotic zone are the *littoral zone* and *pelagic zone*. These zones are areas of *primary productivity* where macrophytes or any of the various microorganisms like phytoplankton and cyanobacteria can convert light energy (or chemical energy) into organic matter. In deeper areas below effective light penetration (i.e., PAR < 1%) is the *profundal zone*.

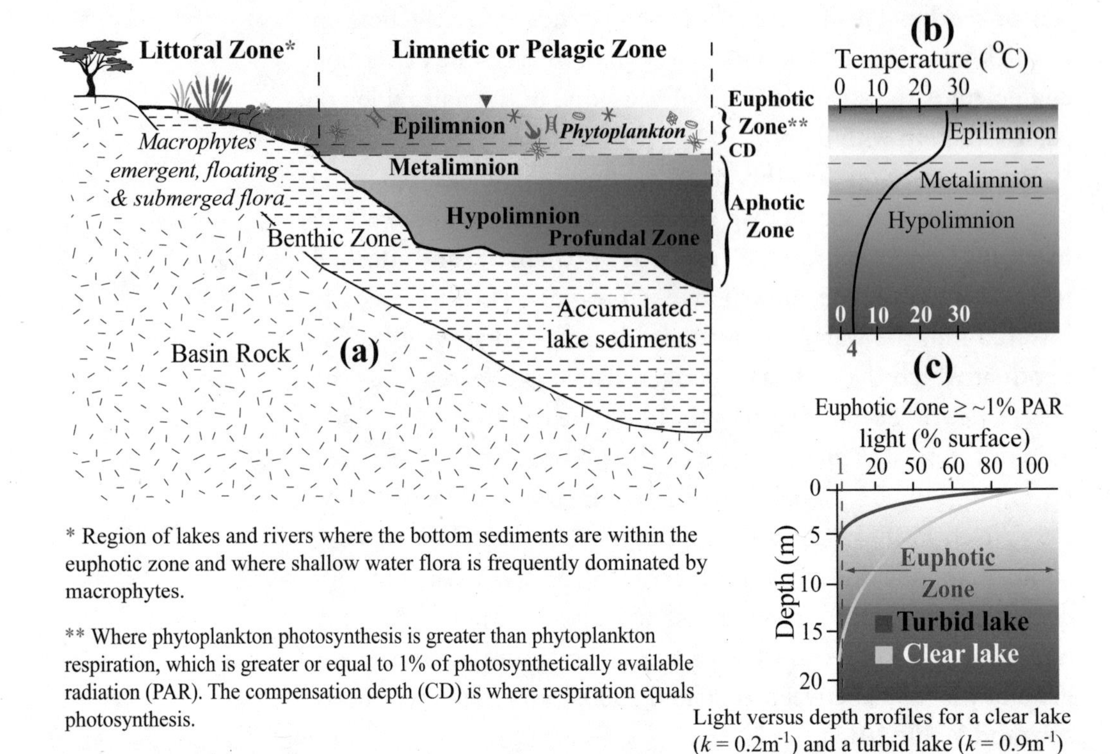

Figure 3.8. Cross section of a freshwater lake (a), showing summer thermal stratification (b) and zones based on light gradient, (a) and (c). Graphs (b) and (c) modified with permission from WOW (2004).

Littoral zones are the nearshore area where the lake-bed sediments are within the photic zone and the shallow water flora is predominately macrophytes. Rates of transpiration in macrophytes are extremely high and result in a quantitative efflux of water vapor from the leaves that is much greater than evaporation from the equivalent area of open water (Wetzel 1975). The littoral zone often has high species diversity and is commonly the area of a water body where fish reproduction and development occur (Cooke et al. 2005). This area is also usually important waterfowl habitat. Plant biomass in the littoral zone typically replaces itself two or more times per summer in productive lakes and reservoirs, leading to inputs of nonliving dissolved and particulate organic matter called *detritus* to the water column and sediments. Detritus, whether from the watershed or from in-lake productivity, is a stable source of energy and nutrients for lake and reservoir autotrophic and heterotrophic production (Green, Robertson, and Wilde 2015).

Pelagic zones are the open water regions beyond the littoral zone and consist mainly of phytoplankton and some floating macrophytes. Plankton (phytoplankton and zooplankton), and the fish grazing on them, dominate the pelagic zone (Cooke et al. 2005). The phytoplankton includes algae that can produce unsightly "blooms" and low water clarity. The pelagic community obtains energy from sunlight and from detritus transported to it from stream inflows and the littoral zone. The phytoplankton of most enriched lakes and reservoirs are often dominated by one or a few species of highly adapted algae and bacteria (e.g., green algae and the nuisance blue-green algae called *cyanobacteria*).

Profundal zones are the deep zones of lakes or ponds, located below the range of effective light penetration (the aphotic zone). This is typically below the *thermocline*, the vertical zone in the water through which temperature drops rapidly. The temperature difference may be large enough to hamper mixing with the littoral zone in some seasons that cause a decrease in oxygen concentrations. Based on the description by Thienemann (1925), the profundal zone is often defined as the deepest, most vegetation-free, and muddiest zone of the lake benthos, where the benthic zone refers mainly to the animal communities that live there.

3.3.1b. Nutrient Concentration and Trophic Classification

Eutrophication and other ecological processes that occur in one zone directly or indirectly affect processes in other zones. For example, nutrients that can cause algal blooms may come from lake sediments and decomposition of littoral zone plants, as well as from external loading (Cooke et al. 2005).

Nutrient-poor lakes are described as *oligotrophic* and are typically clear with low concentrations of plant life. *Mesotrophic* lakes have good clarity and an average level of nutrients (mainly phosphorous and nitrogen) and plant growth. *Eutrophic* lakes are nutrient enriched, resulting in good plant growth and potential algal blooms, and *hypertrophic* lakes have excessive nutrient concentrations and generally low clarity, crowded macrophyte growth, and overwhelming algal blooms resulting in decreased dissolved oxygen and destruction of lake habitat. Hypertrophic lakes are usually due to human activities such as urbanization or overfertilization due to agricultural practices. See Wetzel (1975); Kalff

(2002); and Green, Robertson, and Wilde (2015) for more detail on trophic classifications.

3.3.2. Thermal Stratification of Lakes

3.3.2a. Thermal Radiation in Lake Water

The net amount of solar radiation affecting a lake is referred as *net radiation surplus* (Q_B) (Wetzel 1975):

$$Q_B = Q_S + Q_H + Q_A - Q_R - Q_U - Q_W \tag{3.5}$$

where

Q_S = direct solar radiation,
Q_H = indirect and reflected radiation from sky and clouds,
Q_A = longwave thermal radiation (infrared radiation) from the atmosphere and surrounding topography,
Q_R = radiation reflected from the lake,
Q_U = radiation scattered upward and lost, and
Q_W = emission of longwave radiation.

As most components are negligible at night, the net radiation surplus is equal to the longwave thermal radiation of the atmosphere minus that emitted from water:

$$Q_B = Q_A - Q_{W.} \tag{3.6}$$

Empirical studies have shown that Q_B = –11 (temperature of the water minus the temperature of air) in cal cm^{-2} day^{-1} (Hutchinson 1957). If the mean value of Q_B is positive (the lake is gaining energy), the lake could still be losing heat through evaporation, transpiration, and convective heating of the air (Hutchinson 1957).

Driven by infrared radiation, thermal stratification strongly influences the flora and fauna associated with lakes as well as the overall ecology of individual lake systems, including distribution of dissolved and suspended material, sedimentation, nutrient cycling, water chemistry, and mass balance as it applies to evaporation and transpiration (chapter 4).

Most compounds become denser as they transform from a liquid to a solid, as the molecules become more tightly packed. As discussed in section 2.1, water is unique in that it is most dense at 4°C and becomes less dense at both higher and lower temperatures (see Figure 2.2). Water in its solid state of ice floats, while at temperatures just above freezing, water sinks. The density-temperature relationship of freshwater is shown in Figure 3.9. Because of this relationship, lakes tend to stratify; that is, they separate into distinct thermal layers. This is particularly true for lakes in temperate climates.

In temperate climates, water temperatures in lake waters often decrease to near or below 4°C during the winter season. If ice cover does not occur, the density differences between different temperatures around 4°C are small, allowing complete (wind) mixing of the water column. Upon the arrival of spring air temperatures and as daylight hours increase, the water at the surface (in deeper lakes) heats more quickly than it can be mixed by the wind with the deep, cooler (denser) waters below. As water near the surface warms, the density difference between two successive warm water temperatures, for example 29°C and 30°C, is greater than cooler water, for example 4°C and 5°C. The density difference between 29°C and 30°C is 37.25 times greater than the density difference between 4°C and 5°C (Green, Robertson, and Wilde 2015). The wind energy needed to mix the warmer water would be proportionally greater than the energy needed to mix the cooler water. Relative thermal resistance to mixing (Wetzel 2001) is the phenomenon that allows a thermocline to set up and the water body

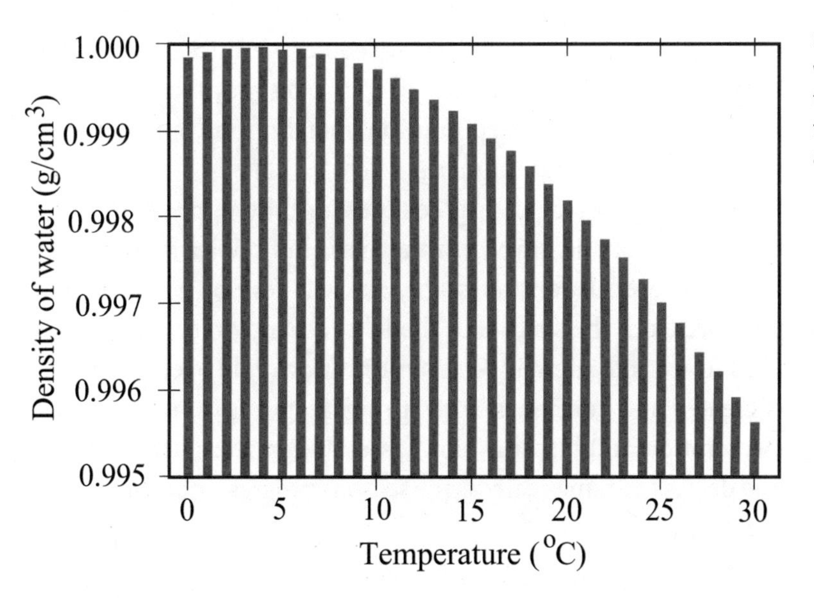

Figure 3.9. Density of water distribution by water temperature. Modified from Green, Robertson, and Wilde (2015).

to stratify. At warmer water temperatures, a difference of only a few degrees is enough to prevent complete mixing.

The lake water column absorbs infrared radiation from the lake surface in the form of heat, which increases as the average daily air temperature increases. If wind is negligible, then a layered temperature profile, as seen in Figure 3.8a and 3.8b, would be expected during late spring to summer months in temperate climates.

Near the surface, the least dense warmer water forms the *epilimnion* layer. As the temperature decreases, water density increases and the colder heavier water sinks, characteristically forming the *hypolimnion* layer near the bottom. Over time, the water column in deeper lakes, typically greater than six meters, will be divided into three layers; overlying the hypolimnion is a zone of transition termed the *metalimnion* (middle layer) where the thermocline (the maximum rate of temperature change with depth) exists. The metalimnion ranges from warm too cold between the epilimnion and hypolimnion.

This sequence of stratification and thickness of the layers can vary depending on the season, location, climate, and morphology of the specific lake as well as other factors. For example, to distribute the warmer upper layer water to lower depths, the water column must have an energy source to mix the upper heat deeper, and in most lakes, wind provides that energy. Thus, a lake with minimum wind interaction will generally have a very warm but shallow layer at the surface with cold water below. Shallow lakes exposed to strong winds will have a cooler but thicker upper layer overlying the colder water and deeper lakes may form the three-layered structure of the epilimnion, metalimnion, and hypolimnion throughout the summer.

Eventually, solar heating declines and the upper epilimnion begins to cool and sink with the arrival of cooler seasons such as fall. As the fall continues, the lake will develop a similar temperature from the surface to the lakebed. In winter, ice forms at the surface and a new, inverse stratification (cold over cool water) is created and continues until spring. In terms of mass balance, evaporation and transpiration losses are greater in the summer and decrease dramatically as the seasons progress into winter (Figure 3.10).

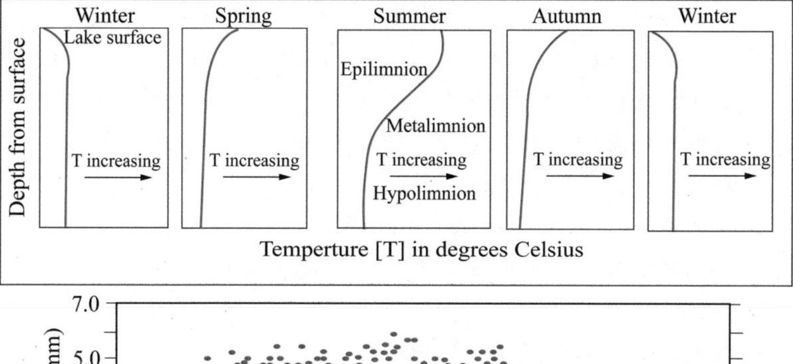

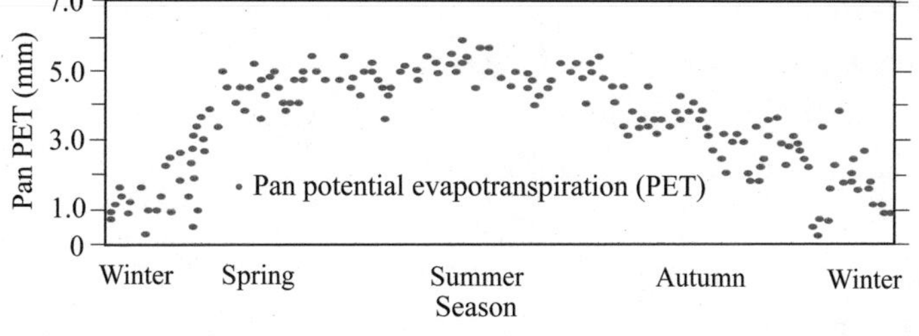

Figure 3.10. Seasonal pan evaporation data with water temperature profiles for a deep temperate-zone lake. Modified from Evans (1996) and Averett and Schroder (1994).

3.3.2b. Thermal Classification

Considered the father of limnology, the German scientist François-Alphonse Forel (1901) was the first limnologist to grasp the importance of thermal layering. This led to his development of a classification of lakes based on thermal stratification. Hutchinson and Löffler (1956) and Hutchinson (1957) later modified and improved Forel's classification system into what is accepted today and briefly summarized here.

A *holomictic lake* is a lake that has a uniform temperature and density from top to bottom at a specific time during the year. This uniformity in temperature and density allows the lake waters to completely mix. Based upon thermal stratification and frequency of turnover, holomictic lakes are divided into *cold monomictic lakes*, *dimictic lakes*, *warm monomictic lakes*, *polymictic lakes*, and *oligomictic lakes* (Wetzel 1975). This classification of lakes by thermal stratification presupposes lakes with enough depth to form a hypolimnion. As a result, very shallow lakes are excluded from this classification system.

Monomictic lakes: Lakes in which water mixes to the bottom in the deepest areas over one extended period throughout late fall and winter. These are deeper lakes that typically do not freeze.

Dimictic lakes: Lakes in which water mixes to the bottom in the deepest areas only during the spring and again in the fall (autumn), referred to as spring and fall turnover. These lakes have extended periods of stratification in summer and are under ice in winter.

Polymictic lakes: Lakes in which water frequently mixes to the bottom in the deepest areas throughout the open-water period.

Oligomictic lakes: These lakes are generally tropical with rare circulation periods at irregular intervals, and temperatures are always well above 4°C.

A number of lake types do not undergo circulation, and the primary mass does not mix with the lower portion. A *meromictic lake* is a lake that has layers of water that do not intermix (Hutchinson 1937, 1938). The deepest stratum is perennially stagnant and termed the *monimolimnion*; this underlies the *mixolimnion*, which periodically circulates. These two layers are separated by a steep salinity gradient called a *chemocline* or *halocline* (Wetzel 1975).

The deepest layer of water in such a lake does not contain any dissolved oxygen. In addition, the layers of sediment at the bottom of a meromictic lake remain relatively undisturbed because there are no living aerobic organisms.

3.4. Lake Morphometry

3.4.1. Introduction

How a lake varies in three dimensions—the *x*, *y*, and *z* directions corresponding to length, width, and depth, respectively, and defining shape and volume—is referred to as *lake morphology*. The morphology of a lake basin has important effects upon most major physical, chemical, and biological parameters, including mass balance in the form of storage.

The morphology of lakes is highly varied and reflects a lake's mode of origin, evolution, water movements, temperature stratification, sedimentation, and degree of loading from the surrounding catchment. Lakes formed from volcanic or meteoric craters and many karst sinkholes may have an almost perfect circle shape while those formed from tectonic faults or scoured glacial valleys will have high length-to-width ratios. *Catchment morphometry* is often related to lake morphology with the size and slope of the drainage basin determining runoff and inflow from other wetland sources depending upon the climatic zone.

The morphology of a lake is best described by a detailed topographic analysis of the lake bottom, known as a *bathymetric map*. These bathymetric contours are created from depth soundings made along transects and connecting the points of a specific depth with a contour line (Håkanson 1981). Originally this was done by a sounding line (marked line with weight attached) measured from a stationary boat (see Welch 1948). This method is subject to error due to wind effects on boat stationarity. Soundings now are created with small and relatively inexpensive single beam sonic or echo sounders (also known as depth sounders or fathometers) that release a sound pulse in a single narrow beam and "listens" for a return echo. Sounding location can be done with handheld global positioning systems from a range of stationary satellites resulting in accurate positioning.

The US Geological Survey utilizes a *multibeam echosounder* that emits multidirectional radial sound waves in a fan-shaped swath to map lake and river bottom bathymetry even in murky waters. The timing and direction of the returning sound waves provides detailed information on water depth and the shape of the lake bottom, river channel, or any underwater feature of interest. This system integrates several individual components, such as inertial navigation, data-collection, and data-processing, to efficiently generate high-resolution maps (Huizinga and Heimann 2018).

Topographic elevations (h) in relation to a known datum such as mean sea level can be used for measuring lake depth and for determining the morphometric parameters discussed in section 3.5.2. This relationship with elevation and basin area is frequently used when investigating reservoir storage (volume) at specific stage intervals or with *ephemeral lakes* that periodically are drained due to climate or sinkhole interaction and topographic elevation contours can be measured when the area is dry.

For an excellent resource for basic lake survey techniques, see the curriculum offered by Water on the Web at http://www.waterontheweb.org/curricula/ws/unit_03/U3mod8_9.html.

3.4.2. Morphometric Parameters

Utilizing bathometry maps, such morphometric parameters as surface area; maximum length, width, and depth; and lake volume can be determined (Hutchinson 1957). The most common parameters are *shoreline* and *shoreline development.* Shoreline is the intersection of land with water of the lake, and this parameter can fluctuate widely in ephemeral lakes in response to variations in precipitation and discharge. Shoreline development (D_L) is the ratio of the length of the shoreline (L) to the length of the circumference of a circle of area equal to that of the lake:

$$D_L = \frac{L}{2\sqrt{\pi A_0}} \tag{3.7}$$

This method of describing lake shape as it relates to the size of the littoral zone is such that lakes with similar volumes or surface area, but with increasing shore length, will result in progressively larger littoral zones (Kalff 2002). Nearly circular lakes, such as crater, kettle, or sinkhole lakes, approach the minimum shoreline development value of unity ($D_L = 1$). However, most lakes are subcircular to elliptical in form with D_L values of approximately 1.4 and less than 3 (Wetzel 1975). Flooded valleys with a number of tributaries will have many branching arms and a dendritic shoreline with D_L values > 3.5 (Kalff 2002). Lake Mälaren, the third largest lake in Sweden has a D_L as high as 10 (Håkanson 1981). A prominent elongation (l) is more important than high irregularity (sinuosity) for producing high D_L values.

Maximum length is the distance on the lake surface between the most distant points on the lakeshore. This length is the maximum effective length for wind to interact across the lake surface without land disruptions (Wetzel 1975). The distance determines the wave height of both surface and internal waves. *Fetch,* related to maximum length, is the distance at which the wind can create turbulence (Kalff 2002). Fetch can be calculated in various ways (Håkanson 1981) and is a useful indicator of thermocline depths as wells as the suspension, resuspension, and transport of particles of varying size and density in a lake system (Håkanson and Jansson 1983).

Maximum width is the maximum distance on the lake surface

at a right angle to the maximum length between the shorelines (Wetzel 1975).

Breadth (b) is defined as the length from shore to shore at right angles to the maximum length (Hutchinson 1957). The mean width ($\bar{b}$) is equal to area divided by the maximum length, or

$$\bar{b} = \frac{A}{l}. \tag{3.8}$$

Area (A) of the surface and each contour at depth (d) is determined by manual or computer-generated planimetry or less precisely by a grid enumeration analysis (Welch 1948; Olson 1960). For computing lake volumes, it is necessary to measure the contour intervals on each map. The area of contour at depth z or elevation h is designated Az or Ah.

Maximum depth (z_m) is the greatest depth of the lake that may vary slightly with lake level and ideally should be referred to independently with some datum such as altitude above mean sea level (h).

Mean depth ($\bar{z}$) is the volume divided by the surface area:

$$\bar{z} = \frac{V}{A}. \tag{3.9}$$

Mean depth is one of the most useful morphometric parameters in terms of biological processes. Water nutrient levels as well as algal biomass tend to decline with increasing mean depth in both freshwater and saline lakes (Chow-Fraser 1991). Algal biomass and subsequent chlorophyll—a major component of energy flow in the aquatic food chain and consequently fish catches—also decline with increasing lake depth.

Mean to maximum depth ratio ($\bar{z}/z_{max}$) provides a useful approximation of lake morphology. For example, Neumann (1959) demonstrates that the average lake shape approximates an elliptical sinusoid to an elliptical cone with a $\bar{z}/z_{max}$ ranging between 0.33 and 0.35.

Shallow lakes with flat bottoms have $\bar{z}/z_{max}$ ratios between 0.50 to 0.67 (Figure 3.11). The ratios greater than 0.50 are found in lakes formed from calderas, grabens, and fjords. For most lakes, the value of $\bar{z}/z_{max}$ is > 0.33 and < 0.50, which is typical of lakes formed in easily eroded lithology. A ratio of 0.33 is the value that

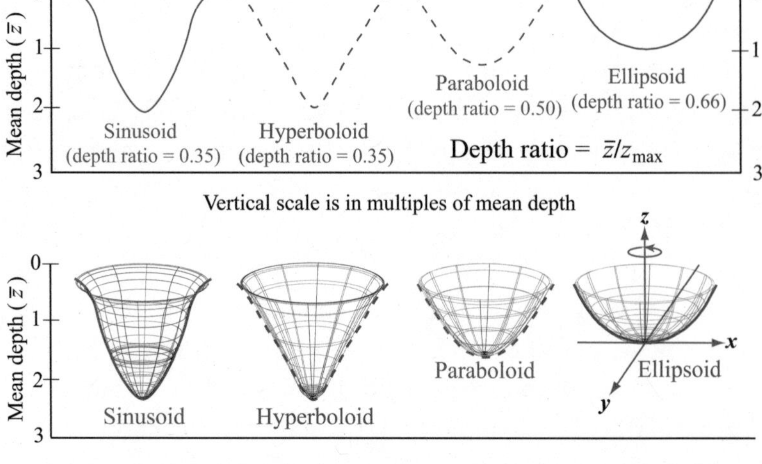

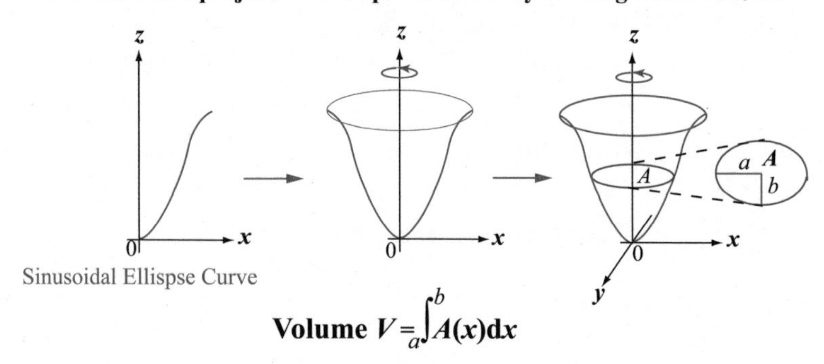

Figure 3.11. Vertical cross sections through four different forms of ideal lake basin shape with equal surface areas and equal mean and maximum depths and volumes. Elliptical data from Carpenter (1983).

best represents a conical depression and these low values are representative of lakes with deep holes, such as solution or kettle lakes. Thus, $\bar{z}/z_{max}$ can be linked to sediment deposition and lake formation as well.

Recall from calculus that curves can be illustrated in three dimensions by rotating the curve about its central axis (see section 1.3.1 in the appendix). Thus, by rotating the elliptical curves from lake cross sections about the z axis, an "ideal" geometric shape of the lake basin can be visualized (Figure 3.11). Although this is not the actual shape of a specific lake basin, from this illustration the relationship of $\bar{z}/z_{max}$ ratios and shallow versus deep lakes can be demonstrated along with conical relationship between lake morphology and volume approximation, discussed next.

Volumes of the basin can be calculated by applying the definite integral to the equation of the elliptical curve. Beginning with the general volume integration equation (3.10) and substituting the depth (or height) z for the traditional y terminology then

$$V = \int_{a}^{b} A(x)dx \tag{3.10}$$

where A is the cross-sectional area

$$A = \pi z^2 \tag{3.11}$$

and

$$dV = \pi z^2 dx \tag{3.12}$$

The equation of an ellipse is

$$\frac{x^2}{a^2} + \frac{z^2}{b^2} = 1 \tag{3.13}$$

where a and b are the major and minor axes of the ellipse, respectively (Figure 3.11).

Substituting $z^2 = b^2[1 - (x^2/a^2)]$ then

$$V = \int_{-a}^{a} \pi b^2 \left(1 - \frac{x^2}{a^2}\right) dx \tag{3.14}$$

and integrating

$$V = \frac{4}{3}\pi ab^2. \tag{3.15}$$

Volume development (D_v) is a measure of lake form utilizing the shape of a cone. As illustrated in the three-dimensional projections in Figure 3.11, the mean to maximum depth ratio ($\overline{z}/zm$) is an expression similar to the ratio of lake volume and that of a cone of basal area A and height z. Since the volume of a cone is one-third the product of the basal area and the height, D_v can be expressed as

$$D_v = \frac{A\overline{z}}{\left(\frac{1}{3} z_{max} A\right)} = 3\frac{\overline{z}}{z_{max}}. \tag{3.16}$$

The ratio of mean to maximum depth provides an estimate of lake volume without a specific reference to an ideal conical form. Lakes with a D_v of ~ 1 are exactly cone-shaped. D_v that are > ~ 2 are lakes with a flat bottom and steep sides (Kalff 2002). Timms (1992) reported that shallow petri dishes and deeper laboratory beakers both have a D_v of 3, demonstrating that the D_v relation to lake depth is limited.

Relative depth ratio (z_r) is defined as the maximum depth as a percentage of the mean diameter (represented by the square root of the lake area in km^2) or

$$z_r(\%) = \frac{z_{max}\sqrt{\pi}}{20\sqrt{A_0}}. \tag{3.17}$$

The majority of large shallow lakes have a z_r of less than 2% whereas deep lakes with small surface areas exhibit greater stability and usually have a relative depth of greater than 4% (Wetzel 1975). High z_r lakes have small surface areas available for wind-induced turbulence and tend to be stably stratified. Inland lakes with z_r of 4%–5% become so stratified that they do not completely mix or circulate on an annual basis, resulting in permanent anoxic hypolimnion (Kalff 2002). In general, lakes with relatively high relative depth ratios are typically nutrient poor and highly transparent.

Depth-area or *depth-volume curves*, also known as *relative hypsographic curves*, are a graphic representation of the relationship between the surface area (or cumulative volume) and its depth. This correlation is expressed in terms of percentage of the lake surface area or volume at a given depth (Figure 3.12). Lake shapes generated by these percentage relationships range between very convex (f (−3)) to concave (f (3)) and express the shape of the hypsometric curve but not necessarily the actual lake basin (Håkanson 1981). These forms were described by Neumann (1995) and also illustrated in Figure 3.11.

By revolving these curves around the z-axis, their idealized geometric shape can be illustrated and the extreme variability between very convex and concave lakes can be demonstrated. Very convex lakes are seen to be very shallow with only about 20% of their area found at depths greater than 10% of the maximum depth (Kalff 2002). These types of lakes have high $\bar{z}/z_{max}$ ratios and one or more deep holes occupying only a small percentage of the area. Due to their extremely shallow natures, they would be expected to have a wide littoral zone and subject to sediment resuspension due to wind-generated turbulence effects on the shallow bottom

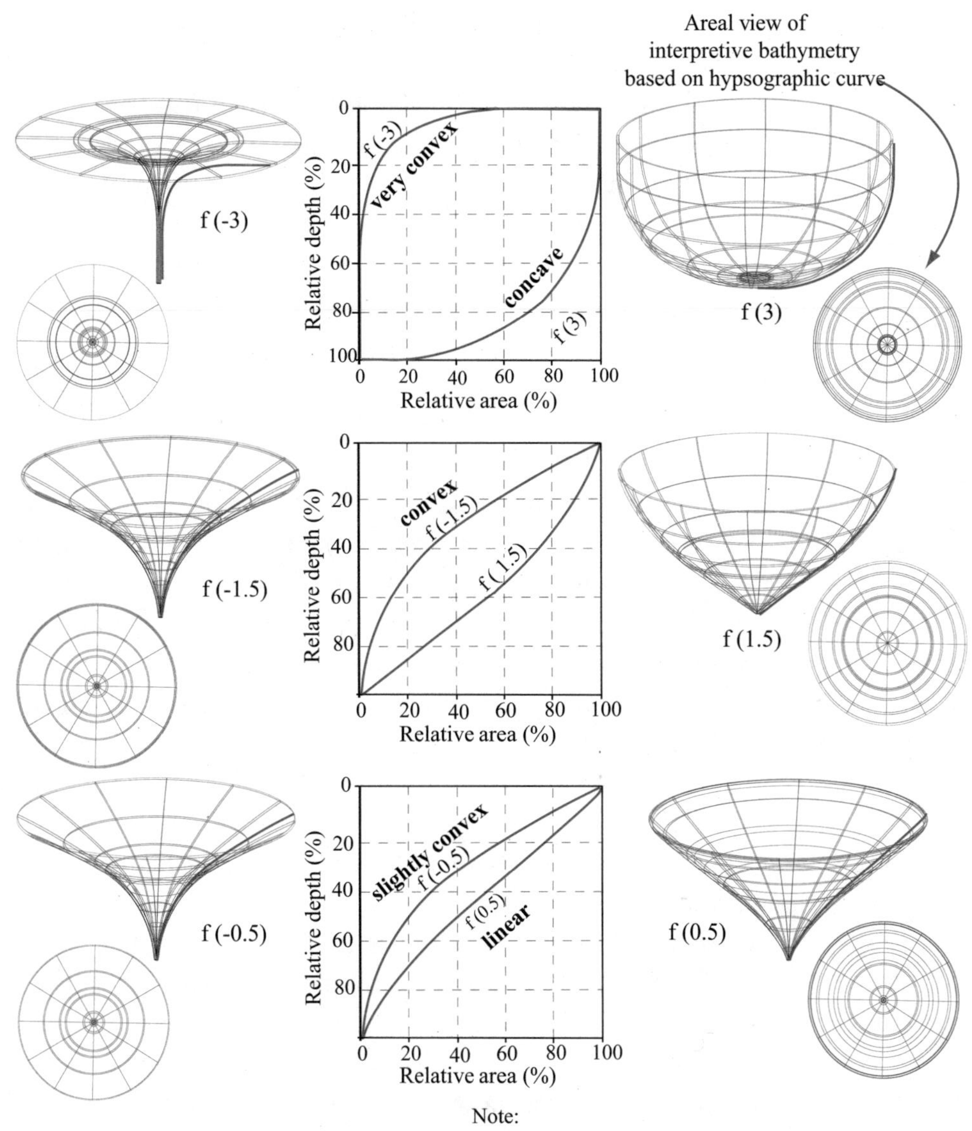

Figure 3.12. Lake form as described by relative hypsographic curves of lake surface area and depth. Lake shape ranges between very convex to concave. Lake shape is expressed three-dimensionally by rotating curves around the *z*-axis. Hypsographic curve data from Håkanson (1981).

sediments. The concave lake (f (3)), in contrast, has a trough-like basin, with steeper walls and a flat bottom, thus these wider, deeper areas will be more protected against wind effects but will have a smaller area for littoral zone development.

Relative hypsographic curves represent the relative proportions of the bottom areas of lakes between the lake stage and strata under examination and are important for investigating the relationship between lake morphology and biological productivity (Wetzel 1975). In general, biological productivity is greater in lakes that have a superposition of zones of photosynthetic production and decomposition (Wetzel 1975). An important factor is the extent of shallowness in a lake available for rooted plant growth as well as light penetration for phytoplankton and algal growth along with the interrelationships of these in developing the littoral zone.

3.5. Lake Volume or Storage

When investigating mass balance to lake systems, the volume (V), or lake storage (S) capacity, is paramount for water supply, flood control, and various relationships between other limnologic properties of a lake system. Volume/storage is the integral of the areas of successive depth determined by planimetry from the lake surface to the point of maximum depth and given by the standard volume integral:

$$V = \int_{z=0}^{z=z_{\max}} A_z \, dz. \tag{3.18}$$

There are several standard procedures for evaluation of the integral and estimating volume. For the best approximation for all methods, the area of contours, as closely spaced as possible, are calculated. These areas (A) are plotted against the depth z, generating a *hypsographic curve* where the area of the curve is then measured planimetrically (discussed in more detail in section 3.6.4). Two alternative methods using the actual elevation (or depth) and area are the *conic method* (Figure 3.13) and the *engineering method* (Figure 3.14).

3.5.1. Conic Method

The volume of a pyramid is one-third its base times width times height, and this is basis of the conic method. In geometric terms, a pyramid or cone with the top portion cut off is called a *frustrum*. The volume of a frustrum is $z/3 \times \left(A_1 + A_2 + \sqrt{A_1 \times A_2}\right)$ where z is the height between the two areas, and A_1 and A_2 are the areas of the two cross sections. The conic method of storage calculation is based on finding the volume of a frustrum (layer of water) of a

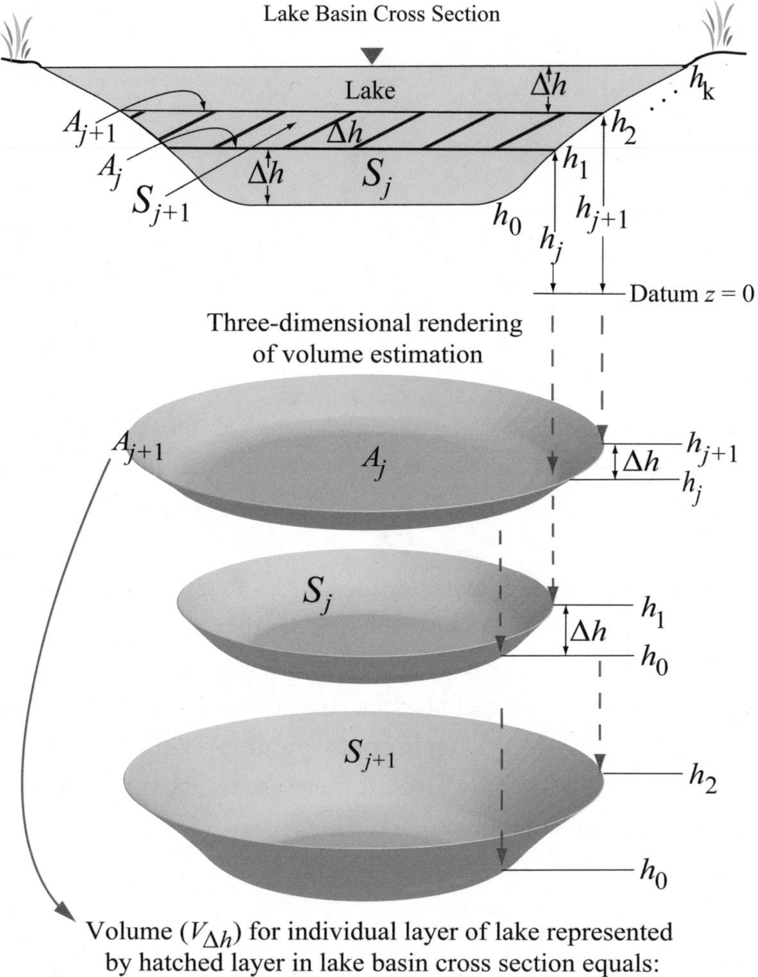

Figure 3.13. Conic method for determining lake volume or storage. Equation from Feldman, Ely, and Goldman (1981).

circular cone (see Figure 3.11). This is done by summation of a series of truncated cones between contoured areas and the change in depth z (or elevation h) between them such that

$$V_{j+1} = V_j + \frac{(z_{j+1} - z_j)\left(A_{j+1} + A_j + \sqrt{A_{j+1}A_j}\right)}{3}. \quad (3.19)$$

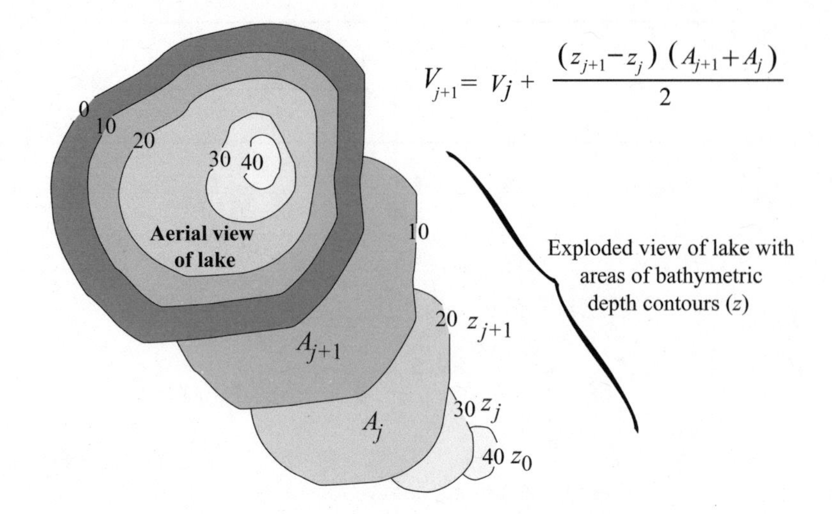

Figure 3.14. Engineering method for approximating lake volume (V).

This procedure consists of determining the volumes of successive layers of water and then summing the volumes to obtain the total volume of the lake. Thus, V_j is the calculated volume below the depth z_j.

More simply stated, consider the lake as an irregularly shaped cone divided into segments by different depth contours. The upper and lower surfaces of each segment are delineated by sequential depth contours. The volume of the lake is calculated by summing the "surface area" of each depth contour and applying the formula for the volume of a cone. Thus, if there are two depth contours, the volume is calculated as

$$V = \frac{h}{3}\left[A_1 + A_2 + \sqrt{A_1 A_2}\right] \tag{3.20}$$

where h is the difference between depth contours, A_1 is the area of the upper depth contour, and A_2 is the area of the lower contour. The last or bottom cone is calculated as

$$V = \frac{\pi}{3} r^2 h \tag{3.21}$$

where r is the radius of the top of the cone and h is the elevation or depth z (i.e., the difference between the last elevation contour (top of the cone) and the maximum depth (z_{max}).

3.5.2. Engineering Method

The method used by engineers for computing reservoir volumes is derived from an *end-area formula* sometimes applied to find the volume of prismoidal forms (Taube 2000). The average end-area formula or method is where the area at one section is added to the area of the next section, divided by 2, and then multiplied by the distance between them: $[(A_1+A_2)/2] \times (z_2 - z_1)$. The equation applied to lakes is

$$V_{j+1} = V_j + \frac{(z_{j+1} - z_j)(A_{j+1} + A_j)}{2}. \quad (3.22)$$

This a one-dimensional equation that gives good estimates of lake storage when the areas change by getting wider (x-axis) or by getting longer (y-axis), but not both. In lakes where both the x and y axes change, the engineering method tends to overestimate the volume.

In general, the shape that more closely approximates that of a lake is a frustrum. The conic method is generally more correct and efficient for determining lake storage and can give significantly different results than the engineering method. Here are two hypothetical examples demonstrating this:

Example 1. A square lake with a surface area of 100 × 100 [L²], a bottom area of 50 × 50 [L²], and a depth of 10 [L] (i.e., a frustrum shape).

Conic method: $V = 10/3 \times (10{,}000 + 2{,}500 + \sqrt{10{,}000 \times 2{,}500}) = 58{,}333$ [L³]
Engineering method: $V = 10/2 \times (10{,}000 + 2{,}500) = 62{,}500$ [L³]

Example 2. A long narrow lake with upper area of 1,000 × 10, a bottom area of 900 × 5, and a depth of 5 (i.e., a frustrum shape).

Conic method: $V = 5/3 \times (10{,}000 + 4{,}500 + \sqrt{10{,}000 \times 4{,}500}) = 35{,}347$ [L³]
Engineering method: $V = 5/2 \times (10{,}000 + 4{,}500) = 36{,}250$ [L³]

For both examples, the frustrum equation is the same as the conic method. The volume is overestimated in both examples with the engineering formula. The engineering method always overestimates volume for an ideal shape, whether square or round (e.g., greater change in x and y directions). In the elongated lake (more one-dimensional change along the x- or y-axis) or where

Table 3.2. Three methods for approximating lake volumes in Ogemaw County, Michigan

		Computed volumes (m^3)		
Lake name	Area (km^2)	Conic	Engineering	Mean depth
Frost	0.24281139	2,404,056	2,421,325	2,438,594
Robinson	0.08093713	78,942	77,709	71,541
Eagle	0.07810433	168,987	175,154	170,220

Source: Data modified (metric conversion) from Taube (2000).

the bathometric contours intervals are reduced (e.g., more precisely defined morphology with increased data), the degree of overestimation is reduced when using the engineering method (see Case Study 3.1). Thus, for the least effort (data collection) and best results, the conic method is preferred.

3.5.3 Mean Depth

A third method, utilized when there is no lake bathymetry map, is to determine mean lake depth and multiply it by lake area. Average depth is calculated by averaging depth sounding. The soundings should be spaced in a uniform grid pattern using regular intervals for reliable estimates and increased accuracy. Lack of soundings in very shallow water close to shore is a common source of error when applying this method (Taube 2000). When mean depth is compared with the conic method and engineering method, using data from three different lakes, Taube (2000) has found that they each gave similar results, with differences attributed to lake basin shape (Table 3.2).

3.5.4. Absolute Hypsographic Curves

Relative hypsographic curves are useful for comparing lake morphology but cannot be used to calculate actual areas or volumes of a specific lake below a specific depth unless the lake area or volume is known (Kalff 2002). Many limnologic parameters, such as primary productivity, oxygen and nutrient concentrations, and temperature stratification, are directly related to lake depth in association with its corresponding area and volume. A graph of the actual depth plotted against the cumulative area or cumulative volume is termed an *absolute hypsographic curve* (illustrated in Case Study 3.1). When plotting depth versus lake area, the absolute hypsographic curve can be used to approximate lake volume. Absolute hypsographic curves are useful for researching phytoplankton primary production measurements per unit volume [L^3] or area [L^2], seasonal changes in hypolimnetic concentrations of oxygen, total phosphorous and other chemical parameters, and investigating oxygen consumption rates (O_2 L^{-1} day^{-1}) in the hypolimnion (Kalff 2002).

In addition to volume calculations, hypsographic curves provide important information on area and volume distribution that is much easier to visualize and utilize and less cumbersome than bathymetric maps. They are used in best management practices for predicting the best times to implement various lake management strategies, such as aquatic plant management, habitat restoration, and muck removal. The curves can be used to predict lake water mixing and the ability to dilute incoming materials that in turn impacts lake biological productivity.

Hypsographic curves are also useful for comparing lakes and differences in lake behavior based on lake volume to surface area ratios. For example, a shallow lake may show a much more dramatic change of lake area versus a deeper one with a similar area during drought or flooding conditions even though the lakes are similar in area and location.

Case Study 3.1

City of Winters, Texas, Elm Creek Dam and Reservoir

Henningson, Durham, and Richardson Inc. 1979. *Preliminary Engineering Report, Winters Elm Creek Dam and Reservoir.* Winters, TX: Submitted to the Farmers Home Administration, US Department of Agriculture.

Using morphemetric depth and cumulative area data from Winters Elm Creek dam and reservoir in the city of Winters, Texas, Henningson, Durham, and Richardson Inc. (1979) created an absolute hypsographic curve by using the engineering method to determine reservoir volume (Figure 3.15).

The conic method was used on the deepest portions of the reservoir for comparison with the engineering method for the same interval. The conic method calculated a volume of 12,212,704 cubic meters versus 12,115,259 cubic meters using the engineering

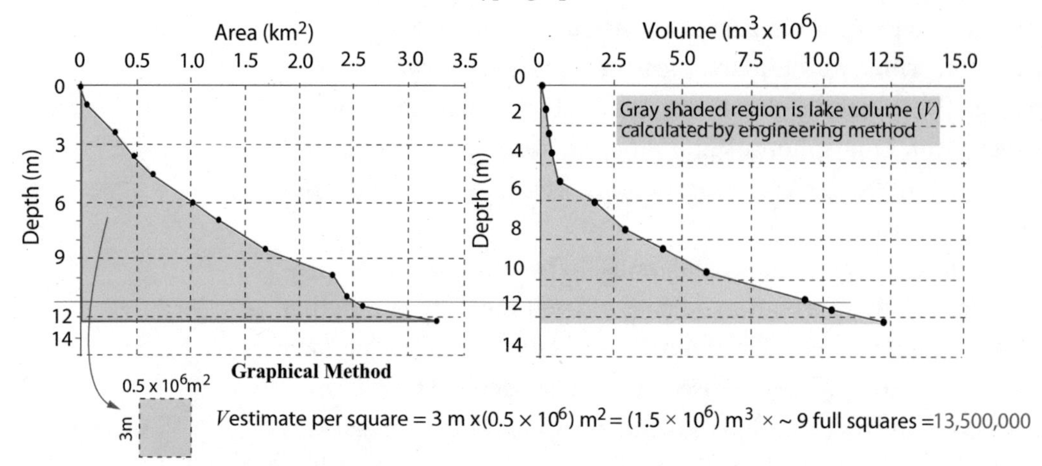

Volume (storage) calculations for Winters Elm Creek Reservoir

Area (km^2)	Depth (m)	Volume (m^3)
0	0	0
0.004	1.2	2,467
0.08	2.4	51,806
0.31	3.7	288,634
0.59	4.9	838,766
0.92	6.1	1,758,942
1.10	7.3	2,965,286
1.40	8.5	4,452,863
1.70	9.8	6,354,889
2.31	11.0	8,829,250
2.60	11.6	10,329,152
3.25	12.2	12,115,259

Volume ($V_{\Delta z}$) calculations using below methods for comparison:

Conic Method

$$V_{j+1} = V_j + \frac{(z_{j+1}-z_j)\,(A_{j+1}+A_j+\sqrt{A_{j+1}A_j}\,)}{3}$$

$$V_{j+1} = 10{,}329{,}162 + \frac{(12.2-11.6)\,(3.3+2.6+\sqrt{8.6}\,)}{3} = 12{,}212{,}704$$

Engineering Method

$$V_{j+1} = V_j + \frac{(z_{j+1}-z_j)\,(A_{j+1}+A_j)}{2}$$

$$V_{j+1} = 10{,}329{,}162 + \frac{(12.2-11.6)\,(3.3+2.6)}{2} = 12{,}115{,}259$$

Figure 3.15. Methods for determining lake storage (volume) using absolute hypsographic curves (graphical), conic, and engineering methods. Data on Winters Elm Creek Reservoir from Henningson, Durham, and Richardson Inc. (1979).

equation. These two results represent a less than 0.8% difference in storage capacity between the two methods that is probably a result of elongated lake morphology and the detailed contour data reducing the overestimation usually seen with the engineering method.

The mean depth method was not applicable in this study due to lack of sounding data, but a *graphical method* can be applied here by looking at the relationship between area and depth on the hypsometric graph (Florida LAKEWATCH 2001). By utilizing one square grid under the hypsographic curve, total lake volume can be approximated. For example, using the Winters Elm Creek hypsographic curve at the area of 0.5 km^2 (= 500,000 m^2) and a depth of 3 m, the volume per square was approximated as

Table 3.3. Comparison of volume approximations

Method	Computed volume (m^3)	Difference from mean volume (%)
Conic	12,212,704	3.15
Engineering	12,115,259	3.92
Graphical	13,500,000	7.06
Mean Volume	12,609,321	

1,500,000 m^3 per square. This product was then multiplied by the number of squares within the hypsographic curve. As a general method, the rectangles that are more than halfway inside this area are to be counted and those that are more than halfway outside the area should be not be counted. In the case of Elm Creek, nine can be included in the count, which results in a volume of or 13,500,000 m^3.

In conclusion, there is a difference of 10.4% (approximately 1.3 million cubic meters) when comparing the graphical method to the conic and engineering methods, and a 7% difference when comparing the graphical method with the overall average volume from the three methods (Table 3.3). Of the methods discussed, the conic is preferred followed by the engineering. When lake bathymetry data is lacking, an absolute hypsographic curve can be generated to approximate lake storage.

3.6. Summary

Regardless of origin, a lake's surface shape, surface area, varying shorelines, bathymetry, and depth all influence such lake behavior as sedimentation, resuspension, thermal and photic stratification, and development of a littoral zone for macrophyte growth and estimates of transpiration. Drainage basin morphology (e.g., valley types) often is the principle control of lake morphometry, with the size and slope of watersheds along with climate determining volume and rates of runoff, stream, and groundwater flow to the lake system. Knowing total lake volume allows for the determination of the epilimnion or hypolimnion layers and subsequent ecosystem parameters for fish spawning habitat and as well as dilution capacity and calculation of oxygen concentrations.

By investigating and monitoring the change in lake stages and volumes, impacts and estimates of runoff, stream, and groundwater flow and seepage can be used to calculate and model the various processes of mass balance to and from lake systems. These processes are particularly important when lakes are used as reservoirs. A reservoir may be used as a single purpose structure such as water supply or flood control, or it may be a multipurpose reservoir with zones of storage identified for different purposes, such as a combination of flood control, water supply, or hydroelectric power. Hence, knowledge of a lake's morphometric, hypsographic, and hydrological characteristics is essential for estimating storage capacity for current or projected consumptive use or as means of flood control. Location and elevation of dams and spillway control structures should be designed as a component of a thorough lake mass balance study done on the scale of the watershed impacting the lake system.

CHAPTER 4

Evapotranspiration

4.1 Introduction

The effects of *evaporation* and *transpiration* on a lake system or wetland water balance are regularly combined into a single estimate of water loss termed *evapotranspiration* (*ET*). In general, the rate of *ET* is proportional to the difference between vapor pressure at the water or leaf surface and the vapor pressure in the overlying atmosphere and can be described by *Dalton's law of partial pressures*:

$$ET = C(vpw - vpa) \tag{4.1}$$

where *ET* is the rate of evapotranspiration, *C* is a coefficient that includes meteorological factors such as wind speed and solar radiation, *vpw* is the vapor pressure at the wet or leaf surface, and *vpa* is the vapor pressure of the overlying atmosphere.

As there are almost always adequate amounts of moisture in a lake/wetland system, the meteorological factors affecting *ET* are similar. Meteorological factors that increase *ET* include solar radiation (increased temperature at the lake surface), a decrease in humidity, and increased wind speed.

In endorheic (closed) lake basins where there is no other significant outlet other than seepage, and in wetlands where the water table is often close to ground surface, *ET* may be the most significant factor in removing water from the system. Because of the significant influence of *ET* on lake mass balance, understanding the physical components of evaporation and transpiration is necessary to thoroughly understand lake hydrodynamics. In general, the majority of the approaches used for calculating *ET* predominately rely on direct empirical observations of evaporation using pans or basin structures pan evaporation (section 4.5.2).

These methods often do not adequately account for the influence of emergent aquatic plants and are a source of error when computing a lake mass balance.

4.2. Evaporation

Evaporation of water consists of the change of state from liquid to vapor and the net transfer of this vapor into the atmosphere. The process occurs when some molecules of the liquid attain sufficient kinetic energy to overcome the forces of surface tension and escape from the surface of the liquid (Dunne and Leopold 1978). To do this, energy from an outside source is required. This energy comes from solar radiation, sensible heat transfer from the atmosphere (e.g., air flow), or heat advected into the water body by inflowing warm water (Dunne and Leopold 1978) (Figure 4.1). Because of the dominant control of solar radiation, evaporation is primarily a function of latitude, season, time of day, and cloudiness.

To evaporate, water molecules must be located near the atmosphere/water surface, with motion in the appropriate direction, and have sufficient kinetic energy to surmount liquid-phase inter-

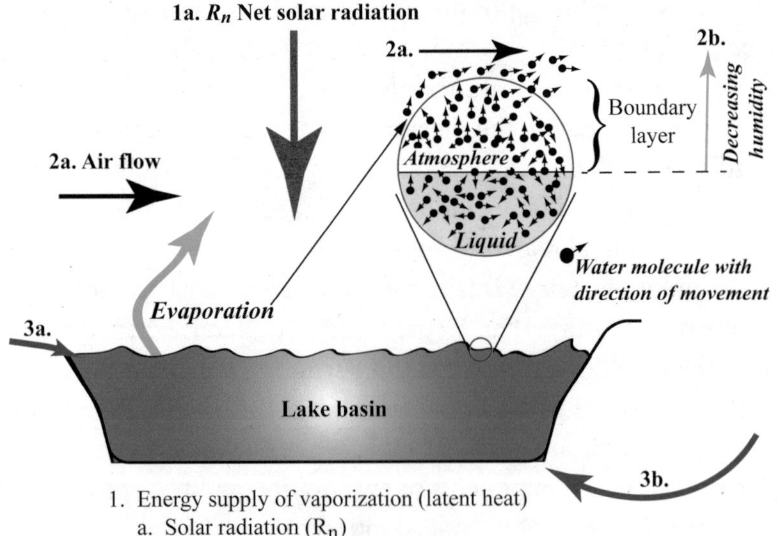

Figure 4.1. Processes and factors affecting lake evaporation.

molecular forces (e.g., hydrogen bonding). Since only a small proportion of the molecules meet these criteria, the rate of evaporation is limited.

Kinetic energy of a molecule is proportional to its temperature hence evaporation proceeds more rapidly at higher temperatures. As the more rapidly moving molecules escape, the remaining molecules have an overall lower kinetic energy and the temperature of the remaining liquid decreases, resulting in evaporative cooling until the liquid is reheated by the ambient temperature and direct sunlight.

If evaporation takes place in a closed vessel, the escaping molecules accumulate as a vapor above the liquid. As the density and pressure of the vapor increases, the number of molecules returning to the liquid increases. When the process of escape and return reaches equilibrium, the vapor is said to be saturated and no further change in either density or vapor pressure will take place (Giancoli 1988). In an open system such as a lake, this accumulation of molecules forms a boundary layer above the evaporative surface and when saturation is reached no further evaporation will occur. This equilibrium state is directly related to the vapor pressure of the substance, as stated by the *Clausius-Clapeyron equation*:

$$ln\left(\frac{P_2}{P_1}\right) = -\frac{\Delta H_{vap}}{R}\left(\frac{1}{T_2} - \frac{1}{T_1}\right) \tag{4.2}$$

where P_1 and P_2 are the vapor pressure at temperatures T_1 and T_2, respectively; ΔH_{vap} is the enthalpy of vaporization; and R is the universal gas constant. If the liquid is heated, it will come to a boil when the vapor pressure arrives at the ambient pressure. At lower pressures water will boil at lower temperatures and vice versa (Figure 4.2).

The ability of molecules to change from a liquid to a gas state is due to the kinetic energy of the individual particle. Kinetic energy (KE) is defined by the mass (m) and velocity (v) of the particle such that

$$KE = \frac{1}{2}mv^2. \tag{4.3}$$

The higher the kinetic energy, the easier it is to change states. Even at lower temperatures, a molecule of water can evaporate if

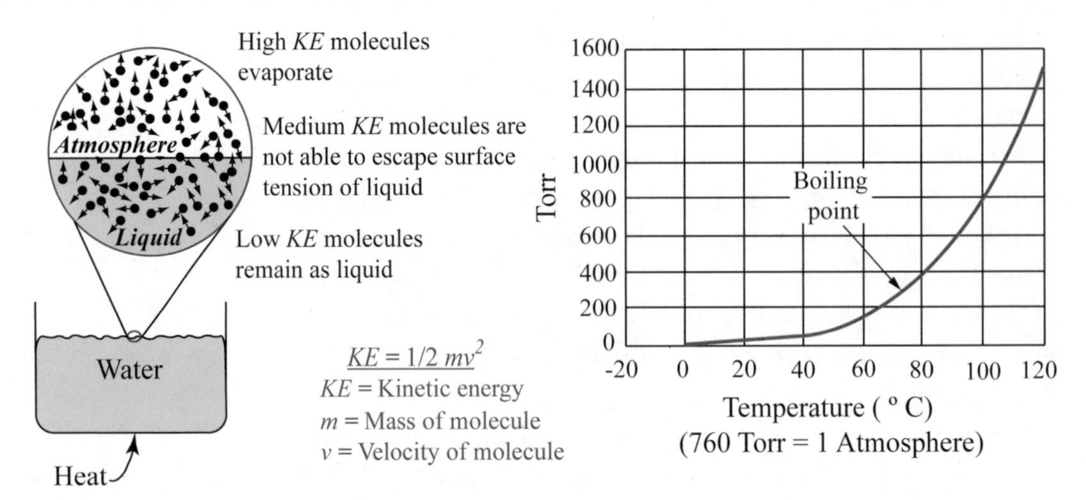

Figure 4.2. Vapor layer above evaporative surface as it approaches equilibrium and its linear relation with pressure and temperature.

it possesses the minimum amount of kinetic energy required for vaporization.

Evaporation is also influenced by relative humidity and the equilibrium gradient between the evaporating liquid and the atmosphere, wind speed immediately above the evaporative surface, and water availability. For example, the higher the humidity, the closer to equilibrium between the atmosphere and the evaporating surface, and thus the lower the evaporation rate. During evaporation, the water molecules form a vapor layer a few centimeters thick above the evaporative surface. When this layer becomes saturated (e.g., 100% humidity) then evaporation ceases. However, wind can remove the layer and replace it with drier air, thus increasing the equilibrium gradient and evaporation potential.

4.3. Transpiration

Transpiration is the evaporation of water from plants. Water is absorbed by the plant roots from the soil, transported up the plant stem, and evaporated (transpired) into the atmosphere from the leaves via tiny pores in the leaf called *stoma* (Figure 4.3). When water enters the roots and into the plant *xylem*, hydrogen bonding links each water molecule such that the molecules are pulled up the thin xylem vessels, analogous to a string of pearls. The water moves up the plant, enters the leaves, moves into air spaces in the leaf, and then evaporates through the stomata as water vapor.

Approximately 90%–99% of the water absorbed by roots is lost

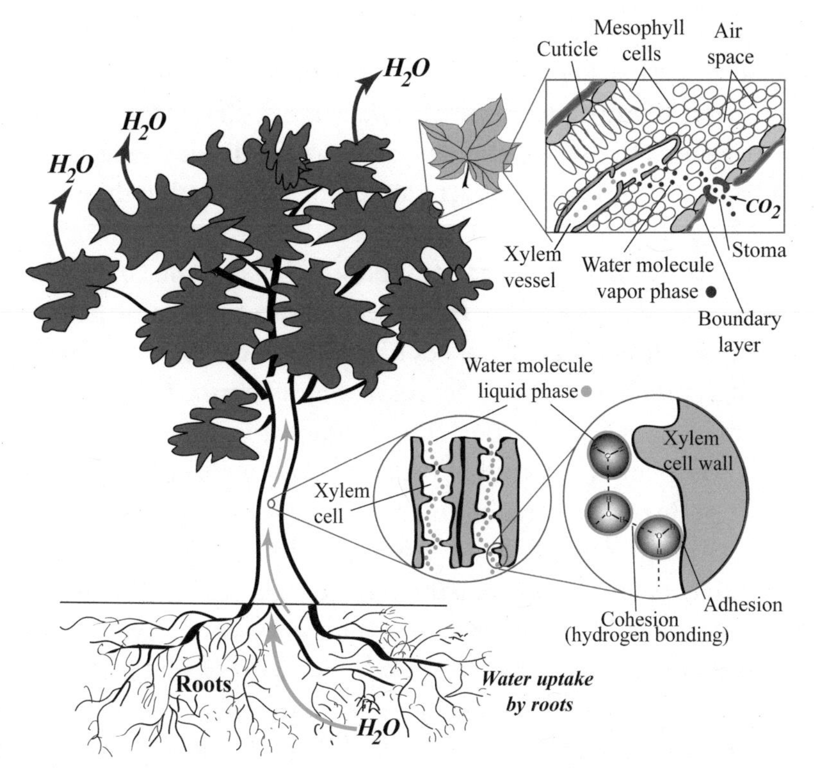

Figure 4.3. Transpiration and water movement through macrophytes (plants).

through the stomata in plant leaves (Nobel 1991). There are hundreds (up to 400 per mm^2) of stomata in the epidermis of the leaf. To conserve water, most stomata are located in the underside of the leaf away from direct solar radiation. The primary function of the stoma is to allow the intake of carbon dioxide (CO_2) for *photosynthesis*. This requires the stoma to allow the water to escape as vapor. Of all the water absorbed, less than 10% remains in the plant for growth (Taiz and Zeiger 2002). Therefore, plants require large amounts of water to maintain basic metabolic functions. For example, it is estimated that 2.59 square kilometers of corn (*Zea mays*) can transpire 12,000–15,000 liters of water per day. A large oak (*Quercus virginianus*) can transpire 151,000 liters per year (Nobel 1991).

Plants transpire for a variety of reasons. Evaporative cooling allows the plant to release heat during periods of high temperatures. The water that enters the roots contains dissolved nutrients vital for plant metabolism and growth. During plant transpiration, the stomata are open, allowing gas exchange between the atmosphere and leaf. Carbon dioxide is essential for photosynthesis.

The gradient between CO_2 uptake and water loss, however, is disproportionate as much more water is transpired then CO_2 is absorbed. This difference in CO_2 absorption and transpiration is primarily due to three reasons: (a) water molecules are smaller and move faster to the leaf stomata than carbon dioxide molecules; (b) approximately 0.036% of the atmosphere is CO_2, thus the availability is much less than that of water; and (c) carbon dioxide must travel to the plant *chloroplasts* for photosynthesis while the water molecules only move from the leaf cell surface to the atmosphere. Because of the large differences in water and carbon dioxide exchange, problems in water conservation can occur. The larger or the more numerous the stomatal openings are, the more CO_2 can enter the plant structure, but the more water escapes the system.

The rate of transpiration in plants is dependent on two major factors: the difference in the pressure gradient between the soil and the atmosphere surrounding the plant, and the resistance of water movement within the plant system. Loss of water due to evaporation from the leaf creates a negative pressure gradient or suction. Water moves from the higher gradient to the lower gradient, and the greater this potential the faster the rate of transpiration. For example, transpiration is greater during periods of low humidity and high soil saturation and more reduced when the humidity is higher or when the soil is unsaturated.

Resistance to transpiration occurs at the leaf-atmosphere interface (Figure 4.3). There are three main areas of resistance: the cuticle, stoma, and boundary layer. The cuticle is the waxy layer on the outer tissues of the plant and functions as a barrier of water movement out of a leaf. The thicker the cuticle layers, the slower the transpiration rate. In general, plants in hot, arid climates have a greater need for water conservation and have a thicker cuticle than plants from moist cool environments.

As already mentioned, the stomata are the pores in the leaf that allow for gas exchange where CO_2 enters and water exits. When stomata are open, transpiration increases; when closed, transpiration rates decrease. The stomata are the primary control mechanism for release and conservation of water as well as for the absorption of CO_2 needed for photosynthesis. Stomata are sensitive to a variety of environmental factors that trigger them to open or close. Two specialized cells called *guard cells* make up each stoma. The amount of water in the guard cells creates the

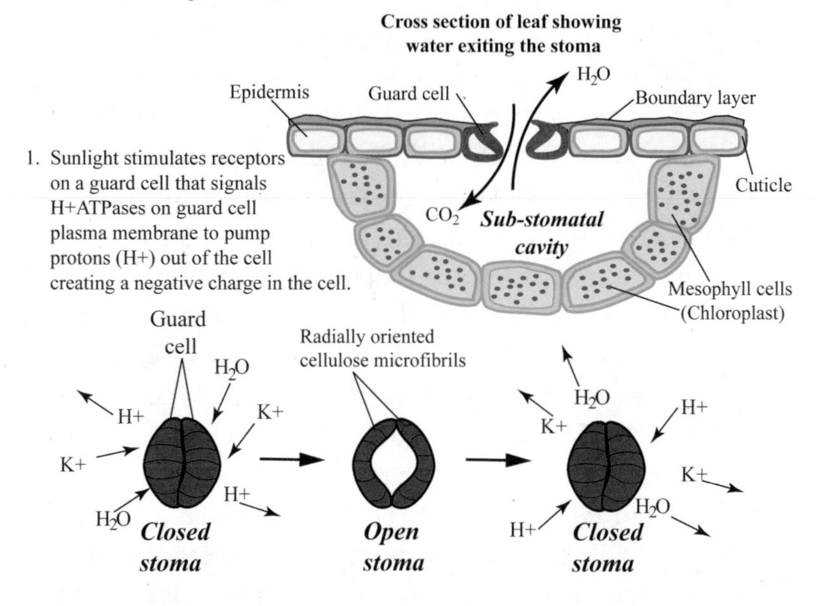

Figure 4.4. Function of stomata in transpiration and respiration.

pressure needed to allow the stoma to open (Figure 4.4). Thus, if the plant is lacking in water, the guard cells will not have enough water to maintain the proper pressure and the stoma will close and conserve water (Nobel 1991).

Environmental cues that affect stomata opening and closing are light, water, temperature, and CO_2 concentration. Stomata will open in the light (photosynthesis) and close in the dark. Stomata can close in the light when there is limited water, CO_2 concentrations in the leaf are greater than needed for photosynthesis, or the temperature is too hot, which causes increased loss in water due to increased evaporative cooling.

4.3.1 Xylem Transport

The *xylem* is the longest part of the pathway from the youngest roots to the newest leaves before water exits a plant. The issue of the processes by which water is raised through the xylem to the leaves (of considerable height at times) has been studied and disputed for years by botanists. The current explanation of this process is the *cohesion-tension theory* (Figure 4.5).

The cell walls contain cellulose and lignin, making them extremely rigid, and since xylem cells contain no membranes, they are open from top to bottom. These cells overlap each other as a series of connective pathways to the leaf. There is no single column of xylem cells transporting water (Taiz and Zeiger 2002).

Water movement from the roots via the xylem to the upper leaves is primarily a function of the negative pressure (suction) generated by the leaves during evaporation (Figure 4.5a). To maintain its water, a leaf cell must also exert a greater suction than evaporation by osmosis (see section 4.4.2). Suction coupled with the cohesive properties of hydrogen bonding between the water molecules and adhesive bonding between the xylem walls and the oxygen molecules can pull water up into the canopies of large trees. However, experiments have confirmed that if water is pumped up a tube by suction similar to that generated by leaf evaporation, it can only raise the water column to 9.75 meters before cohesion is lost and the water column breaks (*cavitation*) (Figure 4.5b). If the experiment is duplicated using a very fine tube, the water can be transported to vertical distances much greater than 9.75 meters (Figure 4.5c). This is due to the capillary action of the fine tube that increases the adhesive bonds between the water and the walls that help maintain the cohesive properties of the water column. This theory was proposed by Dixon and Joly (1894) and has been supported by Curtis and Clark (1950) and Levitt (1956).

When soil water is at *hydrostatic pressure* (h) greater than the atmosphere (*superatmospheric*), its pressure potential is considered positive. When the pressure is lower than atmospheric (*subatmospheric*) the pressure is considered negative, commonly known as *suction* or *tension*. Therefore, water under a free water surface is considered *positive pressure potential*, while the water at the surface is at zero pressure and water that has risen in the capillary tube is described as a *negative pressure potential*, often termed *capillary potential* (Figure 4.5d).

The height h ($-h_2$) of the liquid column in the *capillary tube* is given by Jurin's law (Giancoli 1988):

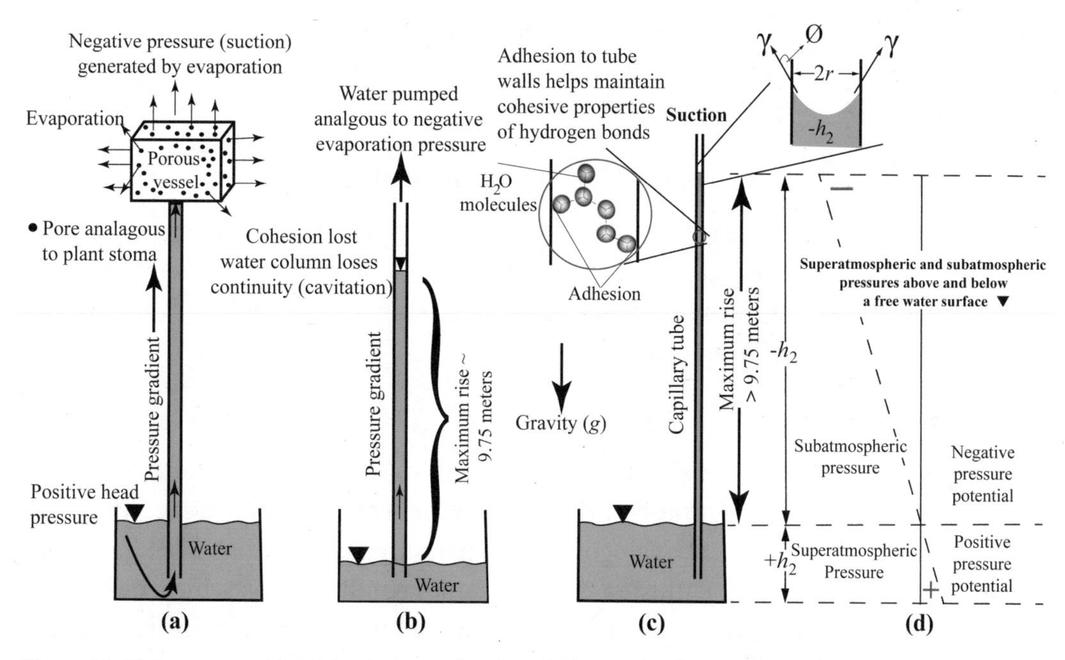

Figure 4.5. Water movement in a tube demonstrating the *cohesion-tension theory* of flow in the xylem and suction due to negative pressures potential due to capillary action.

$$h = \frac{2\Upsilon \cos \varnothing}{\rho g r} \quad (4.4)$$

where ϒ is the *liquid-air surface tension* (force/unit length), Ø is the contact angle, ρ is the density of the liquid, *g* is the acceleration due to *gravity* [Lt^2], and *r* is the tube radius. For water, the thinner the tube or smaller the radius, the higher it can move.

Xylem is made up of capillary tubes of this kind and the cell walls consist of cellulose and other polysaccharides that have an affinity for water increasing the adhesive properties. Xylem contains two types of water conducting cells: *tracheids* and *vessels*. Vessels are wider and longer (up to several meters) than tracheids and allow for rapid water transport. The larger widths, however, increase the probability of cavitation. When cavitation occurs, an air bubble can form, creating a temporary barrier to further transport. Tracheids are short, narrow cells that allow for increased upper transport but at a slower rate and reduced volume due to the restrictive nature of the lesser widths. But if cavitation occurs, the problem is restricted to one smaller tracheid cell. Vessels are an evolutionary product of the flowering dicot plants. Conifers and other gymnosperms only have tracheids and are less competitive in environments where water availability is limited (Taiz and Zeiger 2002).

4.4. Molecular Movement of Water

To complete the examination of hydrodynamics of evaporation and transpiration it is important to have a basic understanding of water movement at the molecular level. This mainly consists of *diffusion* and *osmosis*.

4.4.1. Diffusion

Diffusion is the movement of molecules from an area of high concentration to an area of lower concentration along a diffusion gradient analogous to the pressure gradient discussed in section 4.3.2. Diffusion continues until equilibrium is achieved. Rates of diffusion are affected by temperature and the density gradient of the molecules involved. In plants water diffuses out via the stomata into the atmosphere. In a lake system evaporation is diffusion of water molecules to the atmosphere. Lake diffusion also contributes to the dispersion of nutrients and solutes as well as potential pollutants.

Diffusion can be readily understood based on *kinetic theory* and the random motion of molecules. Consider a vessel of cross-sectional area A containing a uniform concentration of one type of molecule, referred to as *background molecules*, with a second molecule introduced at an initial higher concentration—for example, the placement of colored dye into a container of water (Figure 4.6). The background molecules (in this example, the water) are not shown to focus on the diffusion of the dye molecules. Since we are observing the effects of diffusion only, the pressure and temperature gradients are uniform, thus eliminating hydrodynamic and convection flow, respectively. For simplicity, it is also assumed that the concentration gradient of the dye molecule changes primarily only along the x-axis and not along the y or z axes.

Even though the molecules are in random motion, there will be a net flow of dye molecules to the right because of the difference in concentration. To perceive why this is, consider the small section in vessel of length Δx. Molecules from both regions I and II will cross into this central area as a result of their random motion.

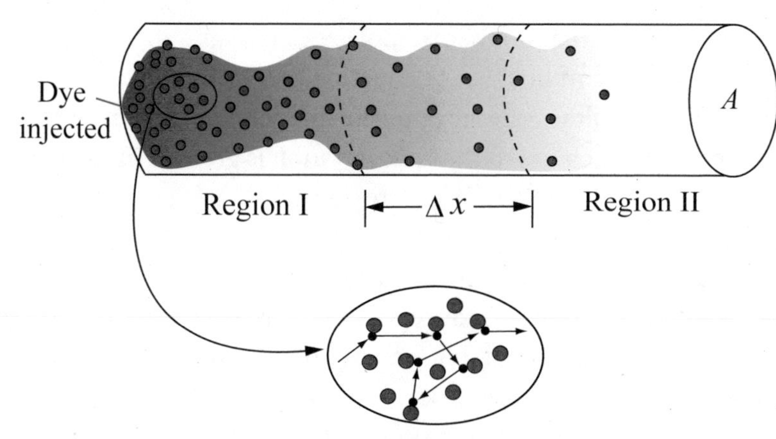

Figure 4.6. Diffusion and the movement of molecules from regions of high concentration to low concentration as it applies to Fick's law. Modified from Giancoli (1988).

Table 4.1. Diffusion constants, D (20°C, 1 atmosphere)

Diffusing molecules	Medium	D (m^2/s)
H_2	Air	6.3 {times} 10^{-5}
O_2	Air	1.8 {times} 10^{-5}
O_2	Water	100 {times} 10^{-11}

Source: Modified from Giancoli (1988).

Since there are more molecules in region I than in region II, more molecules will cross into the central section of the vessel from region I, and eventually from the central area to region II. Hence there is a net flow of dye molecules from left to right, from high concentration to low concentration. The flow stops only when the concentrations become equal.

The greater the difference in concentration, the greater the flow rate. This is known as *Fick's law*. In 1855, physiologist Adolf Fick determined experimentally that the flow rate across unit area J is directly proportional to the change in concentration per unit distance $(n_{v2} - n_{v1})/\Delta x$. This is called the *concentration gradient* and in terms of derivatives

$$J = -D\frac{dn_v}{d_x} \tag{4.5}$$

where D is the *constant of proportionality* or *diffusion constant* with representative values given in Table 4.1. If x is given in meters and the concentration n is the number of molecules per cubic meter, then J is the number of molecules crossing the unit area per second. The minus sign in equation (4.5) is by convention to remind us that flow is in the opposite direction of the concentration gradient such that if the concentration is greatest in region I of Figure 4.6 then flow is toward region II.

The statistical distribution of molecules in space is time-dependent and is given by the diffusion equation (4.5). The theory of diffusion is tied to the concept of mass transfer and driven by a concentration gradient, but because of the random motion of molecules, diffusion can still occur when there is no concentration gradient (though there will be no net flux). In three-dimensional space the diffusion equation can be derived in a straightforward way from the *continuity equation*, which states that a change in density in any part of the system is due to inflow and outflow of material into and out of that part of the system. Effectively, no material is created or destroyed:

$$\frac{\partial n_v}{\partial t} + \nabla \times J = 0 \tag{4.6}$$

Where n_v is the concentration of molecules (number per unit volume); $\nabla$is the vector operator representing diffusion in the x, y, and z directions; and J is the flux of the diffusing material, which in the example of Figure 4.6 are the dye molecules.

Example: Estimating the mass of water vapor that passes through a stomatal pore

In the cross-sectional view of the leaf in Figure 4.3, water passes from the liquid phase to the vapor phase at the walls of the mesophyll cells. The water vapor then diffuses through the intercellular spaces and eventually exits the leaf through the stomatal pores. The diffusion constant D for water vapor in air is $D = 2.4 \times 10^{-5}$ m^2/s. An average stomatal pore has a cross-sectional area of approximately $A = 8.0 \times 10^{-11}$ m^2 and a thickness (dx) of 2.5×10^{-5} m. The concentration of water vapor on the interior side of the pore is about $C_2 =$ 0.022 kg/m^3 versus the outside atmospheric concentration of 0.011 kg/m^3 (Cutnell and Johnson 2004). With these data the mass of water vapor that passes through a stomatal pore in one hour can be determined by modifying equation (4.5) to

$$\text{mass} = \frac{(DAdC)t}{dx}$$

or

$$\text{mass} = \frac{\left(2.4\times10^{-5}\,\text{m}^2\right)\left(8.0\times10^{-11}\,\text{m}^2\right)\left(\frac{0.022\,\text{kg}}{\text{m}^3} - \frac{0.011\,\text{kg}}{\text{m}^3}\right)}{2.5\times10^{-5}\,\text{m}}$$

$$= 3.0\times10^{-9}\,\text{kg}.$$

A single leaf may have millions or more stomatal pores, hence transpiration from a single plant can be considerable.

4.4.2. Osmosis

Osmosis is the net movement of water across a selectively permeable membrane driven by a difference in solute concentrations on the two sides of the membrane. In its most simple definition, a selectively permeable membrane is one that allows unrestricted passage of water but not solute molecules or ions (Figure 4.7).

Osmosis in plant cells is the diffusion of water and preferential solute molecules through a semipermeable membrane or differentially permeable membrane, such as the stomatal pores on a leaf. It is movement of water across a partially permeable membrane from a region of higher solute concentration to a region of lower solute concentration.

The application of pressure can prevent osmosis from occurring. Osmotic potential is the minimum pressure required to prevent fluid from moving as a result of osmosis. Fluid will enter the cell via osmosis until the osmotic potential is balanced by the cell wall's resistance to expansion. Any water gained by osmosis may help keep a plant cell rigid or turgid. The turgor pressure develops against the cell walls as a result of water entering the cell's vacuole. This pressure is referred to as the *pressure potential* (Nobel 1991). The osmotic potential and pressure potential combined make up the water potential of a plant cell. If there are two cells next to each other of different water potentials, water will move from the cell with the higher water potential to the cell with the lower water potential. Water enters plant cells from the environment via osmosis. Water moves because the overall water potential in the soil is higher than the water potential in the roots and plant parts. If the soil is desiccated, then there will be no net movement into the plant cells and the plant will die.

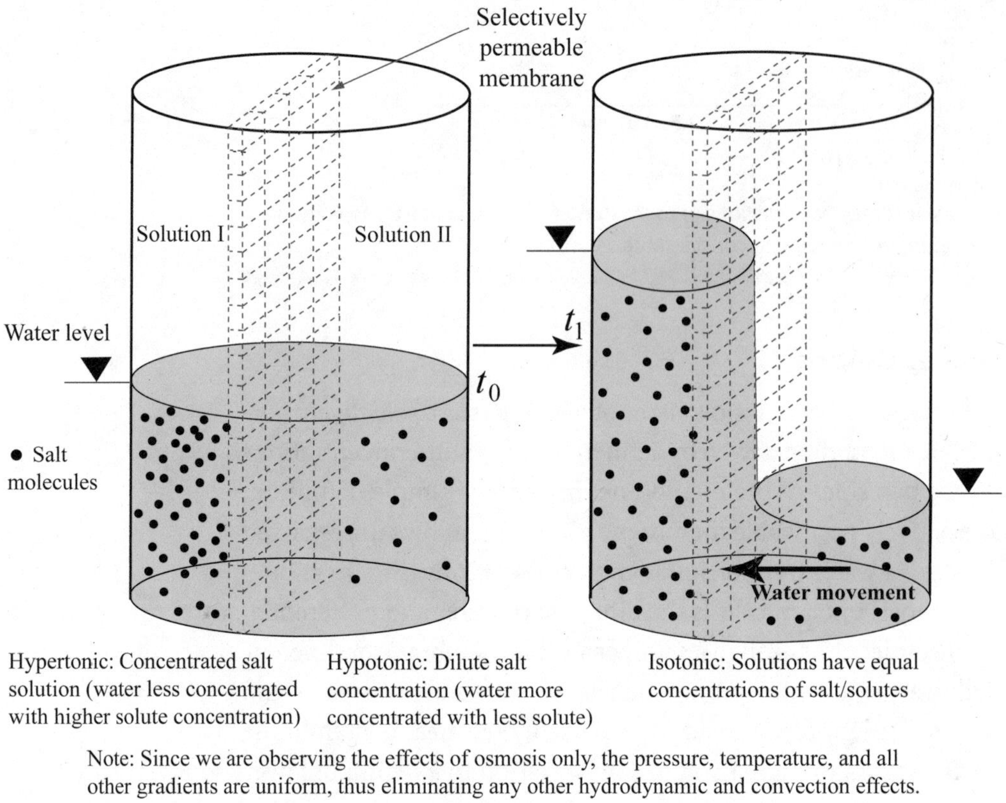
Selectively permeable membrane
Solution I
Solution II
Water level
Salt molecules
t_0
t_1
Water movement
Hypertonic: Concentrated salt solution (water less concentrated with higher solute concentration)
Hypotonic: Dilute salt concentration (water more concentrated with less solute)
Isotonic: Solutions have equal concentrations of salt/solutes
Note: Since we are observing the effects of osmosis only, the pressure, temperature, and all other gradients are uniform, thus eliminating any other hydrodynamic and convection effects.

Figure 4.7. Osmosis is the movement of water molecules from a hypotonic to hypertonic solution, resulting in isotonic equilibrium.

4.5. Estimates of Evapotranspiration

Evapotranspiration (*ET*) and *potential evapotranspiration* (*PET*), which is pertinent to lake systems, can be estimated by a number of empirical and theoretical methods using equations that incorporate such meteorological variables as solar radiation, humidity, and wind speed as well as such hydrogeological variables as precipitation and seepage in and out of a lake. *PET* is defined as the amount of water that would be evapotranspired if there were always enough water available for plants to transpire at the maximum rate possible (e.g., no soil water deficiency). Methods of estimating *ET* and *PET* fall into four main categories, and some approaches use a combination of these:

1. *Water budget method.* This approach estimates *ET* by estimating overall lake storage from surface and subsurface

inflow as well as direct precipitation, seepage, and stream outflow where *ET* will be the difference in lake storage over some time interval (see Case Study 4.1).

2. *Direct empirical observations of evaporation using pans or basin structures. ET* is estimated as the fluctuation in water level as measured with precise water-level gauges. These structures incorporate other meteorological measurements such as wind speed and air temperature. Pan methodology varies in size and depth of the pan. Pan placement can be above ground, sunk below ground, or floated on the lake or reservoir of interest.
3. *Mass transfer method.* This approach calculates *ET* by assuming that *ET* is controlled by wind speed and the difference in vapor pressure between the water surface and atmosphere; it incorporates either the water budget or the energy budget methods.
4. *Energy balance method.* To convert one gram of liquid water to vapor at normal lake temperatures (596 calories/gram at 0°C, 580 calories/gram at 27°C and 540 calories/gram at 100°C) requires approximately 590 calories of heat energy (Dunne and Leopold 1978). This approach utilizes estimates of incoming solar radiation received by the lake that can be used for evaporation.

4.5.1. Water Budget Method

The water budget method uses the basic mass balance approach in a lake system for estimating evaporation over time given by

$$ET = \text{Inflow} - \text{Outflow} - \Delta\ \text{Storage}. \qquad (4.7)$$

Inflow parameters include direct precipitation, indirect surface and subsurface inflow from the catchment basin from the precipitation event, and stream flow and potential groundwater flow from underlying aquifers. Outflow consists of stream flow out and seepage to underlying aquifers (see Figure 11.3). Equation (11.10) from chapter 11 can be modified to include stream (or ponor) flow (Q) and rearranged algebraically to expand statement (4.7) such that

$$ETA = PA + Q_{in} - Q_{out} - \left[\frac{K_A A}{L} + \frac{K_p a}{L}\right] h + \frac{K_A h_a A}{L} - K_A A + \frac{K_p a h_a}{L} - K_p a - A\frac{\partial h}{\partial t}. \tag{11.10}$$

Accurate measurements for each of these variables require detailed instrumentation and field methodology for monitor wells and water-level gauges, as well as weather parameters. Detailed knowledge of the hydrogeology of the system is needed to determine how all of these components interact. Seasonal variations must be included to determine how they affect lake water bodies, such as heat storage. For example, during summer months in a deep lake, heat may be stored to a depth that would otherwise be used for evapotranspiration, resulting in lower estimates based on solar radiation. In the winter, when the lake water is warmer than the surrounding atmosphere, the heat difference may result in higher *ET* than expected. Because of complexity of the system as well as budget and time constraints, the water budget method is not feasible for routine measurement of *ET*.

Case Study 4.1

Lake Jackson Seepage Model

Evans, W.L. 1996. "Modeling the Hydrodynamics of Closed Lake Basins: A Treatise on Lake Seepage." Master's thesis, Department of Geology, Florida State University.

Evans (1996) used a seepage model developed to estimate potential evapotranspiration for Lake Jackson in Florida. Seepage derived from the model was subtracted from the daily change in lake level or stage where

$$PET = \frac{dh}{dt} - S. \tag{4.8}$$

When compared to 1983 pan evapotranspiration data, the modeled *PET* is approximately two times greater (Figure 4.8). This indicates that the pan method significantly underestimates lake *PET* and probably does not adequately take into account the con-

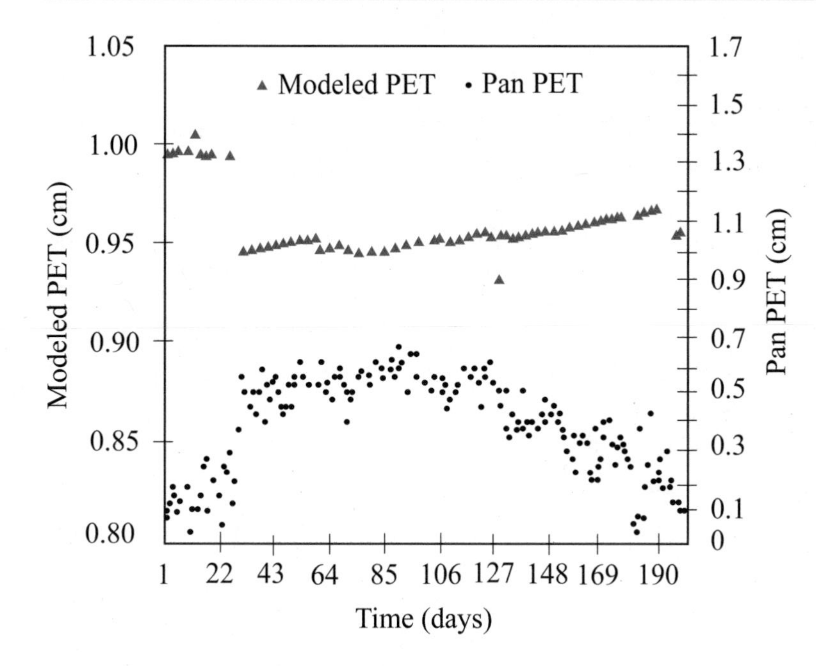

Figure 4.8. Comparison of modeled potential evapotranspiration (PET) with 1983 pan PET estimates at Lake Jackson, Florida. Modified from Evans (1996).

siderable effects of transpiration. Seasonal variation in the modeled *PET* is not observed because of the resolution of the model (e.g., cm versus mm). In addition, because the model incorporated noncontinuous declining lake-level data, the seasonal variations may have been filtered.

Between 1987 and 1993 there has been approximately 5.0 square kilometers of emergent hydrilla (*Hydrilla verticillata*) and 0.05 square kilometers of water hyacinth (*Eichhornia crassipes*) in Lake Jackson (Van Dyke 1994). This represents over 30% of the total surface area of Lake Jackson. In addition to hydrilla and water hyacinth, pickerelweed (*Pontederia cordata*), water lily (*Castalia sp.*), American lotus (*Nelumbo lutea*), spatterdock (*Nuphar luteum*), arrowhead (*Sagittaria sp.*), and other aquatic plants are abundant at Lake Jackson. The quantitative effect of transpiration at Lake Jackson is unknown, but it is probable that it has a significant impact on the lake hydrodynamics. Thus, the potential evapotranspiration derived from the seepage model is more consistent with the vegetative effects and more representative of the Lake Jackson system.

4.5.2. Direct Empirical Observations Using Pan Measurements

Because of the difficulty in measuring *ET* from a mass balance approach, pan or tank methodology offers a relatively inexpensive, mobile, and easily managed system that has numerous advantages for estimating open water evaporation at specific locations. These basic estimates can then be used for preliminary studies or be incorporated into other methods or equations to more accurately estimate *ET* for a specific lake or reservoir system.

The relationship between pan evaporation (*PE*) and *PET* can be expressed as

$$PET = k_2 PE \tag{4.9}$$

where

PET = daily potential evapotranspiration in mm/day $[Lt^{-1}]$
k_2 = pan coefficient (usually taken as 0.7 for Florida conditions but variable throughout the year); and
PE = pan evaporation from a US Weather Service Class A pan where E is the change in water level of the pan over time (dh/dt) (Figure 4.9a).

However, in contrast to lake systems, pan systems receive large amounts of solar energy and conduction through the pan structure, hence evaporation from different pans will vary significantly from the same lake under the same meteorological conditions. In addition, differences in wind, heat storage within lake waters, temperature variations, and seasonality also contribute to inaccuracies. To accommodate for these deviations, empirical coefficients (Table 4.2) that vary seasonally and geographically must be applied to the measurements. In the United States, average evaporation data and empirical pan coefficients have been determined over a wide range of areas and seasons and are available from the National Weather Service as well as from other agencies and the literature.

A great number of evaporation devices of various types have been used in experimental studies of evaporation or evapotranspiration. There is hardly any need to review them since most of the devices are no longer in use at present since they fail to provide

Table 4.2. Examples of pan coefficients for class A pans

Location	Mean pan coefficient	Seasonal range
Auburn, Alabama	0.81	0.72–0.90
Lake Elsinora, California	0.77	0.63–0.97
Salton Sea, California	0.50	0.31–0.83
Fort Collins, Colorado	0.70	0.60–0.82
Lake Okeechobee, Florida	0.81	0.69–0.91
Lake Hefner, Oklahoma	0.69	0.35–1.32
Pretty Lake, Indiana	0.70	0.50–0.90

Source: Data from Dunne and Leopold (1978) and Boyd (1985).

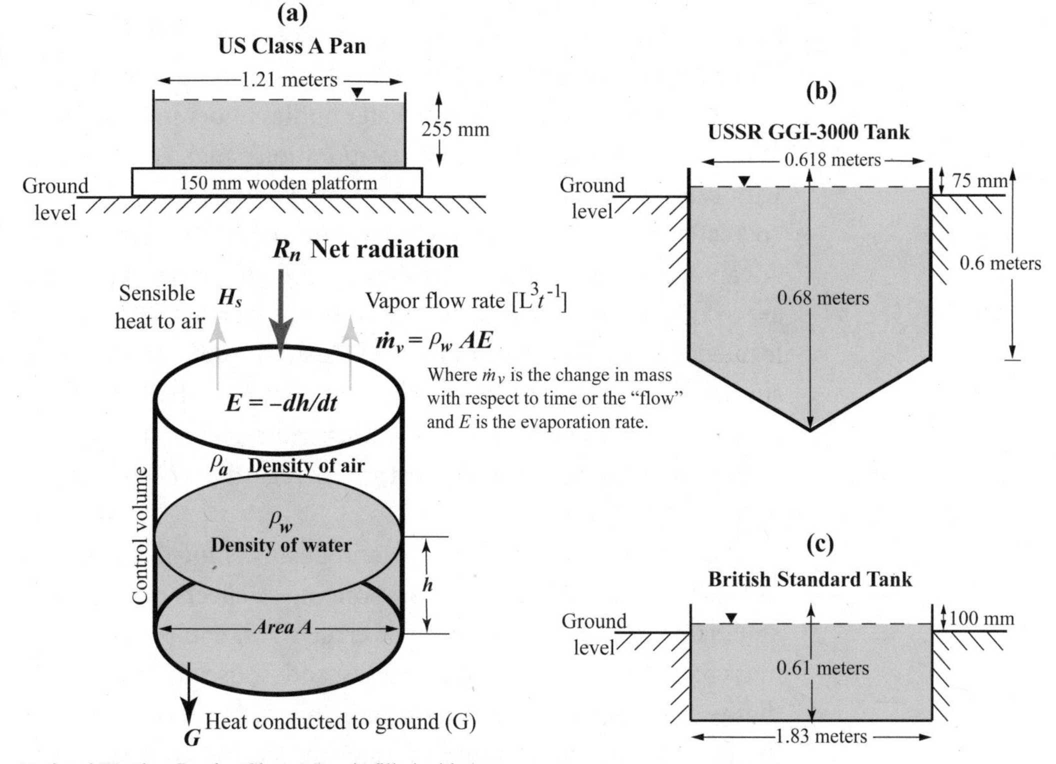

Figure 4.9. Basic pan evaporation process and three common types of pan evaporation devices: (a) Class A pan, (b) GGI-3000 tank, (c) standard concrete basin.

accurate results. The three that remain widely in use are shown in Figure 4.9. The pans of each type are installed either on the shore of a water body or on a special raft. The Class A pan is mounted on a wooden frame on the shore, a little above ground, whereas the GGI-3000 pan is lowered into the water or into the soil in such a manner that the water surface in the pan is level with the soil surface and 75 mm of the upper part of the pan remains above water or ground. The standard concrete basin tank system is sunk into the soil.

Evaporation data estimated from the three empirical pan-type techniques using the local meteorological, geophysical, and actinometrical observations for measuring solar radiation are commonly known and have been standardized. Review of the data for comparison has demonstrated that the determination of reduction coefficients of the three types of installations does not appear to be as essential as site- or area-specific knowledge of those coefficients for each type of installation in obtaining accurate evaporation values from surfaces of the water bodies under investigation. Winter (1981) states, "Data from evaporation pans should be used with caution, even when pans are located adjacent to the lake, if corrections are not made for temperature and wind effects. A disadvantage of using pans is that they need daily service if the data are to have any meaning; data from National Weather Service pans located a long distance from a lake have little value for balance studies." Lack of statistical correlation between spatially different *PET* data sites in north Florida further indicates the importance of proximity for accurate *PET* estimates regardless of the system of interest (i.e., a lake).

The use of experimental pans and basins is for net evaporation without a transpiration component, so effects of emergent aquatic plants are also a source of error when computing a lake mass balance via pans. Benton, James, and Rouse (1978) indicate that water hyacinth, a floating aquatic plant, can transpire more than three times the amount of water than is lost by open water evaporation. Shih (1980) states, "The water budget computation in shallow lakes is complicated because marsh vegetation can transpire large quantities of lake water." Aquatic plants of all types, if they cover a significant portion of the lake, add complications to estimates of evaporation.

4.5.3. Mass Transfer Method

This approach assumes that evaporation (E) is a function of wind speed and the difference in vapor pressure between the water surface and atmosphere such that

$$E = Nf(u)(e_s - e_a) \tag{4.10}$$

where

N = a constant known as the *mass transfer constant*;
$f(u)$ = a function of wind speed;
e_s = vapor pressure of water surface; and
e_a = vapor pressure of air.

Evaporation is determined by incorporating either the water budget or the energy budget methods. E is then plotted against the product of $u_2(e_s - e_a)$ where the subscript 2 is indicative of a two-meter observation height. The slope of this plot results in the mass transfer coefficient N (Figure 4.10).

4.5.4. Energy Balance Method with the Effects of Transpiration and Mass Transfer

As seen from the Lake Jackson data, evapotranspiration has a major influence on lake-level fluctuations. To include the effects of evapotranspiration on the system, the *Penman method* (Pen-

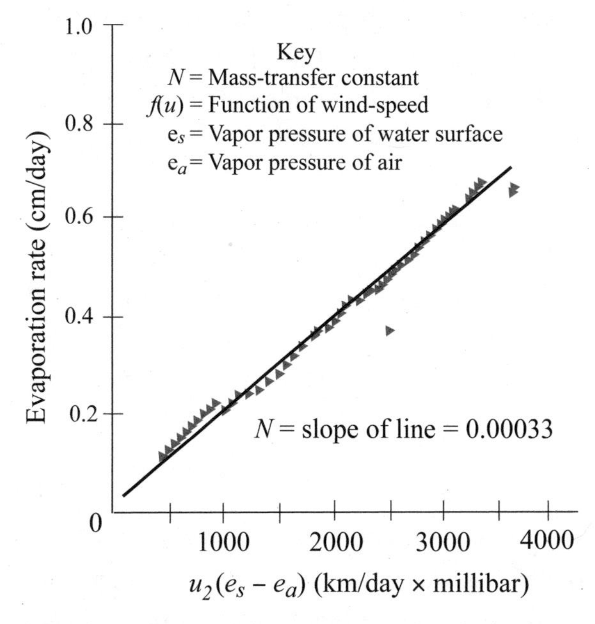

Figure 4.10. Theoretical plot of measured evaporation rate versus $u_2(e_s - e_a)$ to determine the mass transfer coefficient (N). Modified from Dunne and Leopold (1978).

man 1948) as expressed by Allen (1986) can be used. The Penman method is based on four major climatic factors: net radiation, wind speed, air temperature, and vapor pressure deficit (relative humidity).

Penman's method is one of the better energy balance methods for estimating evapotranspiration (Ward 1971) and is used by Florida's Water Management Districts, the US Environmental Protection Agency, the US Department of Natural Resources, and the National Weather Service. Penman's method consists of estimating the energy received by a surface and determining how this energy will be divided between evaporation, heating of the surface, and heating of the air. Unlike other models, Penman's model has a theoretical basis, and the prediction of potential transpiration is based on the factors that cause it; thus, the trends it indicates will have more significance (Takhar and Rudge 1970).

Allen (1986) expressed the Penman equation as

$$PET = \frac{\frac{\Delta Rn}{\lambda} + \gamma E_a}{\Delta + \gamma} \tag{4.11}$$

where

PET = daily potential evapotranspiration in mm/day [Lt^{-1}];
Δ = slope of saturated vapor pressure curve (2.56 mb/°C);
Rn = net radiation, based on incoming radiation, net outgoing thermal or longwave radiation, and surface coefficients (see section 3.3.2a);
λ = *latent heat of vaporization* of water (58.4 calories/cm/mm);
γ = *psychrometric constant* (0.67 mb/°C for elevation 99.0 m NGVD); and
Ea = vapor pressure deficient (based on corresponding temperatures).

4.6. Summary

When describing and managing lake and wetland systems where there is no other significant outlet other than seepage, and in wetlands where the water table is often close to ground surface, evapotranspiration may be the most significant influence in removing water from the system. Meteorological elements, which vary seasonally and spatially in response to regional and local climate, are the most important factors when describing the rate of evapotranspiration. Hence, all formulas used for estimating *ET* are dominated by such meteorological factors as energy input, net radiation (and the ability of the atmosphere to transport water vapor from leaf and lake surfaces via vapor pressure), air temperature, and wind velocity.

Many of the methods used for calculating *ET* do not adequately account for the influence of emergent aquatic plants and are a source of error when computing a lake mass balance. Research (e.g., Benton, James, and Rouse 1978; Shih 1980) has shown that aquatic plants that cover significant portions of a lake can result in water loss up to three times greater than the amount of water lost to open water evaporation. Evans (1996) demonstrated that *PET* derived from the seepage model is more consistent with the vegetative effects and more representative of the lake system under study (Case Study 4.1). Therefore, a combination of seepage modeling and available *ET* methodology in conjunction with the amount and type of vegetation cover influencing a lake system will result in better understanding of lake system hydrodynamics and management.

CHAPTER 5

Rainfall and Surface Flow to Lakes

5.1. Introduction

Surface water is the water system stored on the earth's surface that continuously interacts with the atmospheric and subsurface water systems as part of the hydrologic cycle. When describing these interactions in terms of lake and wetland mass balance, drainage basins are the key component of analysis for water resource management because the concept of water passing through a stream cross section at the basin outlet originates as upstream precipitation (as discussed in section 3.1).

The relationship between rain- or snowfall (as snowmelt) and water transference in a watershed is at the very core of wetland hydrology and mass balance. The total amount and distribution of a precipitation event include its numerous flow paths and the *residence time* within the entire drainage basin as it flows to the lake. Estimates of rainfall amounts to streams and runoff, along with the hydrological character of the catchment, are required to accurately define mass balance interactions to wetlands and lakes within the catchment.

In many areas of the world (e.g., high to mid latitudes where precipitation is marginal), seasonal melting of snow produces critical water resources to lakes, reservoirs, and rivers, and infiltrating meltwater recharges soil moisture and groundwater. Most of the flow processes discussed here are directly applicable to snowmelt. There are many models for forecasting snowmelt runoff. Generally, these incorporate either an energy balance or temperature-index method to compute melt (Maidment 1993). Because of the basic nature of this text, the focus is on precipitation as rainfall. For an excellent discussion on snow and floating ice hydrology, please see Gray and Prowse (1993).

5.2 Precipitation

Precipitation is the major factor controlling the hydrologic cycle of a region and subsequent drainage basins. The ecology, geography, and land use of a region are all dependent upon the function of the hydrologic cycle. Precipitation amounts, temporal and areal distribution, flow paths, and residence time of water in the catchment delineates the mass balance of surface and groundwater discharge, as well as water quality, erosion, and sedimentation impact to rivers, streams, lakes, and other wetland systems.

When rainfall occurs, for example, the water is deposited on many different surfaces and travels by a large number of conveyances (Figure 5.1). It flows into local wetland systems such as lakes and streams, is intercepted by vegetation, or localizes into *depression storage* (e.g., puddles or ponding). Ultimately it moves back into the atmosphere via evaporation and transpiration or to the local groundwater table through *infiltration*. The mechanisms of water movement within the hydrologic cycle are discussed in section 2.2 and shown schematically in Figures 2.2 and 2.3. These

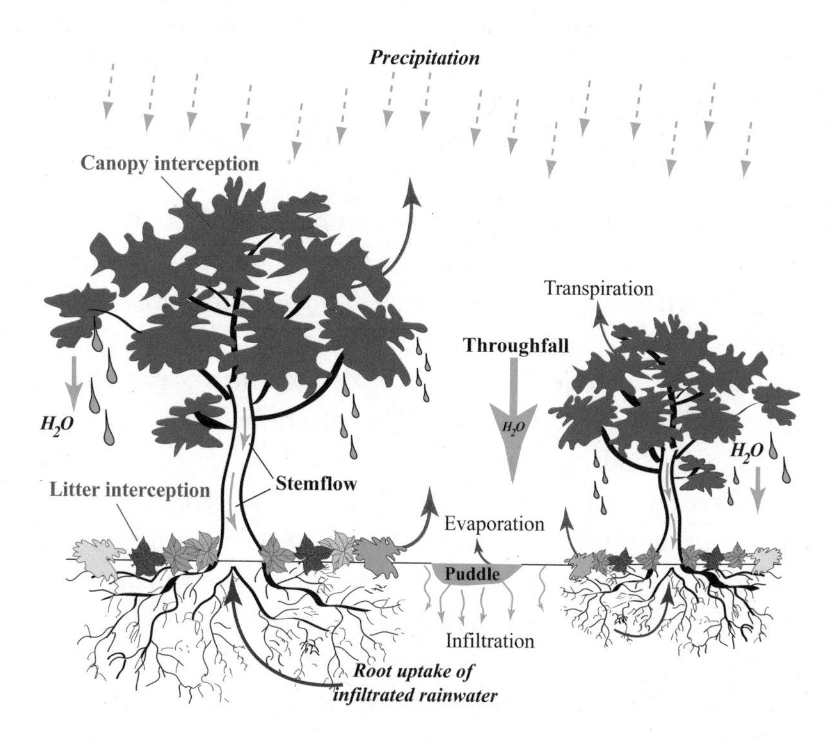

Figure 5.1. Illustration of interception and infiltration. Water reaches the surface by throughfall and stemflow. Water stored in the canopy, leaf litter, or depressions can return to the atmosphere via evaporation.

flow paths depend upon climate, topography, soil, geology, vegetation, and land use. And not only are these processes different for different geographic regions, but they can also vary in the same watershed or catchment basin, depending upon the areal, temporal, and intensity factors of the rainfall event.

Interception is when precipitation is caught by vegetation before it reaches the ground, making it unavailable for flow into wetlands or groundwater. Vegetation, including the canopy and leaf litter, stores the water temporarily before it is subject to evaporation or transpiration back to the atmosphere. Dependent upon the ecosystem flora—such as conifer versus deciduous forest, prairie grassland, farmland, tropical rainforest, or desert—*interception* can range between 8% to 35% of total annual precipitation (Dunne and Leopold 1978). For example, conifer needles can store more intercepted water and generally have more foliage throughout the year than deciduous trees, thus conifer forests appear to have more interception than deciduous forests.

After a period of time during a rainstorm event, intercepted water from gross precipitation (measured in rain gauges) reaches the holding capacity of the vegetation and becomes insignificant. The water reaches the surface either directly or due to a dripping canopy, known as *throughfall*, or down the trunks of trees or underbrush as *stemflow*. Root uptake and subsequent transpiration also reduces the amount of water conveyed to the surficial aquifer or wetlands.

Likens (2013) reported that approximately 82% of precipitation to a forested drainage basin reaches the ground as throughfall. The remaining 18% is evaporated or absorbed directly from the surfaces of vegetation. Once the holding capacity is exceeded, interception has little effect upon the development of runoff and flooding (Dunne and Leopold 1978), which can flow directly to streams, wetlands, and lakes.

Infiltration usually percolates vertically to the water table if the vadose or unsaturated zone is uniformly permeable. The water stored in the surficial aquifer system is known as *baseflow* and is in constant movement from a high to low hydraulic gradient, often discharging to a lake or stream where the water table intersects the surface. Generally, the amount of discharge is directly proportional to the slope of the drainage basin and hydraulic gradient to the lake or wetland. Drainage basins with lower elevations and slopes, thick sandy soils, and permeable to semipermeable underlying geology will have a greater capacity to store groundwater than upland catchments with steeper slopes, shallow soils, and more impermeable geology. The water table will rise as the infiltrated rainwater reaches the surficial aquifer system, which can result in increased discharge to nearby wetlands. Thus, streams, lakes, and wetlands in lowland areas are more influenced by groundwater as opposed to upland wetlands, which tend to be dominated by surface and subsurface runoff.

5.2.1. Mechanisms for Rainfall

Water is carried throughout the atmosphere as clouds and vapor. The amount of water vapor that the atmosphere can carry is a function of temperature (discussed in section 3.2). Having water vapor in the atmosphere is a necessary but insufficient condition for precipitation. There must also be a process to cause uplift cooling and condensation. Three mechanisms for rainfall generation that meet these criteria are convective, cyclonic (or frontal), and orographic (Figure 5.2).

Condensation is the change of water vapor into a liquid. For condensation to occur, the air must be at or near saturation in the presence of condensation nuclei. Condensation nuclei are small particles or aerosol upon which water vapor attaches to initiate condensation, such as dust, smoke, salt, sulfur, or nitrogen oxide. Droplets that coalesce around these nuclei grow as a result of additional condensation, freezing, or collision of other water droplets. The droplets must be of adequate size and gravitational velocity to exceed the cloud updraft force and to not evaporate before reaching the ground. For snowfall, solid ice crystals form by diffusion or collision to form snowflakes when temperatures are close to freezing. In terms of meteorology and the water cycle, precipitation can be defined as the deposition of water in liquid form and as ice particles formed in the atmosphere.

Precipitation varies spatially and temporally according to the general pattern of atmospheric circulation, pressure, and a variety of meteorological factors and local influences. For example, higher rainfall amounts occur near coastal areas since the oceans provide the bulk of atmospheric moisture needed for precipitation.

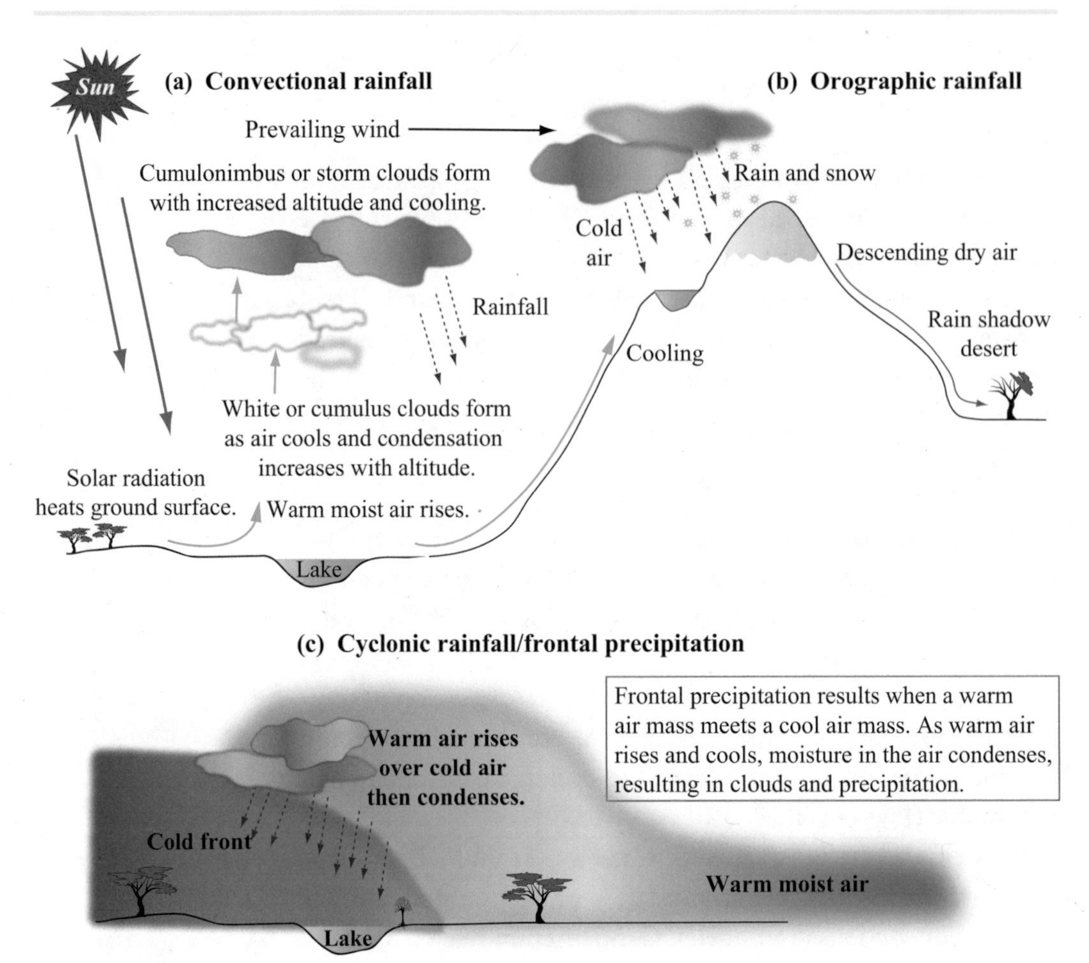

Figure 5.2. Three common mechanisms for rainfall generation.

As maritime air mass travels over mountains, the rising air cools and precipitates on windward mountainsides and is known as orographic rainfall. Corresponding areas leeward of the mountains have less moisture and thus less rainfall and can produce a *rain shadow*, resulting in *xeric* or desert conditions. Orographic influences can modify all types of rainfall systems, and accurate measurements of rainfall can be problematic in mountainous regions (Smith 1979).

5.3. Measuring Precipitation

5.3.1. Rainfall Gauges

Rainfall gauges provide data on rainfall depths observed at a point location over different periods of time. A variety of rain gauges and rainfall recorders and methodology are available that measure rainfall very accurately and are discussed in detail in most of the hydrological textbooks listed in the references at the back of this book (see especially Ball et al. 2016). Rain gauges are of two types, recording and nonrecording. Recording gauges measure automatically in a chronological resolution down to one minute or less and often with digital systems transmitting in real-time for water management applications.

Rainfall gauges are subject to some measurement errors, but they typically provide relatively accurate data. Since they measure rainfall over a small area, the areal distribution and placement of rain gauges as well as the methods used in rainfall calculation are critical when determining estimates of input into a lake system and corresponding drainage basin, watershed, or catchment area. For example, the patterns and intensity from convective (windy) thunderstorms common to coastal areas can be extremely variable and thus subject to considerable error depending on the number and placement of rain gauges. In areas where seasonal rainfall events are relatively uniform in precipitation and wind velocity, accurate rainfall measurement can be estimated with little variation.

Due to modification of the storm's *wind field* and wind-generated turbulence near the top of essentially all rain gauges, errors in measurement occur. Other errors occur from rain splashing out of or into the gauges as well as evaporation prior to measurement. Measurements during periods of snow, high winds, or low-intensity rainfall events can result in underrecording. Tipping-bucket rainfall gauges can be subject to errors due to tipping when partially full or failing to tip. Errors in manual recording or electronic transmission of data from telemetered gauges are also possible.

Errors in measuring precipitation are generally of the order of several percent for a single storm and range up to 30% for poorly distributed gauges in a large windy storm event (Dunne and Leopold 1978). Linsley, Johler, and Paulhus (1975) examined data from a network of rainfall gauges each utilizing a different method for estimating rain volume over a drainage basin that resulted in measurements that varied between 9-18%, depending on the method used to calculate the volume. The most severe errors occur at high latitudes and over water and sometimes exceed 80% of the true value (Aguado and Burt 2001).

Siegel and Winter (1980) compared rainfall input into Williams Lake, Minnesota, using recording-gauge data from three different sources. These were gauges at the lake, a network of gauges in closer proximity to the lake, and a larger regionalized network of rain gauges further from the lake. They found that rainfall could vary as much as 25 mm between the larger network and the small-scale network within 5 km of the lake. As expected, these data show that the closer the rainfall devices are to the lake system the more accurate the measurements.

Uncertainty around precipitation estimates is especially relevant to large lake systems. For example, the surface area of the Great Lakes range is 23%–39% of their respective drainage basins. *Over-lake precipitation* is a major component of the water balance in the Great Lakes and is greater than all the water that enters the lakes through runoff (Croley, Hunter, and Martin 2001). However, most of the Great Lake precipitation estimates are primarily from land-based gauge measurements (Coordinating Committee on Great Lakes Basic Hydraulic and Hydrologic Data 1977).

Because of their large areal extent, the Great Lakes can impact local weather conditions, resulting in variability of rainfall between the lakes and adjacent land mass (Eichenlaub 1979).

The potential uncertainty of land-based measurements and lack of direct measurements of over-lake precipitation can lead to an error rate likely to be more than 15% and as high as 45% during winter months (Neff and Nicholas 2005). The higher uncertainty in the winter and fall is mainly due to a significant difference in temperature at the boundary of the warmer lake water and colder land mass. For example, cold air masses frequently move over the area during the fall and winter. As the cold air travels across the lakes, the warm water heats the air, causing an increase in lake evaporation and subsequent moisture transfer to the overlying air mass. As the air mass moves inland, the cooler land surface causes condensation, which precipitates as *lake-effect snow*. Thus, downwind land gauges can result in significant overestimation of over-lake rainfall, whereas gauges on the upwind locations would be comparable to the over-lake rainfall. For a discussion of other meteorological phenomenon affected by the Great Lakes area see Eichenlaub (1979).

Lake Superior (the northernmost and largest of the Laurentian

Great Lakes) accounts for 39% of its total basin surface area and derives over one-third of its water budget from rainfall directly onto the lake surface. Holman et al. (2012) compared Regional Climate Model output (Version 4, ICTP RegCM4; Giorgi et al. 2012) to historical precipitation estimates from the National Oceanic and Atmospheric Agency's Great Lakes Environmental Research Laboratory (GLERL) hydrologic database for the period 1980 to 2000. Holman et al. simulated average overland and over-lake precipitation (929 mm and 870 mm, respectively) and found both were higher than the corresponding land-gauged values from a Thiessen polygon estimate (829 and 805 mm, respectively). The average monthly precipitation values for overland values from the RegCM4 are higher than the Thiessen polygon calculations.

There is no accepted method for determining the space-time pattern of rainfall that is not influenced by the resolution and accuracy of networks of rainfall observations. Because of logistical and financial reasons, weather stations are mostly located on land. For smaller lakes a network of nearby rain gauges provides an accurate estimate of precipitation. Larger lakes necessitate the use of a number of gauges distributed around the lake system for accurate determination of rainfall. For very large lakes that can locally impact weather conditions, lake gauges over the water body are needed to get correct estimates of rainfall.

Rainfall data for many water budgets is commonly determined from national network data. In the United States, the spacing of National Weather Service gauges is approximately 650 square kilometers per gauge (Winter 1995). These gauges are rarely located in the immediate vicinity of the lake under investigation, thus this method of estimating precipitation is subject to error.

5.3.2. Spatial Interpolation: Methods of Estimating Areal or Spatial Patterns of Rainfall with Deterministic and Computer-Generated Models

Storm rainfall depth is defined as the quantity of rain falling within a storm of a specific duration distributed uniformly over a watershed area. Rainfall depth [L] can be conveyed in millimeters, when using SI units, or inches, when using the English measurement system. *Rainfall distribution* is defined as the quantity of rain falling in successive time increments of a total storm duration. The cumulative fraction of precipitation at consecutive times up to the storm duration is used to develop rainfall distribution. Hence the rainfall distribution begins at a value of zero at the beginning of the storm and ends at a value of 1.0 (US Department of Agriculture 2015).

Although a rain gauge measures one point within the area of the storm, the overall area impacted by the storm can be estimated by using various methods of interpolation of rainfall data within a rainfall gauge network. These interpolation estimates can be used to produce a space-time pattern for estimating precipitation in nongauged areas.

One method is where the storm event over a watershed is represented by isohyetal maps. An isohyet is a contour of constant rainfall that has been interpolated from various rain gauges in the area of the storm event (see section 5.3.2b for details). These types of rainfall data are important when investigating the impacts of storm and stormwater discharge to lake systems. Increases of storage to lake systems from storm events are a result of direct precipitation to the lake, associated surface flow, and increases in the water table. Estimation of rainfall over the entire catchment area is essential to determine rainfall inputs to a lake system.

Despite the various errors discussed earlier, rainfall gauges typically provide relatively accurate measurements of rainfall over time recorded at a point location. The primary uncertainty when working with a network of rainfall gauges is in accurately observing the space-time pattern of rainfall across an area because they cannot observe variations in rainfall patterns between the gauges (Seed and Austin 1990; Barnston 1991; Bradley et al. 1997).

Spatial interpolation as applied to rainfall data is a procedure or method for estimating rainfall at nongauged locations throughout an area based on the available observations. Spatial interpolation is based on the *first law of geography* (Tobler 1970), which is the assumption that points closer together in space are more likely to have similar values than more distant points.

For the human species, the neurotypical perception of space consists of four dimensions: three spatial and one temporal. In addition, spatial correlations describing geological systems often exhibit skewed directional behavior in three-dimensional space (*anisotropy*) or lack *stationarity*, the condition in which the data do not exhibit a trend. A rainfall event is a prime example of these concepts.

The conventional approach for approximating the space-time pattern of rainfall from gauge networks has been to estimate the spatial pattern of rainfall for the whole rainfall event and then to disaggregate the rainfall accumulation for each part of the spatial domain, often a model subarea or subcatchment, using the temporal pattern observed at a rainfall gauge (Ball et al. 2016).

Ideally, the spatial pattern should be constructed using rainfall totals from daily rainfall gauges and, where available, from continuous rainfall gauges. Gauges should be obtained from both within the catchment or study area and for a region around the catchment. As an indicative value, the region used for constructing the spatial pattern should extend to include gauges that are within at least 10 km of the catchment or study area boundary or further if internal catchment gauges are further from the boundary (Ball et al. 2016).

Computer-generated spatial interpolation techniques generally involve interpolation between rain gauge data onto a geographic grid or projected Cartesian coordinate system. The grid resolution should be sufficiently fine to observe the spatial variability in the precipitation field at a meaningful scale for the drainage basin. Verworn and Haberlandt (2011) recommend resolutions of 1 km for a projected grid or 0.01 degree or finer for a geographic grid.

There are many potential approaches that have been developed for spatial interpolation of rainfall estimation. These tend to be characterized as either *deterministic* (or traditional/conventional) and *stochastic* (or *geostatistical*) (Eberly et al. 2004). Stochastic methods incorporate the idea of randomness into the interpolation process. These methods allow for the uncertainty of the predicted values to be calculated. Deterministic methods do not incorporate statistical probability theory into the prediction models.

Deterministic methods use mathematical formulas or other relationships to interpolate values. Common "traditional" or "conventional" approaches include the *nearest* (or *natural*) *neighbor*, *arithmetic mean*, *isohyetal maps*, *weighted mean* (e.g., Thiessen polygon), and *inverse distance weighting*. Methods range from assigning the value of a measured point to neighboring unmeasured locations (nearest-neighbor approach) to estimating ungauged values by use of a weighted average of nearby points or locations (inverse distance weighting). The latter uses distance as the only factor influencing the calculation such that the closer the measured location, the more significant the weight.

There is no preferred technique for constructing a spatial pattern of rainfall for an event. Both computer-generated interpolations and hand-drawn rainfall contours from rainfall totals at the gauges remain valid approaches that will produce acceptable results for many rainfall events. The more traditional deterministic approaches are presented here, along with one example (i.e., *kriging*) of the possible variation in data presentation and values. Basic geostatistical modeling is discussed in section 11.3.2a.

5.3.2a. Arithmetic Mean

The *arithmetic mean* method is the simplest technique for determining areal average rainfall ($\overline{P}$), the average depth recorded at several rainfall gauges distributed over the area under study, where

$$\overline{P} = \frac{\sum_{i=1}^{n} P_i}{N}. \tag{5.1}$$

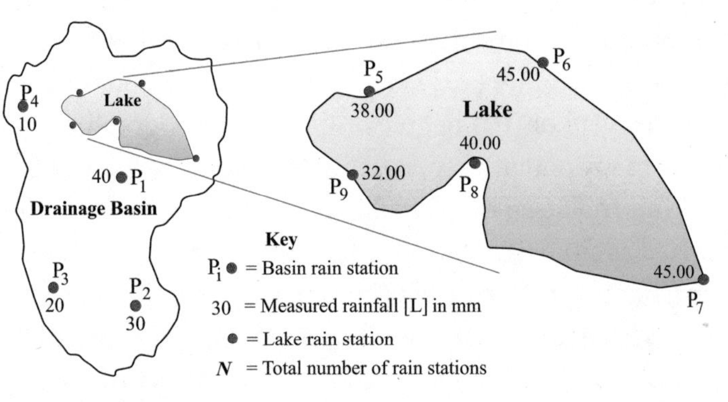

Figure 5.3. Calculation of average areal rainfall by the arithmetic mean method.

Drainage Basin

Station	Measured rainfall (mm)
P_1	40
P_2	30
P_3	20
P_4	10
Total =	100

$\overline{P}_{DB} = {}^{100}/_{4} = 25.0$ mm*

*Drainage basin (DB) without lake rain data.

$$\overline{P} = \frac{\sum_{i=1}^{n} P_i}{N}$$

Mean drainage basin precipitation (25 mm) with lake gauge mean (40 mm) = 32.5 mm

Lake

Station	Measured rainfall (mm)
P_5	38.00
P_6	45.00
P_7	45.00
P_8	40.00
P_9	32.00
Total =	200.00

$\overline{P}_{Lake} = {}^{200}/_{5} = 40.00$ mm

This method is satisfactory if the gauges are uniformly distributed over the area and the individual measurements do not vary greatly about the mean (Figure 5.3).

5.3.2b. Isohyetal Maps

The gauges in catchment basins are rarely distributed in an areal network sufficiently uniform to allow an accurate estimate of rainfall distribution using a numerical average of gauge rainfall measurements. Isohyetal maps are often used with rain gauge networks of any spatial configuration to estimate area averages for studies of rainfall distribution

The isohyetal map is a contour map of constant rainfall quantity (isohyets) representing the spatial variability of rainstorm events over an area. The isohyetal method is one of the more accurate methods but is dependent on knowledge of rainfall characteristics of the drainage basin / catchment area. The isohyetal method is flexible and can resolve some of the difficulties encountered in the weighted average methods by constructing isohyets using the observed rainfall depths and interpolations between adjacent gauges (Figure 5.4).

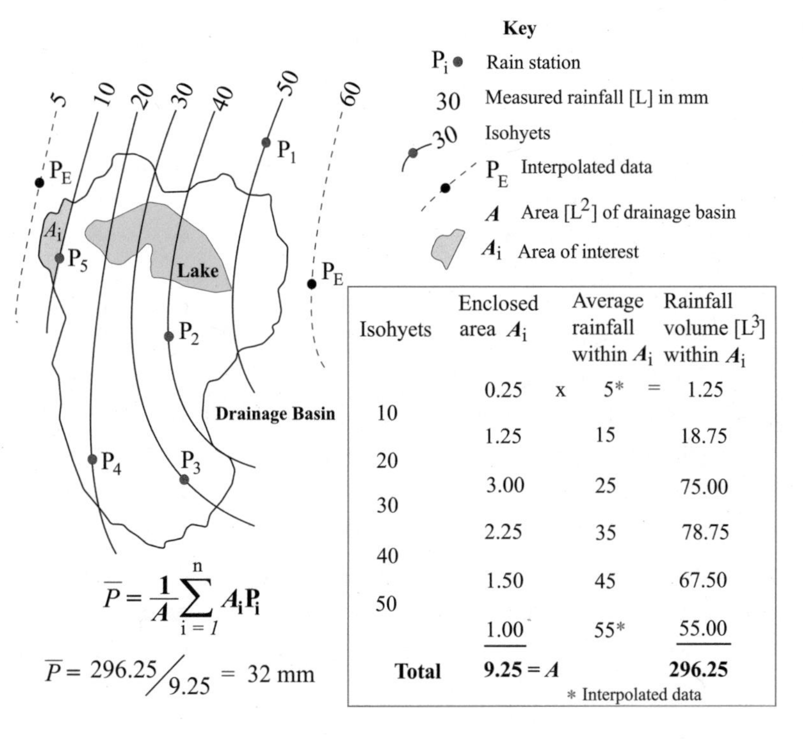

Isohyets	Enclosed area A_i		Average rainfall within A_i		Rainfall volume [L^3] within A_i
	0.25	x	5*	=	1.25
10					
	1.25		15		18.75
20					
	3.00		25		75.00
30					
	2.25		35		78.75
40					
	1.50		45		67.50
50					
	1.00		55*		55.00
Total	**9.25 = A**				**296.25**

* Interpolated data

Figure 5.4. Calculation of average precipitation over a hypothetical drainage basin by the isohyetal method.

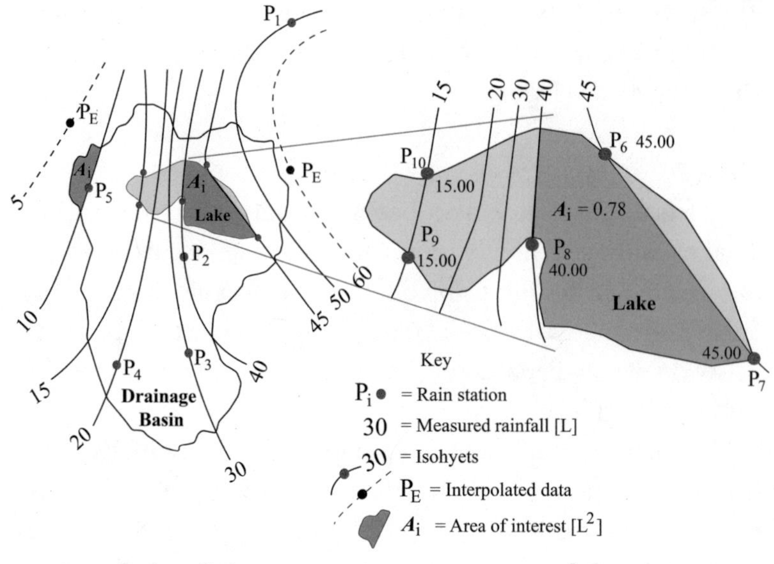

Figure 5.5. Isohyetal calculation of average precipitation over a hypothetical watershed with lake rain gauge data to illustrate the difference between watershed rain average with average precipitation to lake when gauges adjacent to the lake are used.

Drainage Basin

Isohyets	Enclosed area (A_i)	Average rainfall	Rainfall volume
	0.25	5.0*	1.25
10			
	1.83	12.5	22.88
15			
	1.08	17.5	18.9
20			
	1.93	25.0	48.25
30			
	1.03	35.0	36.05
40			
	1.31	42.5	55.68
45			
	0.94	47.5	44.65
50			
	0.88	55.0*	48.40
Total	**9.25 = A_{DB}**		**276.06**

* Interpolated

$$\overline{P}_{DB} = 276.06/9.25 = 29.84 \text{ mm}$$

Lake

Isohyets	Enclosed area (A_i)	Average rainfall	Rainfall volume
	0.05	12.5 *	0.63
15			
	0.17	17.5	2.98
20			
	0.13	25.0	3.25
30			
	0.08	35.0	2.80
40			
	0.78	42.5	33.15
45			
	0.09	47.5*	4.28
Total	**1.30 = A_{Lake}**		**47.09**

* Interpolated

$$\overline{P}_{Lake} = 47.09/1.30 = 36.22 \text{ mm}$$

Where there is a larger network of rain gauges, isohyetal contours can be constructed using computer programs designed to create contour maps (Figure 5.5). Once the map has been constructed, the area A_i between the isohyets within the basin is measured and multiplied by the average P_i of the rainfall depth between the two isohyets (Figures 5.4 and 5.5) to calculate the rain volume per A_i. The average areal precipitation is determined using equation (5.2). To accurately estimate rainfall during a complex storm event, a relatively dense network of rain gauges in conjunction with computer contouring software is required to properly construct an isohyetal map. Note the difference in the two isohyetal maps and rainfall calculation in Figures 5.4 and 5.5 when introducing a lake system with increased rain gauges into the catchment basin.

As demonstrated, precipitation will fall directly onto the lakes and reservoirs during a rainfall event and the average can be significantly different than the overall watershed area. Therefore, when measuring lake precipitation, it is important to place rain gauges as near the lake as possible, if not within the water body.

Case Study 5.1

Hastings, Nebraska, Isohyetal Map

US Department of Agriculture, 2005, *National Soil Survey Handbook*, Title 430-VI (Washington, DC: USDA), http://soils.usda.gov/technical/handbook.

Figure 5.6 is an example of the basic three-step process for creating an isohyetal map used to estimate the average rainfall for a drainage basin in a small research watershed in Hastings, Nebraska, by the US Department of Agriculture (2015). Intially this basin had four permanent rain gauges associated with it, including two within the basin, one on the boundary, and a fourth just outside the basin (Figure 5.6a). After estimating the rainfall amounts between gauges using interpolation (Figure 5.6b), the isohyetal map contours can be created (Figure 5.6c). However, due to the limited number of rainfall gauges, defining the orientation of the contours as well as the overall rainfall average amounts for the watershed can be subjective.

One method for increasing an areal precipitation network coverage is to place temporary rain gauges in the area of interest by employing a *bucket survey* (Figure 5.6d). A bucket survey is a field survey sometimes used in anticipation of an unusually large storm by various federal and state agencies to collect additional rainfall data. As the name suggests, rainfall during a storm event is collected and measured in impermeable narrow-bore tubes such as a small portable home rain gauge, hardware-store bucket, bottle,

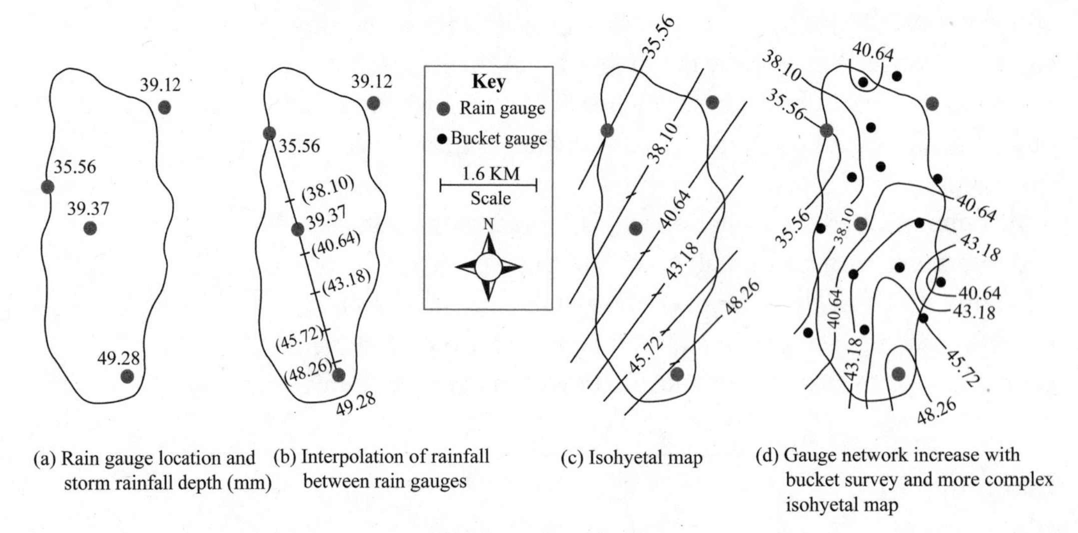

(a) Rain gauge location and storm rainfall depth (mm) (b) Interpolation of rainfall between rain gauges (c) Isohyetal map (d) Gauge network increase with bucket survey and more complex isohyetal map

Figure 5.6. Construction and application of isohyetal maps for determining average rainfall within a Nebraska watershed during a storm event. Modified from US Department of Agriculture (2015).

or similar container. If the location and rainfall volume data are carefully evaluated in conjuction with the existing gauges, these surveys can significantly increase the detail and accuracy of the isohyetal maps and average watershed rainfall estimates in areas with sparse rain gauge coverage. The fourth step is determining the overall average rainfall for the watershed by weighting the rainfall volume based on the percent area between each isohyetal (Tables 5.1 and 5.2). In this example, the area between the adjacent isohyet is measured manually by using a *dot counter*. A dot counter is a transparant sheet with dots placed in a grid center of equal horizontal and vertical spacing (e.g., a square) representing the total area of the watershed divided into parts where one part is one grid square. For this watershed, the total area is the sum of its individual parts, or a total of 174 dots. The percentage of the total area in each isohyet (column 4) is the number of dot points (column 3) divided by the total 174 parts. The weighted volume of rainfall in each isohyetal area (column 5) is then calculated by multiplying the mean rainfall (column 2) by the fraction of area (column 4).

Average rainfall was calculated this way for the isohyetal maps in Figures 5.6c and 5.6d. Although there was variation in the rainfall depths in parts of the two isohyetal maps, the averages of 40.94 mm and 41.43 mm are not significantly different.

Table 5.1. Estimate of watershed average rainfall from isohyetal map in Figure 5.6c

Rainfall limits (mm) (1)	Mean rainfall (mm) (2)	Number of points[a] (3)	Fraction of area (4)	Rainfall weighted by area (mm)[b] (5)
<35.56	35.56	5	0.03	1.07
35.56–38.10	36.83	38	0.22	8.10
38.10–40.64	39.37	47	0.27	10.63
40.64–43.18	41.91	37	0.21	8.80
43.18–45.72	44.45	28	0.16	7.11
45.72–48.26	46.99	11	0.06	2.82
>48.26	48.26	8	0.05	2.41
	Totals	174	1.00	40.94
			Average rainfall[c]	40.94

Source: Modified from US Department of Agriculture (2015).
[a] Number of points was determined by dot method for defining area.
[b] Column (2) × column (4).
[c] The sum of the weighted amounts gives the average rainfall for the entire watershed.

Table 5.2. Estimate of watershed rainfall from isohyetal map in Figure 5.6d

Rainfall limits (mm) (1)	Mean rainfall (mm) (2)	Number of points[a] (3)	Fraction of area (4)	Rainfall weighted by area (mm)[b] (5)
<35.56	35.56	3	0.02	0.71
35.56–38.10	36.83	25	0.14	5.15
38.10–40.64	39.37	58	0.33	12.99
40.64–43.18	41.91	31	0.18	7.54
43.18–45.72	44.45	34	0.20	8.89
45.72–48.26	46.99	17	0.10	4.70
>48.26	48.26	6	0.03	1.45
	Totals	174	1.04	1.43
		Average rainfall[c]	41.43	

Source: Modified from US Department of Agriculture (2015).
[a] Number of points was determined by dot method for defining area.
[b] Column (2) × column (4).
[c] The sum of the weighted amounts gives the average rainfall for the entire watershed.

The more gauges or rainfall depth data there are, typically the more accurate the estimates of mean rainfall over an area. However, spatial arrangement of the gauge network, size of the watershed, relative and areal extent of a significant storm event with respect to the watershed area, rainfall duration and frequency, and other such factors as topography and orographic influences can all influence data accuracy.

5.3.2c. Thiessen Polygon Weighted Average

Alfred H. Thiessen (1911) developed this method assuming that rainfall at any point in the catchment area of interest is the same as that at the nearest rain gauge such that the depth recorded is applied out to a distance of halfway to the next rain station in any direction. The relative weights for each station are a result of the corresponding area and the amount of rainfall that the gauge represents in the overall Thiessen polygon network. The boundaries of the polygon are formed by the perpendicular bisectors of the lines joining the rain gauges (dashed in Figure 5.7). The polygon areas A_i corresponding to the rain gauges P_i are determined (i.e., planimeter- or computer-generated), and then using the rainfall

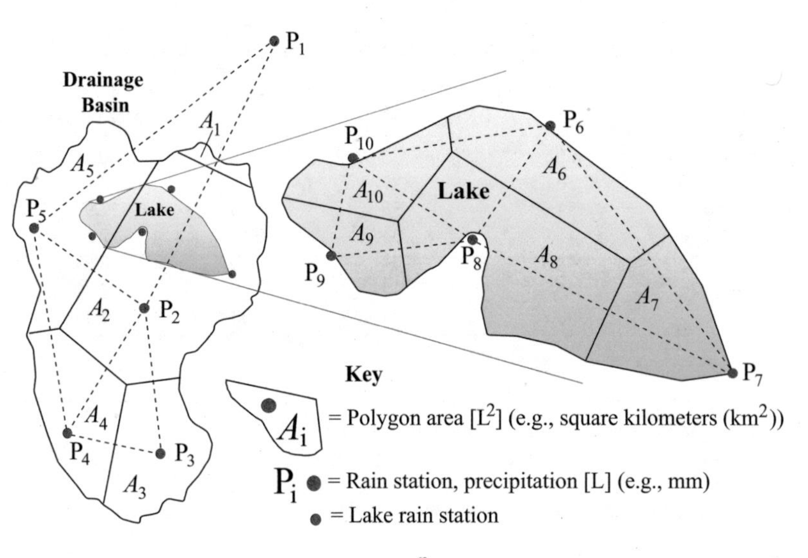

Figure 5.7. Calculation of average areal rainfall using the Thiessen polygon method for a hypothetical lake basin watershed.

$$\overline{P} = \frac{1}{A} \sum_{i=1}^{n} A_i P_i$$

Drainage Basin

Station	Measured rainfall		Area (A_i)		Weighted rainfall
P_1	50.00	x	0.30	=	15.00
P_2	40.00		4.00		160.00
P_3	30.00		1.40		42.00
P_4	20.00		1.55		31.00
P_5	10.00		2.00		20.00
	Total =		**9.25**		**268.00**

$\overline{P} = 268.00/9.25 = 29$ mm

Lake

Station	Measured rainfall		Area (A_i)		Weighted rainfall
P_6	45.00	x	0.30	=	13.50
P_7	45.00		0.31		13.95
P_8	40.00		0.43		17.20
P_9	15.00		0.11		1.65
P_{10}	15.00		0.15		2.25
	Total =		**1.30**		**48.55**

$\overline{P} = 48.55/1.30 = 37.35$ mm

measured at each gauge P_i the areal average precipitation for the watershed and lake are calculated by

$$\overline{P} = \frac{1}{A}\sum_{i=1}^{n} A_i P_i \tag{5.2}$$

The watershed or lake area A is found using

$$A = \sum_{i=1}^{n} A_i \tag{5.3}$$

In Figure 5.7, one of the rain gauges (P_1) is outside the watershed boundary. However, the area A_1 within the watershed is closer to P_1 than to adjacent gauges P_2 or P_5 and is a better representative of the rainfall measured at P_1.

Case Study 5.2

Hastings, Nebraska, Thiessen Polygons

US Department of Agriculture, 2005, *National Soil Survey Handbook*, Title 430-VI (Washington, DC: USDA), http://soils.usda.gov/technical/handbook.

Using the same Nebraska research watershed area from Case Study 5.1, Thiessen polygons are constructed by drawing lines (dashed) connecting the three outer rain gauges (Figure 5.8a).

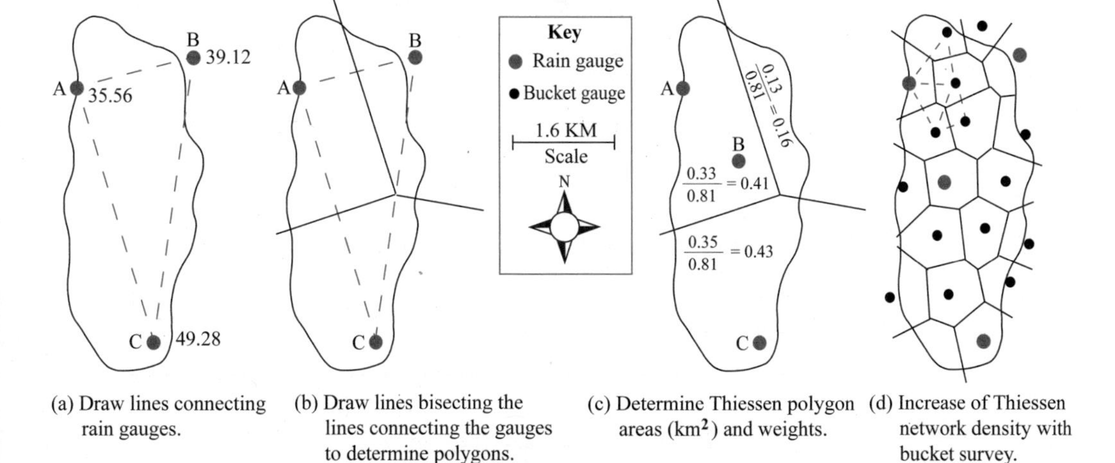

Figure 5.8. Construction and application of a Thiessen polygon map for determining average rainfall within a Nebraska watershed during a storm event. Modified from US Department of Agriculture (2015).

Table 5.3. Watershed rainfall depth from the Thiessen polygon method

Rain gauge	Rainfall (mm)	Polygon area (km^2)	Thiessen weight	Weighted rainfall (mm)
A	35.56	0.33	0.407	14.47
B	39.12	0.13	0.156	6.1
C	49.28	0.35	0.437	21.54
Totals	See weighted rainfall column	0.81	1.04	2.11[a]

Source: Modified from US Department of Agriculture (2015).
[a] The sum of the weighted rainfall gives the average rainfall for the entire watershed.

Lines are then drawn perpendicularly to bisect the lines connecting the gauges (Figure 5.8b), giving the Thiessen polygon area for each gauge. As the final step, the weight of each polygon section within the watershed is determined by measuring the area of each polygon within the watershed and dividing it by the total basin area (0.81 km^2) (Figure 5.8c).

The polygon areas, Thiessen weights, and weighted rainfall (gauge rainfall multiplied by the Thiessen weight) are recorded in Table 5.3. This rainfall average is 42.11 mm (1.66 inches) compared to the 40.94 and 41.43 mm (1.61 and 1.63 inches) estimates from the isohyetal method earlier.

The Thiessen method is generally more accurate than the arithmetic mean. When Thiessen polygon areas (Ai /A), called Thiessen coefficients, have been determined for a stable rainfall network, then rain calculations for an area can be efficaciously calculated. However, if one or more rain stations for the network have been fundamentally altered, then recalculation of Thiessen methodology is required. In the case where data are missing for one gauge, it is simpler to estimate the missing value and retain the original Thiessen polygons and coefficients.

The Thiessen method is a sound and objective procedure to measure areal rainfall but is dependent on a well-established network of representative rain gauges. In mountainous areas this method does not take into account the orographic influences on precipitation and is not recommended.

5.3.2d. Inverse Distance Weighting

When rainfall gauge data are not present near a lake or location of interest, one method for estimating precipitation for a nonsampled location is inverse distance weighting. This is a deterministic, nonlinear interpolation technique that uses a weighted average of the rainfall values from nearby sample points (P_i) to estimate the precipitation at nonsampled locations. The weight of a particular point assigned in the averaging calculation depends upon the sampled point's distance (d_i) to the nonsampled location (Figure 5.9). The weights for this method are reciprocal to the squared distances between the location and the surrounding stations (Bedient and Huber 1992).

The method is called inverse distance weighting because the similarity of two locations should decrease with increasing distance (Tobler 1970). Thus, prediction at a point is more influenced by nearby measurements than by distant measurements. Once distances (d_i) between gauges and location of interest have been determined, precipitation (P_{IW}) at a nonsampled location can be estimated by

$$P_{IW} = \frac{\sum_{i=1}^{n}\left(\frac{P_i}{d_i^2}\right)}{\sum_{i=1}^{n}\left(\frac{1}{d_i^2}\right)}. \quad (5.4)$$

Inverse distance weighting is a deterministic technique; it does not consider the spatial structure (i.e., arrangement) of the sample

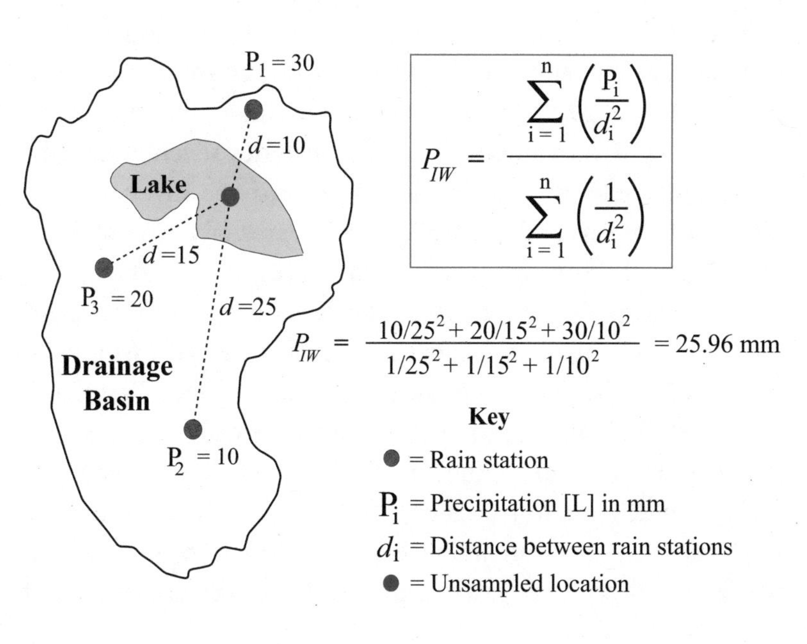

Figure 5.9. Inverse distance weighting for predicting precipitation (P_{IW}) at an ungauged point within a drainage basin and lake system.

points. Therefore, spacing and density of the rain gauges can influence the results and the accuracy of the interpolated values. Because inverse distance weighting computes an average value, the value it calculates for a nonsampled point can never be higher than the maximum value for a sample point or lower than the minimum value of the sample point, thus, if the high and low values of the data are not represented in your sample, this technique may be wildly inaccurate in some locations.

5.3.2e. Computer-Generated Rainfall Interpolations

Case Study 5.3

Stanley River Catchment, Queensland, Australia

Ball, J., M. Babister, R. Nathan, W. Weeks, E. Weinmann, M. Retallick, and I. Testoni, eds. 2016. *Australian Rainfall and Runoff: A Guide to Flood Estimation*. Canberra: Commonwealth of Australia (Geoscience Australia).

Figure 5.10 (modified from Ball et al. 2016) is the Stanley River catchment base map used to demonstrate examples for computer-generated models for both Thiessen polygons and inverse distance weighted average rainfall interpolations (Figures 5.11 and 5.12). For comparison of deterministic with geostatistical methods, a kriging example is included as well (Figure 5.13).

The precipitation data is from Tropical Cyclone Oswald, which generated heavy rainfall in the Stanley River catchment between January 23 and 29, 2013, resulting in flooding in the catchment. The Stanley River catchment, which includes seventy-six subcatchments, is in the upper part of the Brisbane River basin in Southeast Queensland, Australia, with an area of 1,324 km^2 contributing to the Somerset Dam (Figure 5.10).

A significant feature of the Stanley River catchment is the appreciable gradient in rainfall that is typically observed during large storm events. Tropical cyclones, ex–tropical cyclones, East Coast lows, and other rainfall producing systems typically feed moisture into the catchment from the Pacific Ocean. Since the northeastern part of the catchment is only 20 km from the coast but the western side of the catchment is almost 70 km from the coast, the typical direction of storm movement and typical direction of

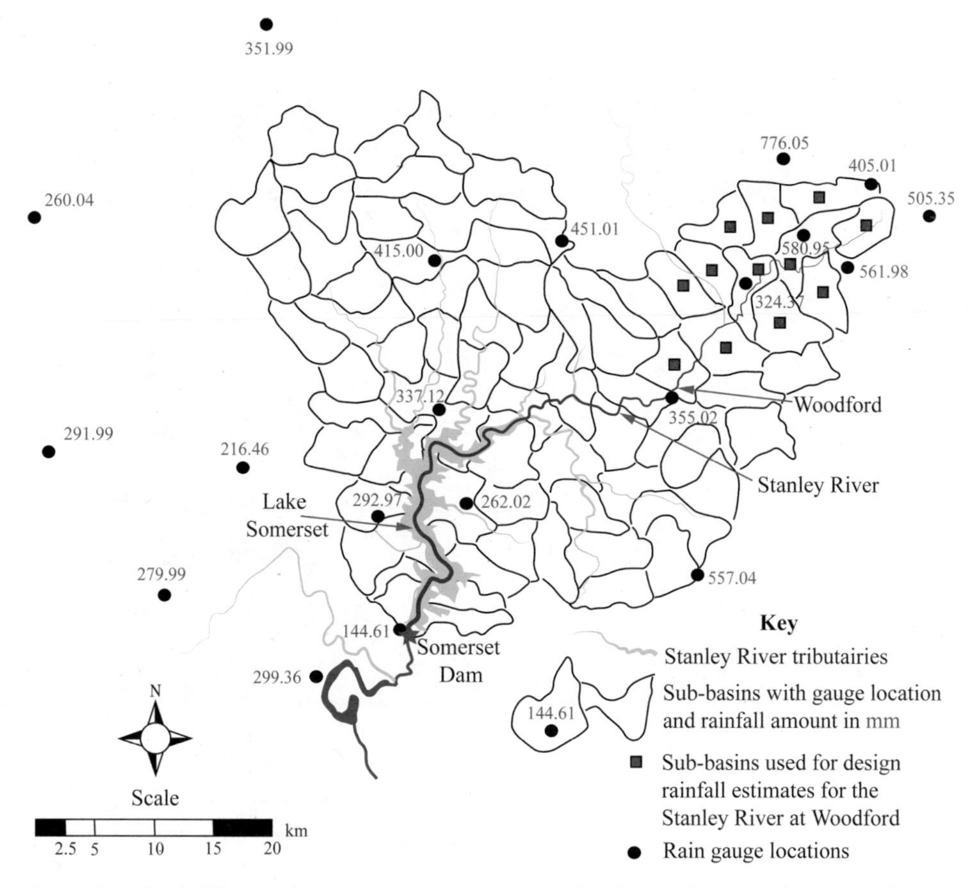

Figure 5.10. Stanley River catchment base map, showing runoff-routing model subcatchments and the locations of daily rainfall gauges, upper part of Brisbane River basin in Southeast Queensland for a January 2013 storm event. Modified from Ball et al. (2016).

flow of warm moist air from the ocean results in a gradient of rainfall totals that reduces from east to west across the catchment in most rainfall events. The strength of the rainfall gradient is enhanced by orographic effects, with the highest totals typically also occurring in the northeastern part of the catchment (Ball et al. 2016).

For all interpolation methods, rainfall totals were observed at twenty continuous rainfall gauges around the Stanley River catchment. The rainfall totals were first interpolated onto a 0.5 km resolution grid over the catchment. Rainfall totals for each of the seventy-six runoff-routing model subcatchments were then computed from the average of the rainfall totals at the grid cells that overlapped each subcatchment (see Figures 5.11–5.13).

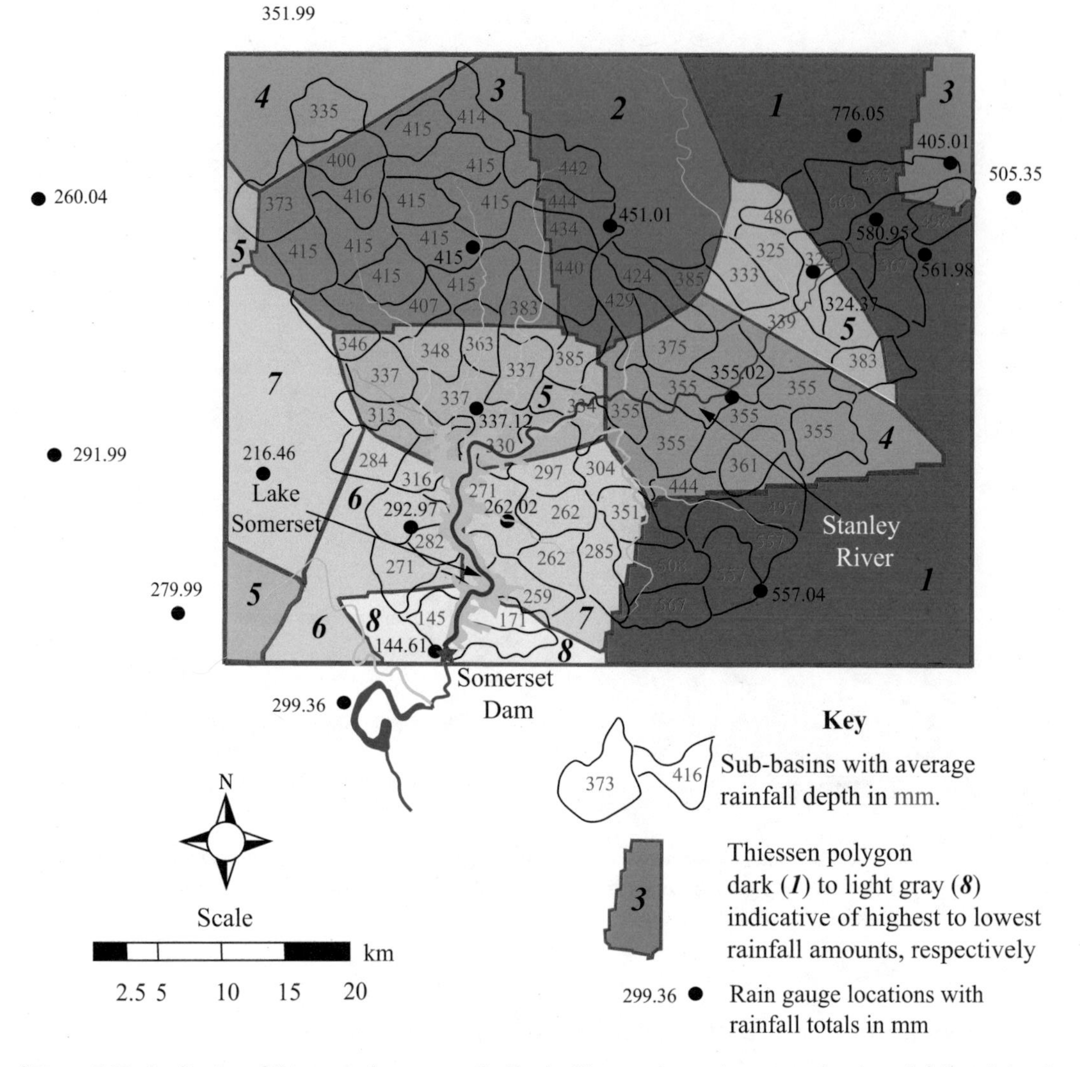

Figure 5.11. Application of Thiessen polygons over the Stanley River catchment base map, showing rainfall totals (mm) for a January 2013 storm event. Modified from Ball et al. (2016).

Geostatistical interpolation methods are stochastic methods, with *kriging* being the most well-known representative of this category. Like the inverse distance weighting method, kriging calculates weights for measured points in deriving predicted values for unmeasured locations. With kriging, those weights are on distance between points and variation between measured points as a function of distance (Figure 5.13). For further discussion, see section 11.3.2b.

The rainfall interpolations were generated for the purpose of

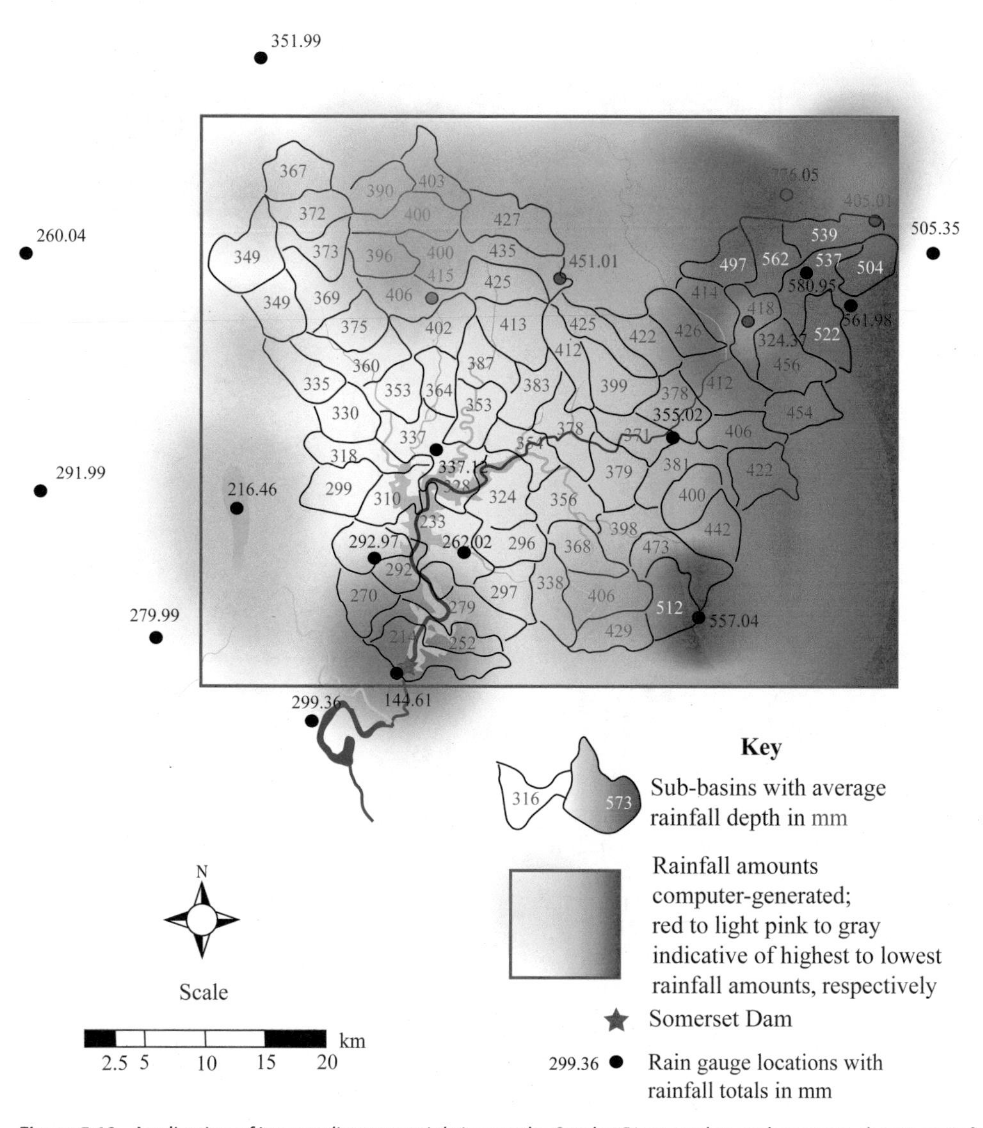

Figure 5.12. Application of inverse distance weighting on the Stanley River catchment base map, showing rainfall totals (mm) for a January 2013 storm event. Modified from Ball et al. (2016).

calibrating design runoff-routing models, which is beyond the scope of this text. These types of models are statistical estimates generally based on some form of probability analysis of flood or rainfall data. *Design rainfall* models can be used for the entire catchment with all seventy-six sub-basins being used to the Stanley River catchment boundary at the Somerset Dam as well as for specific areas of interest within the catchments that are being investigated. For this example, design rainfall estimates are developed for the Stanley River at Woodford, which has a catchment

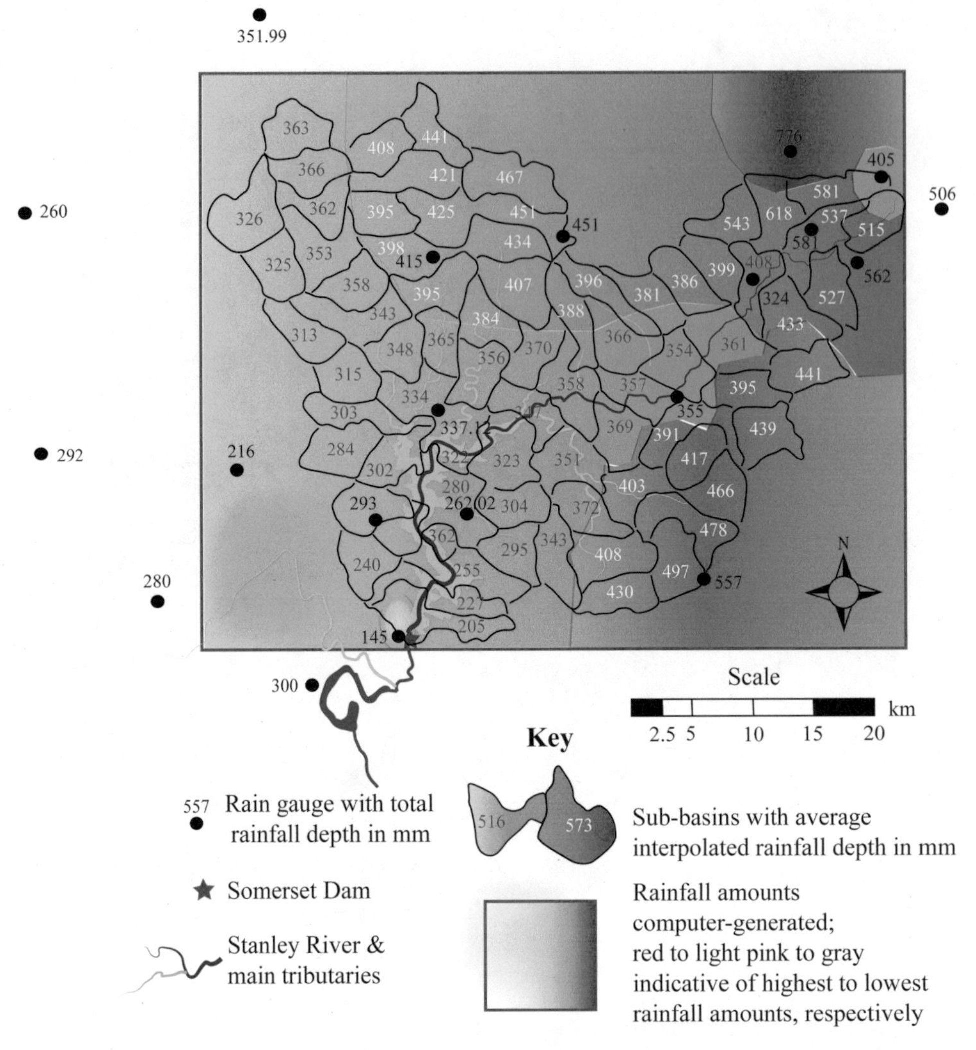

Figure 5.13. Example of the kriging method on the Stanley River catchment base map, showing rainfall totals (mm) for a January 2013 storm event. Modified from Ball et al. (2016).

area of 245 km^2. The catchment area to Woodford is the north-eastern portion of the Stanley River catchment to Somerset Dam and includes fifteen subcatchments in the runoff-routing model (Figure 5.13). See Ball et al. (2016) for details.

5.3.2f. Summary

Other methods of average rainfall interpolation from a network of rain gauge point measurements include *natural neighbors*, *spline interpolation algorithms*, *ordinary kriging*, *variants on kriging* (such as *indicator kriging*, *regression kriging*, and *kriging with external drift*), *reciprocal distance squared method* (Wei and McGuinness 1973; Singh and Chowdhury 1986), and the use of other geostatistics (McCuen and Snyder 1986; Bras and Rodriguez-Iturbe 1985). An introduction to stochastic and geostatistical methods are discussed in section 11.3.2.

Regardless of the interpolation method used, accuracy of any rainfall estimate relies primarily on the distance between gauges and point of application of the estimate. For terrain that is relatively uniform in topography, horizontal distance typically matters. However, in watersheds of relatively flat terrain where rain shadows or lake effect precipitation occurs, rainfall may vary considerably. Conversely, in mountainous areas the vertical distance may be more important than the horizontal. For a rain gauge network, distance between gauges as well as arrangement affect accuracy. Generally, the possible distances between average rainfall and accuracy are generally ignored unless special investigations at gauge sites have been implemented (US Department of Agriculture 2015).

Thus, the spatial pattern produced by mapping should be compared against the point values observed at rainfall gauges. If the mapping reveals anomalies in the interpolation approach, an alternative method should be implemented. Where large gaps in rainfall gauge coverage exist (particularly in mountainous areas), careful review of the simplifying assumptions made in the interpolation procedure should be undertaken to avoid unlikely spatial patterns (Ball et al. 2016).

5.3.3. Remote Sensing Measurement of Rainfall

As discussed so far, the majority of uncertainty in the data collected by a network of rainfall gauges is from the inability to accurately observe variations in the space-time pattern of rainfall between the gauges (Seed and Austin 1990; Barnston 1991; Bradley et al. 1997). Instead, remote sensing methods can be used to estimate rainfall intensity observed on a spatial grid across a wide observation domain for a given period of time. The two most commonly available remote sensing approaches for rainfall estimation are ground-based weather radar (short for *radio detection and ranging*) and satellite observing systems.

5.3.3a. Radar

Radar reflectivity can be used to observe precipitation occurring in the atmosphere over a large geographic area and therefore provides good spatial coverage. Detailed precipitation for large areas can be produced by radar from a single location with high spatial and temporal resolution. With an effective range of 200 km (130 miles) a single radar can cover an area of greater 10,000 km^2 (4,500 mi^2) (Maidment 1993). However, the scanning radar beams leave gaps in spatial coverage, particularly over mountainous regions, where there may be large variation in rainfall and snowfall due to orographic effects.

Weather radars measure the reflectivity returned by raindrops, hailstones, or snow, which are then converted into a rainfall intensity estimate. However, because weather radars do not point at the ground, radar reflectivity does not accurately represent rainfall on the ground. Therefore, there are several different types of errors in this process that degrade the accuracy of the radar rainfall measurement (Joss and Waldvogel 1990; Collier 1996). Areal and point rainfall estimates are often in error by a factor of two or more. Error sources reside in measurement of radar reflectivity factor, evaporation, advection of precipitation before reaching the ground, and variations in the drop-size distribution and vertical air motions (Kessler 1992).

The use of radar data in combination with rain gauge measurements improves precipitation estimates over those based on either form of measurement alone. This improvement is accomplished by adjusting, or calibrating, radar rainfall data with data from rain gauges situated within the radar boundary. The rain gauge data allows forecasters to calibrate the radar data in the form of ground truth, and the radar data allows the gaps between rain gauges to be filled in. Simple techniques that combine sparse gauge reports (one gauge per 1000–2000 km^2) with radar produce smaller measurement errors (10%–30%) than either system alone (Kessler 1992). Thus, radar-derived rainfall accumulations combine the benefits of both these systems: the accurate point data from the rain gauges and the excellent spatial coverage of the radar.

Although radar has been used experimentally for nearly sixty years to measure rainfall, operational implementation has been slow, only just recently increasing rapidly due to computer-generated algorithms and methodology. In 1988, the National Oceanic and Atmospheric Agency (NOAA) and Next Generation Weather Radar (NEXRAD) agencies established the WSR-88D (Weather Surveillance Radar-1988 Doppler) Radar Operations Center in Norman, Oklahoma. The NEXRAD system currently comprises 160 sites throughout the United States and overseas locations. (For more information on weather radar use, see https://www.ncdc.noaa.gov/data-access/radar-data/nexrad).

5.3.3b. Satellite

Meteorological satellites have been operational since the 1970s and are the only observation sources capable of providing unique information about spatial distribution and intensity of precipitation from regions that are inaccessible by ground-based radar and/or gauge techniques (Scofield and Kuligowski 2003). Recent improvements in the ability of satellite-based precipitation retrieval algorithms to produce estimates (with global coverage) at high space and time resolutions makes them potentially attractive for hydrologic forecasting in ungauged basins.

NOAA manages a constellation of geostationary and polar-orbiting meteorological spacecraft. Geostationary satellites help monitor and predict such weather and environmental events as tropical systems, tornadoes, flash floods, dust storms, volcanic eruptions, and forest fires. Polar-orbiting satellites collect data for weather, climate, and environmental monitoring applications, including precipitation, sea surface temperatures, atmospheric temperature and humidity, sea ice extent, forest fires, volcanic eruptions, and global vegetation analysis, as well as search and rescue.

Depending on the type (i.e., geostationary vs. low earth orbit), these satellites can carry a range of visible, infrared, and microwave sensors (Ebert and Manton 1998; Simpson et al. 1996). Available methods for estimating area rainfall include using images of visible (wavelength between 0.4 and 0.7mm), infrared (wavelength between 10.5 and 12.5 mm), and electromagnetic radiation from geostationary satellites. The visible channel measures the shortwave radiation backscattered by the atmosphere and the earth and this is related to the albedo and the brightness of the scattering body. The infrared channel measures thermal radiation emitted by clouds, and this is related to the temperature of the emitting cloud by the *Planck*'s *radiation law* (Planck 1914). The brightness and temperature of precipitating clouds obtained from the visible and infrared channels respectively are an indirect measure of convective rainfall intensity.

Passive microwave frequencies have been used for rain retrieval from low earth-orbiting sensors for about twenty-five years. The techniques that have been developed and refined in time rely on the signal emitted by raindrops over oceans at frequencies at or below 37 GHz, and on the scattering signal of ice particles in the precipitation layer over land at frequencies at or above 85 GHz (Lensky and Levizzani 2008). Depending on the type of satellite sensor and designated goals, different passive microwave precipitation retrieval algorithms have been developed. Each product has its own strengths and weaknesses, but none of them appears to be universally better than the other (Kummerow, Masunaga, and Bauer 2007).

NOAA provides a vast amount of satellite meteorological data and are currently implementing methods of efficiently distributing these data for public use. One such program is the NOAA Big Data Project, which makes satellite data from the newest generation of NOAA's geostationary satellites, the GOES-R series, freely available through cloud infrastructures. The Big Data Project is an innovative approach to publishing NOAA's vast data resources and positioning them near cost-efficient high-performance computing, analytic, and storage services provided by the private sector. (For more information, refer to https://www.ncdc.noaa.gov/data-access/satellite-data.)

5.3.3c. Incorporation of Multisensor Rainfall Estimation

The development of rain gauge, radar, and satellite methods for estimating rainfall developed independently. Incorporation of these three systems is now being recognized to create better estimates of rainfall as well as other meteorological data. The major problem of implementing different systems is defining and solving the joint error uncertainties produced for the different sensor systems.

Radar-derived precipitation estimates are a combination of radar data and rain gauge observations. These estimates may be referred to as *MPE values* because they are derived by using the Multisensor Precipitation Estimator (MPE) program from NOAA's National Weather Service. Incorporation of satellite data with MPE data is the underlying theme in the United Kingdom's FRONTIERS system (Browning 1979) and actively being investigated by NEXRAD.

For example, a NOAA-funded study by Mahani and Khanbilvardi (2009) developed a merging algorithm capable of improving satellite-based rainfall retrieval algorithm by merging its estimates with radar rainfall measurements as well as generating rainfall for the pixels with no radar information. The merging algorithm is also viable for extending the patterns and intensity of the radar-based rainfall to the gap area from the surrounding pixels, and the generating algorithm could generate rainfall over pixels where the radar map shows there was some rainfall but satellite-based algorithm could not estimate any rainfall. (For more information, see https://www.weather.gov/nerfc/MultiSensorPrecipitation.)

5.3.4. Summary

Accurate estimation of rainfall intensity and distribution continues to be challenging. Precipitation intensity can be assessed through a variety of ground and remote sources of observation such as gauge, radar, and satellite. No source of observation or methodology exists that can provide realistic spatial distribution and "true" intensity of rainfall. Observation error and instrument noise as well as limited spatial coverage continue to hinder traditional point gauge, radar, and satellite observations.

Meteorological data are imperative for determining mass balance of lakes and wetlands. Technological and computer-modeling advances are making estimates of rainfall, runoff, evaporation, and other meteorological impacts more accessible as well as more accurate. There are numerous sources in the United States for hydrometeorological data:

Natural Resources Conservation Service:
  National Weather and Climate Center, www.wcc.nrcs.usda.gov
US Geological Survey (USGS): www.usgs.gov

National Oceanic and Atmospheric Association:
  National Weather Service: www.weather.gov
  National Centers for Environmental Information: www.ncdc.noaa.gov
Regional Climate Centers: www.weather.gov/coop/regionalclimatecenters
US Department of Agriculture (USDA): www.usda.gov
  Agricultural Research Service: www.ars.usda.gov/office-of-technology-transfer
  Forest Service: www.fs.fed.us
State climatologists
Other federal, state, and local agencies responsible for water-related projects (e.g., Florida's Water Management Districts)

5.4. Presentation of Rainfall Data

Although computerized digital rain gauges can measure rainfall continuously, a rain gauge record is usually broken down into a set of rainfall depth measurements in successive time increments, such as one-, five-, or ten-minute intervals. Table 5.4 are rainfall data from a single US Geological Survey gauge, 1-Bee in Austin, Texas, from a May 24–25, 1981, storm event that caused flood damage to the Austin area (Massey, Reeves, and Lear 1982 Chow, Maidment, and Mays 1988). The data and graphs here can be easily calculated in spreadsheet software such as Microsoft Excel.

Understanding the distribution of rainfall with time is essential for assessing such hydrologic problems as the design of culverts for watersheds, storm-sewer systems, and stormwater retention structures to prevent flooding and downstream erosion, as well as improving water quality in an adjacent river, stream, lake, wetland, or bay. This information is also useful for studying soil erosion, recognizing the flood potential of various types of storm events, and increasing the understanding of the physics of atmospheric precipitation processes (Huff 1967).

5.4.1. Rain Intensity

Rainstorms vary considerably in space and time, and as rain continues, water reaching the ground infiltrates into the soil un-

Table 5.4. Single-gauge rainfall measurements from gauge 1-Bee, Austin, Texas

			Running totals (mm) over successive increments		
Time (min)	Rainfall (mm)	Cumulative rainfall (mm)	30 min	1 hour	1.5 hours
0	0.00	0.00			
5	0.508	0.508			
10	8.636	9.144			
15	2.540	11.684			
20	1.016	12.700			
25	4.826	17.526			
30	12.192	29.718	29.718		
35	12.700	42.418	41.910		
40	12.700	55.118	45.974		
45	12.954	68.072	56.388		
50	4.064	72.136	59.436		
55	7.874	80.010	62.484		
60	16.764	96.774	67.056	96.774	
65	9.144	105.918	63.500	105.410	
70	9.906	115.824	60.706	106.680	
75	9.144	124.968	56.896	113.284	
80	13.716	138.684	66.548	125.984	
85	19.304*	157.988	77.978*	140.462	
90	12.954	170.942	74.168	141.224*	170.942
95	11.176	182.118	76.200	139.70	181.610
100	6.350	188.468	72.644	133.35	179.324
105	6.350	194.818	69.850	126.746	183.134
110	5.588	200.406	61.722	128.270	187.706*
115	3.810	204.216	46.228	124.206	186.690
120	2.286	206.502	35.560	109.728	176.784
Total	**206.502**				

Source: Data modified from Massey, Reeves, and Lear (1982) and Chow, Maidment, and Mays (1988).
*Maximum depth.

til it reaches a stage where the rate of rainfall (intensity) exceeds the infiltration capacity of the soil. Thereafter, surface puddles, ditches, and other depressions are filled (depression storage), after which runoff is generated. Rate and volume of runoff is related to the rainfall intensity where the peak intensity produces the largest runoff rate.

Intensity is the amount of rain that has fallen over a unit of time. The average storm intensity is calculated by dividing a rainfall depth for a storm event by its duration, which is the amount of time it took for that depth to accumulate:

$$Intensity = \frac{Rain\ Depth}{Duration}. \tag{5.5}$$

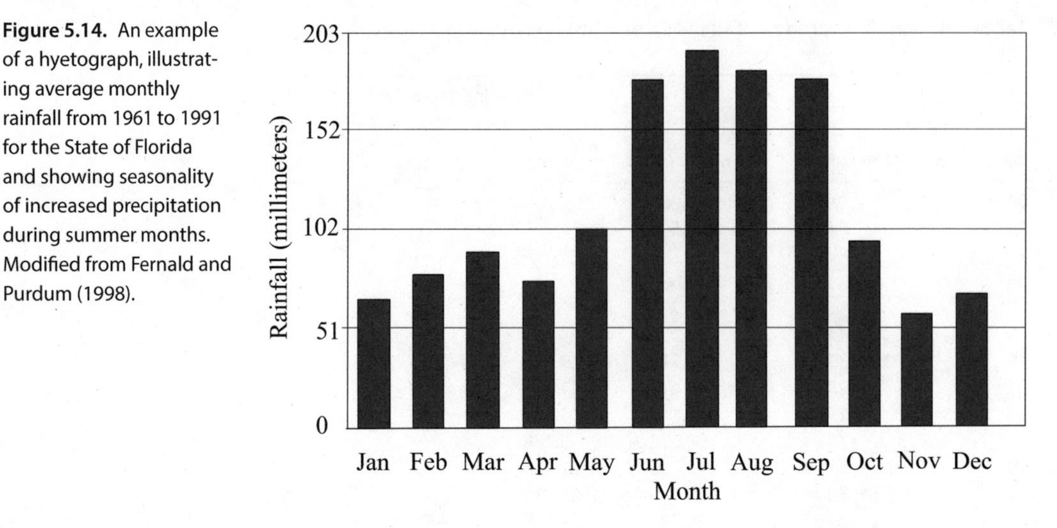

Figure 5.14. An example of a hyetograph, illustrating average monthly rainfall from 1961 to 1991 for the State of Florida and showing seasonality of increased precipitation during summer months. Modified from Fernald and Purdum (1998).

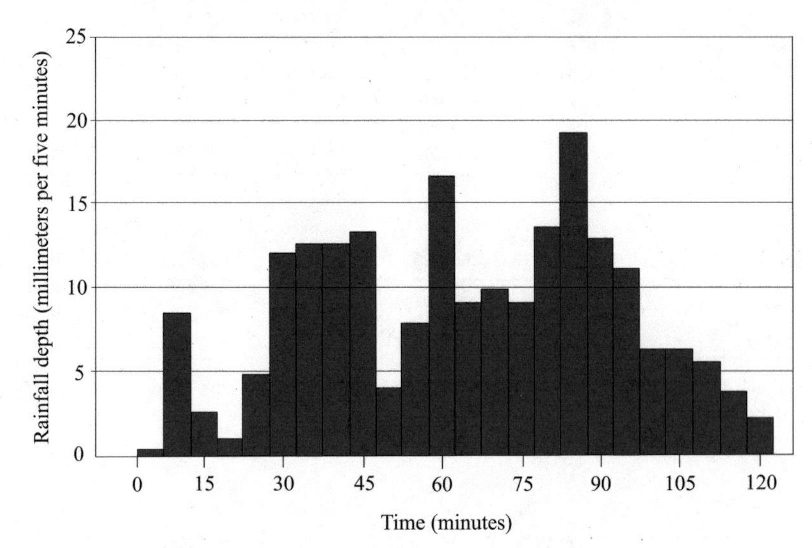

Figure 5.15. Hyetograph showing incremental rainfall from USGS 1-Bee rain gauge in Austin, Texas, from the first two hours of a May 24-25, 1981, storm event. Data from Massey, Reeves, and Lear (1982).

A plot of rainfall depth or intensity as a function of time in the form of a histogram is a rainfall *hyetograph* (Figure 5.14). A hyetograph is a graphical representation of the temporal distribution of rainfall intensity at a point on the ground (rain gauge) or over a watershed during a storm or other long-term rainfall pattern to investigate precipitation trends, such as seasonality (Table 5.4 and Figure 5.15).

For examining storm intensity for a given time interval during a storm—such as fifteen, thirty, or sixty minutes—a *cumulative rainfall hyetograph* or *rainfall mass curve* can be generated by summing the rainfall intervals over time (Figure 5.16). For example,

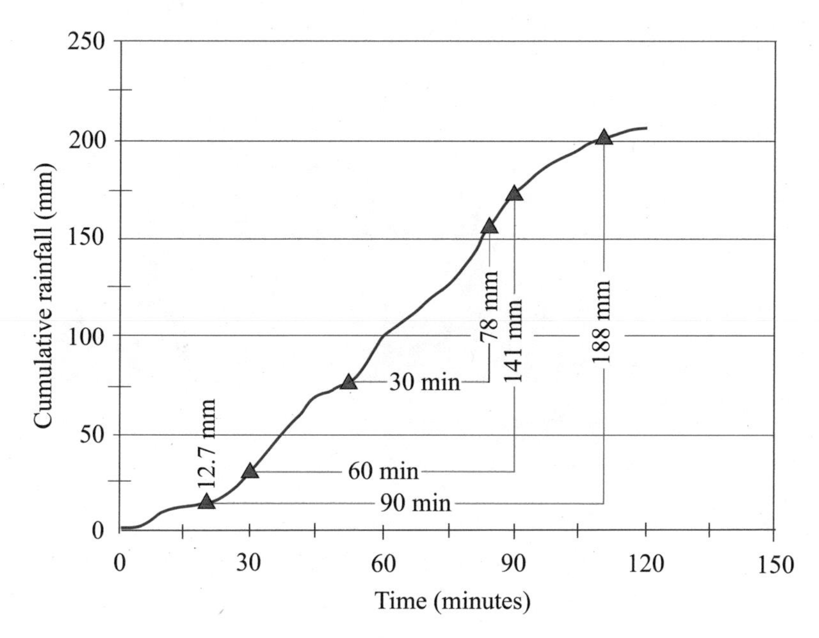

Figure 5.16. Cumulative rainfall hyetograph with maximum storm intensity for thirty-, sixty- and ninety-minute intervals of the May 24-25, 1981, storm event from USGS 1-Bee rain gauge in Austin, Texas. Data from Massey, Reeves, and Lear (1982).

at twenty minutes into the storm, the cumulative rainfall depth is the rainfall data summed over this time interval, or 12.7 mm.

The *maximum intensity* or *maximum rainfall depth* over a given time interval is calculated by a series of running totals. For example, the thirty-minute interval begins with cumulative rainfall at thirty minutes, or 29.7 mm, followed by 41.91 mm at thirty-five minutes, and so on. The maximum intensity for this interval is approximately 78 mm, occurring between fifty-five and eighty-five minutes into the storm (see Table 5.4 and Figure 5.16). The sixty- and ninety-minute intervals are calculated in the same way as the thirty-minute interval.

The maximum rainfall depth can be used to observe the intensity of the storm over time. The thirty-minute interval is equivalent to an average intensity of 78 mm/0.5 hours or 156 mm/hour. The sixty- and ninety-minute-interval average hourly intensities are 141 mm and 188 mm, respectively, which shows the average intensity of the storm is decreasing over time for this storm.

5.4.2. Temporal Patterns of Storms

Using hyetographs and other data, a typical four-component pattern of rainfall is often observed for most storm events (Ball et al., 2016). The importance of temporal rainfall patterns has in-

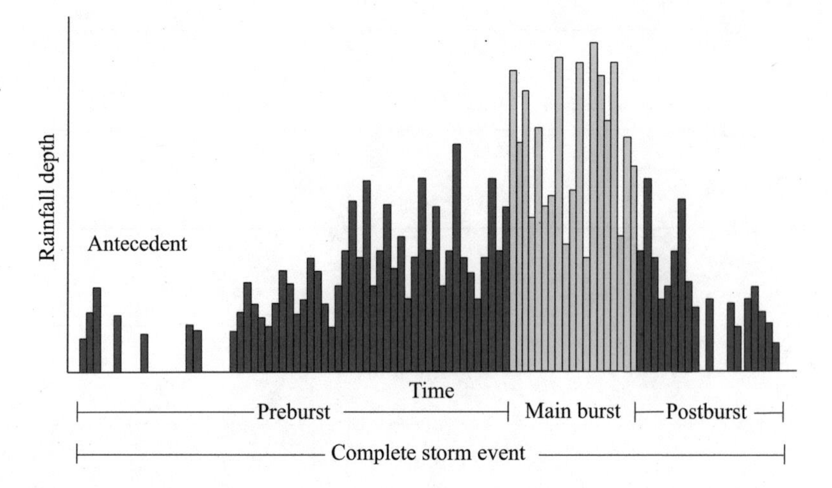

Figure 5.17. Typical components of a storm event recorded on a hyetograph.

creased for investigating catchment response to storm events as it relates to estimating total storm volumes as well as the temporal distribution of rainfall within the storm event necessary for flood management.

To understand the temporal patterns, it is necessary to understand the components of a storm event (Figure 5.17) and how they relate to storm intensity and catchment response. It is important to note the components can be characterized either by *infiltration frequency duration* relationships or by catchment response and are highly dependent on the definitions used. That said, the components of a storm include the following:

1. *Antecedent rainfall*: rainfall that has fallen before the storm event and is not considered part of the storm but can affect such catchment response as decreasing soil infiltration and ponding.
2. *Preburst rainfall*: which occurs before the main burst. With the exception of relatively frequent events, it generally does not have a significant influence on catchment response but is very important for understanding catchment and storage conditions before the main rainfall burst.
3. *Main burst*: the main part of the storm, but very dependent on the definition used. Bursts have typically been characterized by duration. The burst could be defined as the *critical rainfall burst*, the rainfall period within the storm that has the lowest probability, or the *critical response burst*, which

corresponds to the duration that produces the largest catchment response for a given rainfall *annual exceedance probability*, which is a measure of the rarity of a rainfall event (see section 7.2.1).

4. *Postburst rainfall*: rainfall that occurs after the main burst and is generally only considered when aspects of hydrograph recession are important. This could be for drawing down a dam after a flood event or understanding how inundation times affect flood recovery, road closures or agricultural land.

Whereas Figure 5.17 is a generalized depiction of a typical storm pattern, it has been well recognized that temporal patterns exhibit significant variability between rainfall events of similar magnitude and the assumed pattern can have a significant effect on the estimated peak flow (Askew 1975). Pilgrim, Cordery, and French (1969, 82) report, "In nature, a wide range of patterns is possible. Some storms have their period of peak intensity occur early, while other storms have the peak rainfall intensity occur towards the end of the storm period and a large number have a tendency for the peak to occur more or less centrally."

5.4.3. Hyetographs

A *hyetograph* is a specific type of hydrograph, or chart that displays the change of a hydrologic variable over time, for measuring rainfall. Measurements of lake level, streamflow, chemical concentrations, potentiometric surface, temperatures, and numerous other variables with time are other examples of hydrographs (see section 7.4 for a detailed discussion). When investigating the hydrology of a site-specific system, it is often useful to compare different variables, such as rainfall, over a defined time period with lake levels, streamflow, or certain chemical characteristics to observe any related correlations or impacts.

Rainfall that is not infiltrated into the soil, retained on the land surface via depression storage, or intercepted by vegetation is *excess rainfall*, or *effective rainfall.* After flowing across a watershed or basin surface, excess rainfall becomes direct runoff at a watershed outlet, such as a river, lake, or wetland. The graph of excess rainfall versus time is an *excess rainfall hyetograph* and is important for

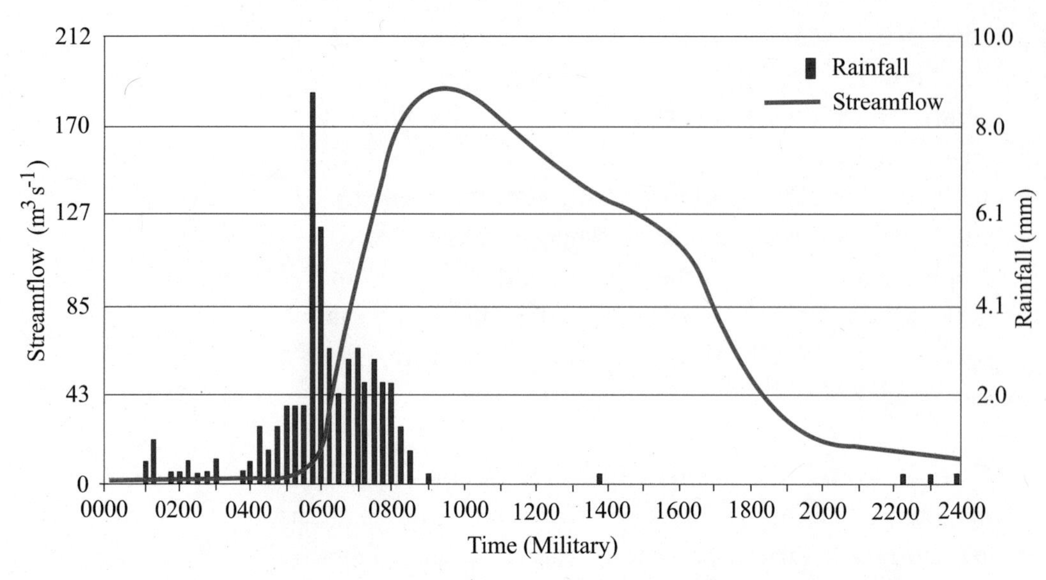

Figure 5.18. Peachtree Creek Watershed hyetograph and Peachtree Creek hydrograph showing how the system was affected by a 51 mm storm event on December 24, 2002, in Atlanta, Georgia. Data from US Geological Survey (2002).

investigating rainfall-runoff relationships (Chow, Maidment, and Mays 1988). The differences between an observed total-rainfall hyetograph and an excess rainfall hyetograph are termed *abstractions*, or *losses*. Abstractions are primarily due to infiltration, with some losses from interception and depression storage.

Hyetographs, in conjunction with other hydrograph profiles, can be used to attain peak discharge and runoff volume estimates to better develop more efficient and cost-effective drainage control systems (Haan, Barfield, and Hayes 1994) and determine the volume of stormwater discharge to a lake, river, or wetland system. For example, on December 24, 2002, about 51 mm (2 in.) of rain fell in the Peachtree Creek watershed in Atlanta, Georgia. Figure 5.18 describes streamflow characteristics during the storm; the rain fell for only a few hours and Peachtree Creek was at baseflow conditions before the rain started.

Note the lag in streamflow after increased rainfall and the continued increase in flow after the storm ended, indicating the sustained flow of water into the stream from surface flow and groundwater generated during the storm. Streamflow was at approximately 1.4 cubic meters per second (50 cubic feet per second) before the creek began to rise. At approximately 0930, peak flow was at about 190 cubic meters per second (6,700 cubic feet

Table 5.5. Comparison of streamflow before and during a December 24, 2002, storm event in the Peachtree Creek watershed, Atlanta, Georgia

Time	Stream stage (m)	Instantaneous streamflow (m^3/s)	Liters per second	Streamflow in liters during fifteen-minute interval
0000 (midnight)	0.86	1.2	1,219	1,093,984
1000	5.3	188	187,756	168,829,366

Source: Data modified from US Geological Survey (2002).

per second), or about 135 times greater than base. This is typical of small streams, especially urban streams where runoff enters a river very quickly (Konrad 2003). Table 5.5 provides calculations of flow volumes before (at base level) and during the rainfall event based on stream characteristics at the gauge location. At baseflow, an estimated 105,234,447 liters (27.8 million gallons) of water will flow by the Peachtree Creek measurement station in one day. Using mean streamflow's for each fifteen-minute period during the storm of December 24, an estimated 16,239,416,553 liters (4.29 million gallons) of creek flow was calculated. That would be about 154 times more water than during a day of baseflow (US Geological Survey 2002).

CHAPTER 6

Stormwater Flow

6.1. Introduction

Once the amount of rainfall in a catchment has been estimated, determining the resulting flow to a lake, wetland, or stream is a fundamental problem for the hydrologist when investigating mass balance and system behavior over time. This knowledge is essential for water resource assessment and management and for flood predictions and control.

As precipitation commences, rainwater is deposited on a variety of surfaces and travels downslope by numerous routes into a stream, wetland, or lake; back to the atmosphere; or to the underlying water table by *infiltration*.

A generalized schematic of a hydraulic budget for a lake and its catchment basin during a rainfall event is demonstrated in Figure 6.1. For the watershed, the input is primarily precipitation coupled with groundwater as it crosses the hydraulic boundary at the surface, usually in the lower topographic regions of the basin. Losses from the basin occur from infiltration, evapotranspiration, runoff, *throughflow*, and streamflow to the lake. Losses from the lake are from seepage/groundwater, evapotranspiration, and streamflow leaving the lake system.

During a storm event, there are four conveyances where water is discharged to a lake, stream, or wetland system: (1) direct precipitation on the lake, (2) surface runoff or overland flow, (3) shallow subsurface flow from the storm, and (4) increased groundwater discharge from the storm event.

The *infiltration rate*, the rate at which water enters the soil, depends on the rate at which water is supplied to the soil surface; and the *infiltration capacity*, which is the maximum rate at which water can enter the soil. If the rainfall rate (*intensity*) [Lt^{-1}] is greater than the infiltration capacity, water will pond at the soil surface;

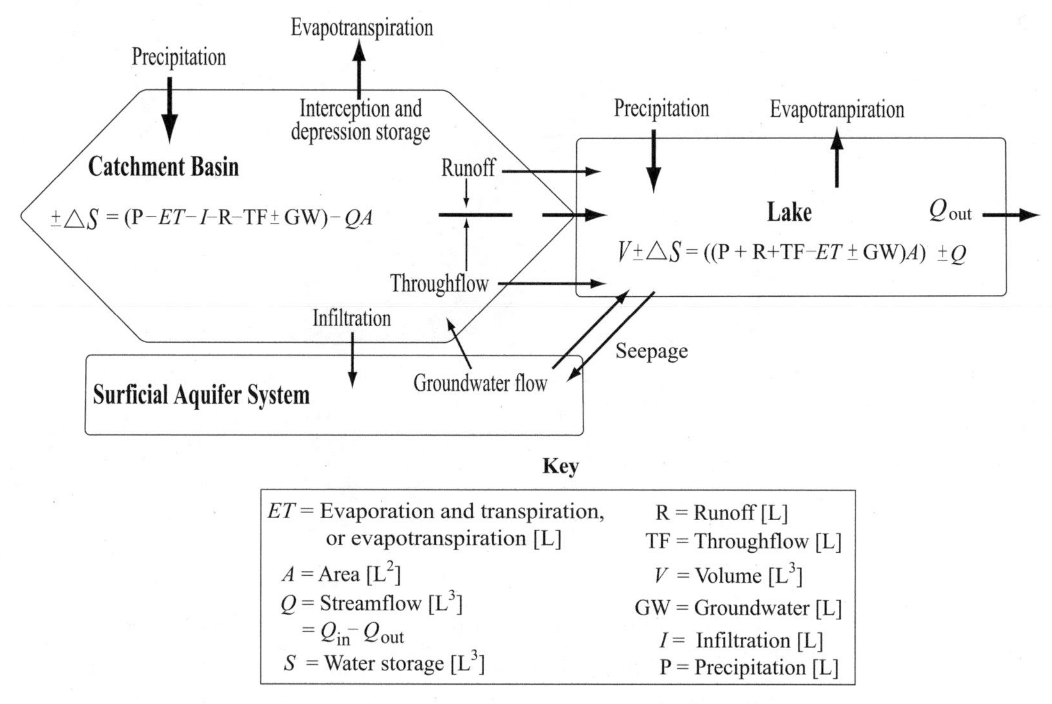

Figure 6.1. A catchment basin as it relates to a lake system mass balance budget within the catchment.

if the ground is sloping, then water will runoff. Runoff produced in this way is called *infiltration-excess overland flow*, or *Hortonian overland flow* who first described this process.

The infiltration capacity of the soil is contingent on its texture and structure, as well as on the soil moisture content (e.g., time of previous rainfall) prior to the storm event. For dry soil, the initial capacity is high but, as the storm continues, it decreases until it reaches a steady value, termed the *final infiltration rate* (see Figure 6.2). The process of runoff generation continues as long as the rainfall intensity exceeds the actual infiltration capacity of the soil, but it stops as soon as the rate of rainfall drops below the actual rate of infiltration (Critchley and Siegert 1991).

There are three primary processes that contribute to runoff: Hortonian overland flow, throughflow, and *saturated* overland flow. Hortonian overland flow, after Robert Horton (1933), is produced during a rainfall event when precipitation intensity exceeds the infiltration capacity in the watershed. Hortonian overland flow can provide a rapid pathway for water to be converted from rainfall to runoff. Hortonian flow is likely to contribute to floods when

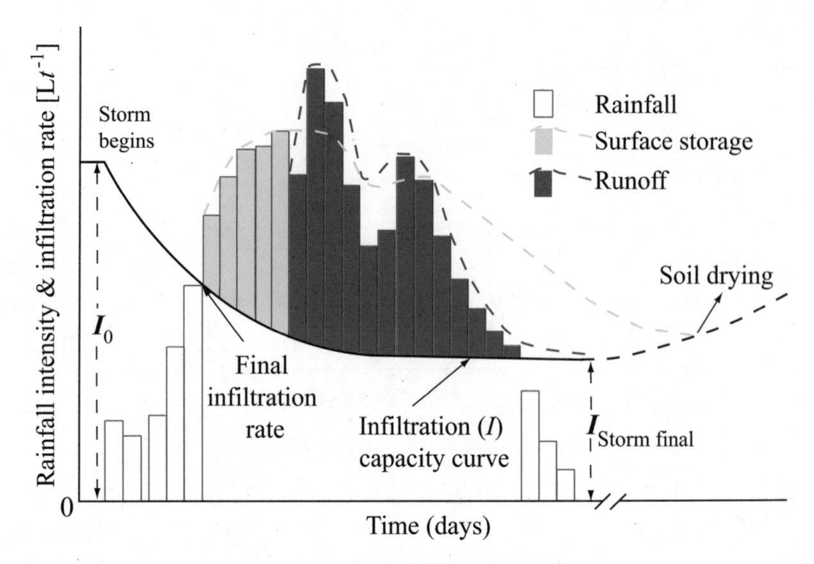

Figure 6.2. Hypothetical schematic demonstrating the relationship between rainfall, infiltration, surface storage, and runoff during a storm event. Modified from Critchley et al. (1991).

catchment surfaces have low infiltration capacity, when there is intense rainfall, where there is a rapid mechanism such as areas with steep slopes, or where there are impervious areas such as rock outcrops or urbanized development resulting in faster and increased runoff to lakes (Figure 6.3).

Ever-increasing urbanization causes catchments to be inundated with residential areas, roads, parking lots, roofs, and other impervious surfaces (Figure 6.3a). Although runoff from impervious surfaces in urban areas is considered Hortonian overland flow, this type of flow requires special consideration because of substantial increase of runoff volume and rapid response of lakes and streams impacted by urbanization. Hydrologic impacts of urbanization are discussed in section 6.6.

Saturated-excess overland flow (Figure 6.3c) occurs when portions of the surface of the drainage basin become saturated as a result of a shallow low hydraulic conductivity zone or rapid rise of the water table to the surface. This usually occurs near the basin bottom where the water table is near the surface, but it can also occur at topographic highs where soils are thin due to erosion (Henderson 1966) or there is an underlying lithology of low hydraulic conductivity such as clays.

6.2. Variable Source Areas

If soil becomes saturated, either from rising soil moisture or because of flow from upslope, then no additional rainfall can in-

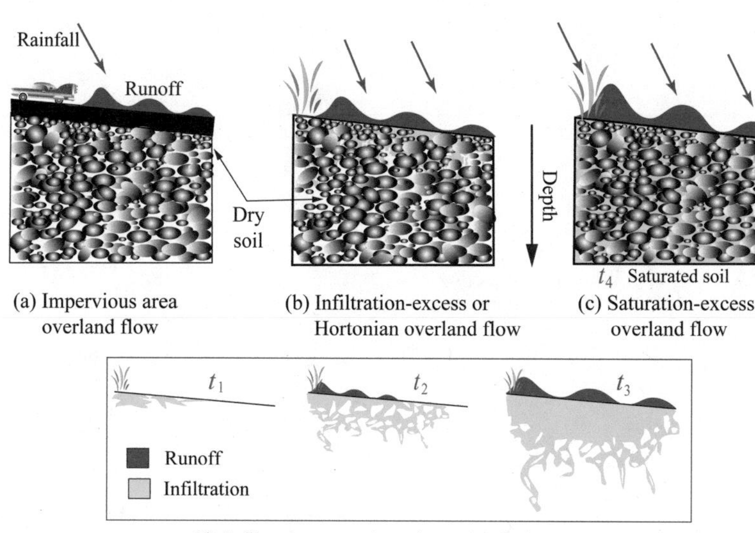

Figure 6.3. Types of surface runoff.

filtrate. Any rainfall striking the saturated soil surface will be converted to saturation excess runoff. These saturated regions of a catchment are referred to as *source areas.* Usually there are some areas within a catchment that are wetter than others. Areas around wetlands, along valleys, and adjacent to streams may remain saturated for long periods with upslope areas being dryer. Saturated areas enlarge and contract with the seasonal wetting and drying of a catchment. Saturated areas may expand during a storm and then shrink once rainfall ceases. As the amount of saturated area changes, so does the source area contributing to runoff. The concept of partial area runoff arises because only part of a catchment may be saturated and this area may be the only contributor to streamflow (Dunne and Black 1970). Saturation excess runoff can contribute to floods when source areas are large and convert intense rainfall to runoff that flows directly to streams or lakes.

Thus, not all the drainage basin area contributes flow to a stream or wetland during a storm event due to the low velocity of subsurface flow and subsequent saturation overflow. Hewlet (1982) used the term *variable source areas* to designate the areas of a drainage basin that contributes flow to an outflow stream in the basin. The variable source area will expand as cumulative rainfall increases over the period of a storm event and decreases as the storm dissipates (Figure 6.4).

The quantity and pathways of precipitation flow within a catchment basin is essential in determining inputs and impacts to lakes,

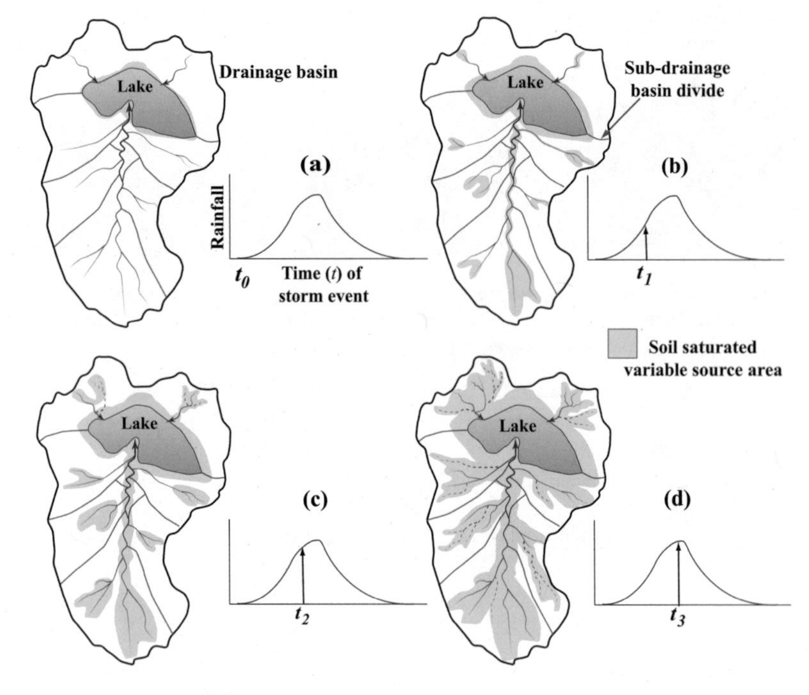

Figure 6.4. Hypothetical example of changing variable source areas within a drainage basin during a rainstorm event. As cumulative rainfall increases, so do source areas, including ponding (b and c) and ephemeral streams (d). The process reverses itself after the rain event, and source areas eventually return to where they started (a).

streams, and wetlands as well as impacts to the underlying aquifer. This knowledge is essential for understanding how rainwater or snowmelt interacts with soil, geology, and biota of the system as well as for flood potential and subsequent planning and best management practices. For example, flow from urban or agricultural areas can transport pollutants directly into a wetland, lowering water quality and causing deleterious effects to the system.

Throughflow (also known as *subsurface stormflow* or *interflow*) is water that infiltrates into the soil rapidly as a result of high permeability soils being underlain by zones of lower permeability; macropores such as fractures, root systems, or animal burrows; or a layer of low hydraulic conductivity that leads to the formation of perched water tables (Figure 6.5a–c). Therefore, throughflow can be an important and rapid source of flow to lakes and wetlands in areas with steep slopes or conductive soils and where the soil profile becomes saturated so that water can move through large pores. In many forested catchments, surface runoff is rare. Soil infiltration rates are never exceeded by rainfall and confined streams limit opportunities for formation of saturated source areas. Instead, given appropriate soil conditions, water may be rapidly transferred downslope as subsurface flow. This process is enhanced where there is an impervious lithology layer that leads to the formation of perched water tables that cause soils to saturate and become highly conductive (Weiler et al. 2005).

Therefore, flow along or above the surface (overflow) is a result of two mechanisms: (a) *infiltration-excess* overland flow (Hortonian) and (b) *saturated-excess* overland flow (Figure 6.5b). Hortonian overland flow is relatively insignificant as runoff in areas of humid vegetated basins, where the soil infiltration capacity typically exceeds all but the most extreme storm events. Infiltration-excess flow is applicable in such areas of low vegetation as arid or semiarid regions, paved urban areas, or areas where the soil surface is thin or has been drastically disturbed, such as in agricultural watersheds (Freeze 1972; Dunne, Moore, and Taylor 1975). Saturated-excess flow is significant in humid regions with dense vegetation and topographic highs sloping down to a flat valley or wetland where the water table is close to the surface (Figure 5.23c).

Rainfall infiltration recharge to the surficial aquifer system occurs at a much slower rate than precipitation and rainwater flow to the lake system. Therefore, a lake can be both a recharge or discharge system in response to seasonal or environmental changes. In Figure 6.5a, for example, the lake system is maintained by baseflow from the water table aquifer. During a storm event the response of the lake system via the various processes depicted in Figures 6.5b and 6.5c is much faster than percolation of rainfall through the vadose zone. As the storm continues (Figure 6.5d), runoff to the lake increases until the lake stage is greater than the surrounding water table. At this point flow reverses itself and the lake is recharging the surficial aquifer system. As the infiltrated water reaches the surficial aquifer system, the water table's hydraulic gradient will increase until it returns to the original conditions of lake discharge (Figure 6.5a).

6.3. Storm Runoff and Baseflow

Storm runoff can be characterized as runoff with high rates of discharge that reach a lake (or stream channel), resulting in increased lake stage and storage after a relatively short period of rainfall (Figure 6.5c). Rainwater that infiltrates into the groundwater moves along much longer pathways and at much slower velocities, intercepting the lake over a longer and more sustained

Figure 6.5. Four-part illustration of storm and runoff processes in a catchment basin as it impacts a lake or another wetland system.

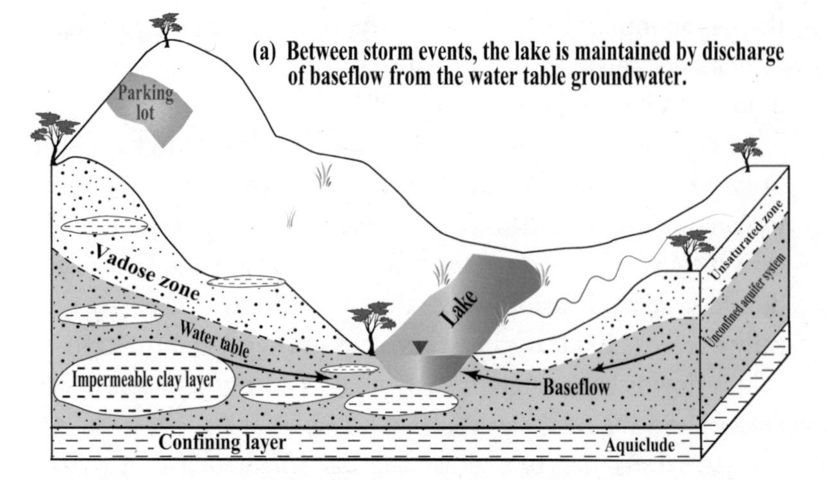

(b) Intitial rainfall causes infiltration to water table, surface flow where litholgy layers of low hydraulic conductivity are near the surface, overland flow near the base of the basin, and direct precipitation to lake.

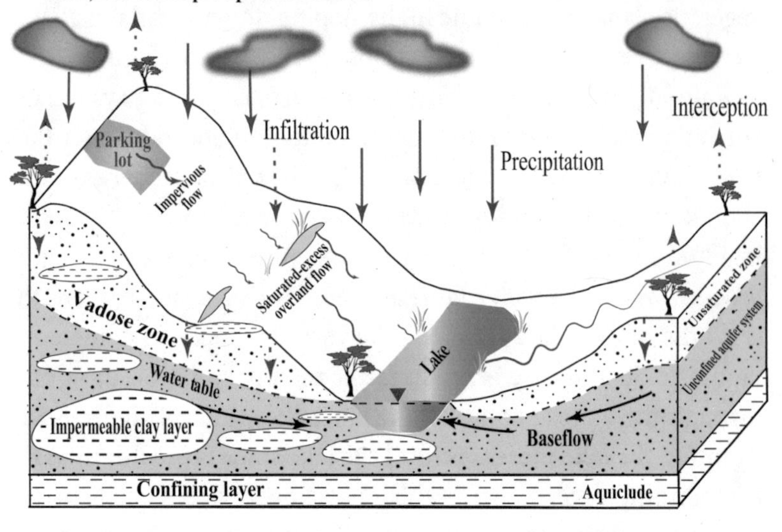

period of time. During periods of little or no rain, lake, streams, and wetlands are sustained by slow-moving groundwater via the surficial aquifer system, known as baseflow or, in some instances, *delayed* or *dry-weather flow*. In response to a rainstorm, lake levels (and storage) increase dramatically due to a "relatively" immediate discharge of storm runoff from the catchment and direct precipitation into the lake system. The terms *quickflow* and *delayed flow* for the components of storm runoff and baseflow were proposed by Hewlet and Hibbert (1967), where quickflow is the characteristic rapid response of a wetland to rainfall and catchment runoff

(c) As the rainfall event increases so does surface flow and throughflow. The lake stage and area will increase correspondingly. The water table will rise gradually as rainwater percolates through soils and lithology.

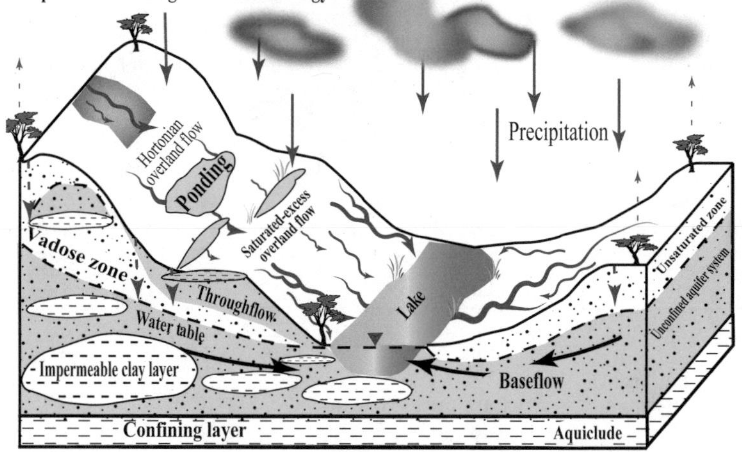

(d) As the lake level increases due to the rainfall processes, the lake may begin to discharge to the water table as a source of recharge when flow reverses from lake to groundwater.

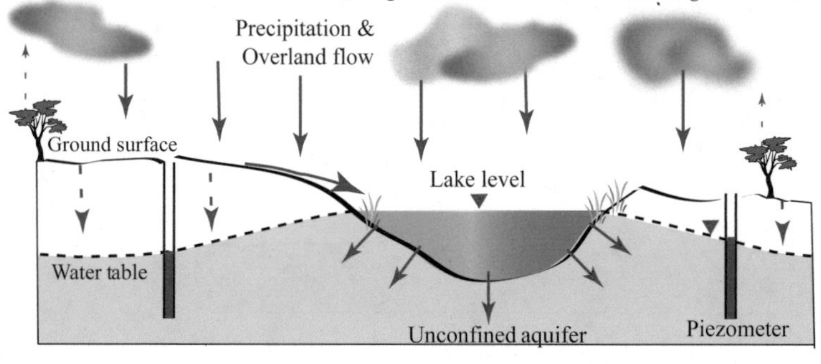

while baseflow is contributed to by slow release of stored water; the initial baseflow represents the contribution from previous events, including quickflow measured from an earlier upstream gauging station.

The concepts of quickflow and baseflow are important when determining the effects of rainfall or storm events in relation to runoff volumes for flood management as well as mass balance estimates to lake systems. Streamflow or lake response to a storm event is also often divided into quickflow and baseflow. Quickflow is often referred to as *direct runoff* or as *surface runoff* but, as noted previously, can include throughflow. During floods, quickflow is of the greatest relevance, but, particularly for modeling, baseflow must be considered where it provides a significant contribution to a stormwater hydrograph (Figure 6.6).

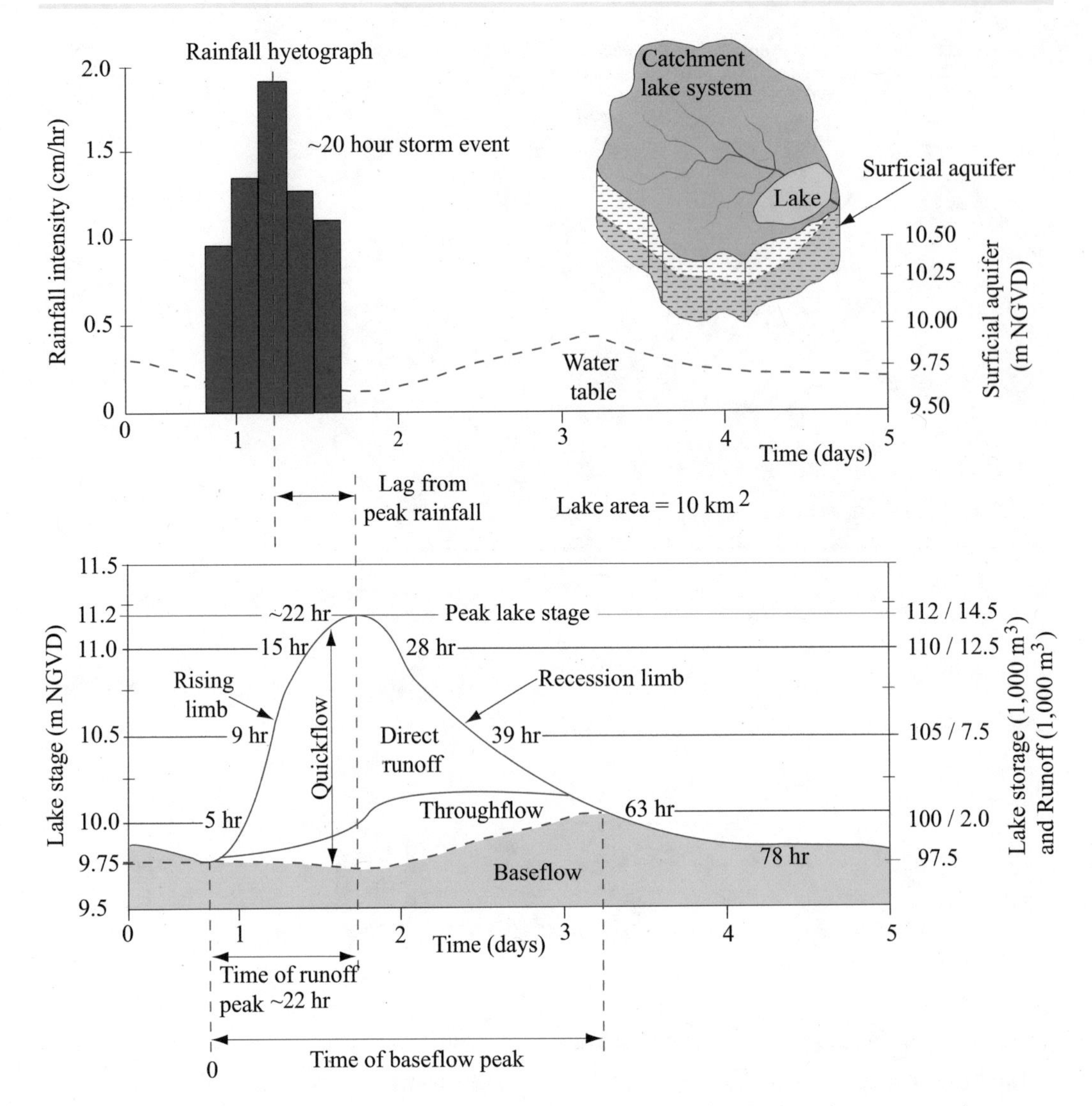

Figure 6.6. Hypothetical hyetograph with corresponding lake hydrograph and surficial aquifer piezometer demonstrating the typical response of a lake system to a storm event. (Note: lake storage is based on the *conical method* of determination).

There are a range of processes that contribute to the conceptual baseflow hydrograph as shown here. The initial baseflow represents the contribution from previous events; then as the hydrograph rises, baseflow can be depleted as water enters bank storage or is removed by transmission loss. Later, baseflow can increase as bank storage reenters the lake, or through other processes such as interflow and discharge from groundwater (Laurenson 1975).

Generally, quickflow will be explicitly modeled, as by a runoff-routing model, for example (section 7.5), or data from the lake stage or stream gauge. Then estimates of baseflow must be incorporated to produce a runoff hydrograph that represents an unbiased estimate of the peak flow. Baseflow provides a significant contribution to peak flows in around 70% of Australian catchments (Ball et al. 2016).

6.4. Separation of Baseflow and Quickflow

In Figure 6.6, the flow components have been separated into direct runoff (quickflow), throughflow, and baseflow (groundwater flow). To accurately quantify these components, intensive field studies are required to distinguish paths by which water has reached a stream or lake. The boundary between quickflow and baseflow is difficult to define and depends on the geological structure and composition of the basin, land use, and climate. Permeable aquifers, such as limestone and sandstone, generally have high baseflow contributions, whereas impermeable lithology, such as clays, or impermeable urban areas provide little or no water to baseflow. Baseflow tends to be higher after times of wet weather and can be low during prolonged droughts. Groundwater use from pumping can also reduce baseflow (see section 9.2.3).

During the extent of a storm event, as the stream and wetland levels increase, as seen in the rising limb, the baseflow continues to fall (Figure 6.6) until the precipitation percolates down to the water table. The baseflow usually increases at its highest level at the end of the surface runoff, rather than at the rise of the hydrograph, resulting in increased flow to streams and lakes after a significant storm event. When the recession limb drops to the level of the water table, groundwater flow provides the total input to wetland systems, as demonstrated by the general recession curve.

In practice, where these data are lacking, separation of hydrograph components is made based on travel time, which matches analytical methods, rather than on the basis of the physical processes (Maidment 1993). A variety of techniques have been developed to separate baseflow, including software from the US Geological Society's Hydrograph Separation Program that performs hydrograph separation, estimating the groundwater, or baseflow, a component of streamflow.

Horton (1933) described the *normal depletion curve* or *master baseflow recession curve* as a characteristic graph of flow recession for a specific stream or lake by compiling the historic recession curves and then superimposing many of them together to generate a curve similar to the unit hydrograph discussed in section 6.5.4b.Recession curves often take the form of an exponential decay function:

$$Q_t = Q_0 e^{-at} \tag{6.1}$$

where Q_t is the flow at some time t after the recession started, Q_0 is the flow at time t_0, a is an exponential decay or recession constant for the basin, and t is the time since the recession began having the dimensions of time (Singh and Stall 1971). The recession constant a can be expressed by

$$e^{at} = \frac{Q_t}{Q_0} \text{ and } -at = ln\frac{Q_t}{Q_0}$$

where

$$a = \frac{1}{t} ln\frac{Q_t}{Q_0}. \tag{6.2}$$

Thus, Q_0 is the baseflow at time = 0 and Q_t is the baseflow at t time units later. Obviously, this constant a will be some number less than one and will approach one for flat recessions and approaching zero for steep recessions. Farvolden (1963) investigated the effects of the hydrogeology on the shape of the recession curve of stream. He found that streams in limestone regions where much of the drainage takes place in the subsurface are characterized by flat recession. Areas of low permeability, such as granite, typically have steep baseflow recessions.

This exponential relationship of baseflow to a lake or stream can be examined by utilizing the conservation of mass that requires the baseflow to be balanced by the change in groundwater storage. During recession the inflow to the groundwater reservoir (i.e., surficial aquifer system) can be assumed to be zero, and based on the discussion in section 9.12, the baseflow to the lake or stream would be a function of the groundwater elevation ($\bar{h}$) (Figure 6.7).

As such the hydraulic head (and hydraulic gradient) will decrease with time, reducing the baseflow (groundwater) discharge to the wetland such that

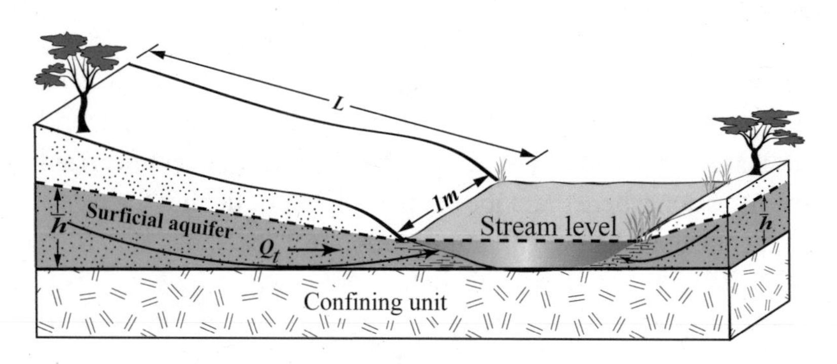

Figure 6.7. Basic illustration of baseflow (Q_t) to a lake or stream system.

$$Q_t = f(\bar{h}) \tag{6.3}$$

where Q_t is discharge [L^3] per meter width of the lake or stream [L^2t^{-1}] (Hornberger et al. 2014). The rate of change of the groundwater stored will be a function of the $\bar{h}$ (average head), the specific yield (S_y) (or storativity S of a confined aquifer; see section 9.2.2), and the length of the aquifer L (Figure 6.7). Therefore, based on the equation of continuity,

$$S_y L \frac{d\bar{h}}{dt} = -Q_t = f(\bar{h}). \tag{6.4}$$

This states that baseflow per unit (meter) width to a stream channel or reservoir is proportional to the current storage in the aquifer. Or using an equation of direct proportionality

$$Q_t = aS_y L\bar{h} \tag{6.5}$$

where a is a proportionality constant having dimensions of time [-t] Substituting equation (6.5) into equation (6.4) gives

$$\frac{d\bar{h}}{dt} = -a\bar{h} \tag{6.6}$$

by integrating

$$\bar{h} = \bar{h}_0 e^{-a} \tag{6.7}$$

where $\bar{h}_0$ is the mean hydraulic head at the initial time or beginning of the recession curve. Then from equation (6.6)

$$Q_t = \left(aS_y\bar{h}_0\right)e^{-at} \tag{6.8}$$

which brings us back to equation (6.1), or $Q_t = Q_0 e^{-at}$.

Meyboom (1961) analyzed the baseflow recession component

of a stream hydrograph of Elbow River to attain the quantity of groundwater recharge of the Calgary area (Alberta, Canada) using Boussinesq's (1877) exponential decay function. Meyboom found discharge versus time plotted on semilogarithmic paper linearized the recession where the slope of the straight line defines the recession constant (Figure 6.8) and equation (6.1) becomes

$$Q_t = \frac{Q_0}{10^{t/t_1}} \tag{6.9}$$

where Q_0 is the discharge at time $t = 0$ and t_1 is the time at one log cycle later ($Q_0 \times 0.1$) and t equals any time of interest when the value of Q is of interest. For example, if $t = t_1$ then

$$Q_t = \frac{Q_0}{10} \tag{6.10}$$

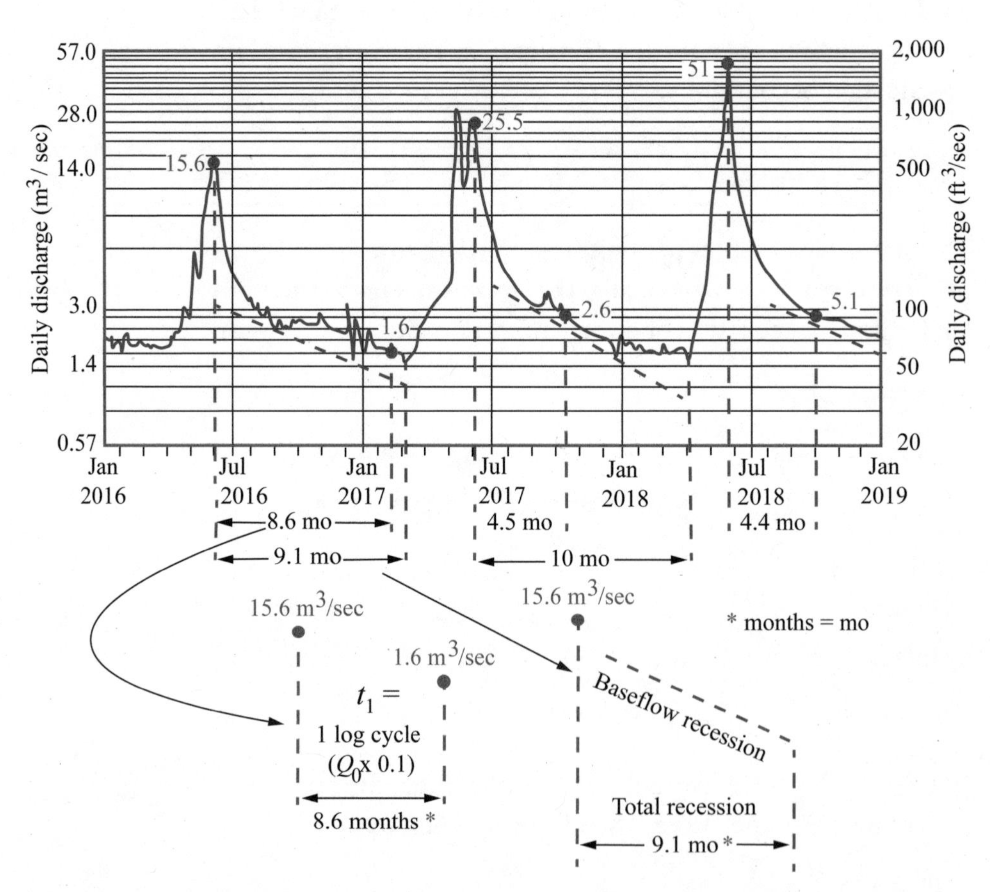

Figure 6.8. Stream hydrograph for Little Bighorn River showing baseflow recession. Stream data are from US Geological Survey water data gauge records for Little Bighorn River station number 06289000 at https://nwis.waterdata.usgs.gov/nd/nwis.

The total potential groundwater volume Q_{tp} is the baseflow amount that would be discharged by a complete groundwater recession (Meyboom 1961) and can be determined by integrating equation (6.9) over the time of interest such that

$$Q_{tp} = \int_{t_0}^{t} Q\,dt = -\frac{Q_0 t_0 / 2.3}{10^{t/t_0}} \Bigg\{ \begin{matrix} t \\ \\ t_0 \end{matrix} . \qquad (6.11)$$

Meyboom (1961) used equation (6.12) to determine baseflow recharge between recessions where he set t_0 to equal zero and t to equal infinity, thus

$$Q_{tp} = \frac{Q_0 t_1}{2.3}. \qquad (6.12)$$

Meyboom (1961) termed this the *total potential groundwater discharge* where the volume is defined as the total volume of groundwater that would be discharged by baseflow during an entire recession if complete depletion takes place uninterrupted. This is the total water in groundwater storage at the beginning of the recession. The difference between the remaining potential groundwater discharge at the end of the recession and the total potential groundwater at the beginning of the recession is a measure of the recharge that takes place between recessions (Meyboom 1961).

Case Study 6.1

Little Bighorn River Groundwater Recharge

P. Meyboom, 1961, "Estimating Ground-Water Recharge from Stream Hydrographs." *Journal of Geophysical Research* 66 (4): 1203–14.

Equations (6.9)–(6.12) were employed by Meyboom (1961) to estimate groundwater recharge between recessions for the Little Bighorn River (Figure 6.8). The first recession total potential recharge (Q_{tp}) is calculated from equation (6.12) as

$$Q_{tp} = \frac{15.6\ \text{m}^3/\text{sec} \times 8.6\ \text{months} \times 30\ \text{days/month} \times 1{,}440\ \text{min/day} \times 60\ \text{sec/min}}{2.3}$$

$= 151{,}192{,}487\text{m}^3$ or about $1.5 \times 10^8\ \text{m}^3$

The value of Q_t at the end of the recession that lasts 9.1 months is

$$\frac{Q_0 t_0 / 2.3}{10^{t/t_0}} = \frac{1.5\times10^8}{10^{9.1/8.6}} = \frac{1.5\times10^8}{11.6} = 1.3\times10^7\,\mathrm{m}^3.$$

Baseflow storage remaining at the end of the first recession can be determined by evaluating equation (6.12) from *t* equals 9.1 to *t* equals infinity, or by simply subtracting actual groundwater flow from the total potential discharge or

$$\frac{Q_0 t_1}{2.3} - \frac{Q_0 t_0 / 2.3}{10^{t/t_0}}$$

$1.5 \times 10^8\ \mathrm{m}^3 - 1.3 \times 10^7\ \mathrm{m}^3 = 1.37 \times 10^8\ \mathrm{m}^3$

The second recession has an initial value Q_0 of 25.5 m^3/sec and takes approximately 4.5 months to complete a log cycle of discharge, thus the total potential discharge is

$$Q_{tp} = \frac{25.5\ \mathrm{m}^3/\mathrm{sec} \times 4.5\ \mathrm{months} \times 30\ \mathrm{days/month} \times 1,440\ \mathrm{min/day} \times 60\ \mathrm{sec/min}}{2.3}$$

129,318,261 m^3 or $1.3 \times 10^8\ \mathrm{m}^3$.

Q_t at the end of the recession that lasts 10 months is

$$\frac{1.3\times10^8}{10^{10/4.5}} = \frac{1.3\times10^8}{166} = 7.8\times10^5\,\mathrm{m}^3$$

and the remaining baseflow storage at the end of the recession is

$1.3 \times 10^8\ \mathrm{m}^3 - 7.8 \times 10^5\ \mathrm{m}^3 = 1.29 \times 10^8\ \mathrm{m}^3$.

The recharge that takes place between the two recessions is the difference between the potential recharge (Q_{tp}) of the second recession and the remaining groundwater potential of the first recession ($1.3 \times 10^8\ \mathrm{m}^3$), or $1.0 \times 10^6\ \mathrm{m}^3$.

In Meyboom's method, baseflow recession curves are derived manually, connecting successive points of minimum stream discharge, and are represented with straight lines by plotting the hydrograph flow data semilogarithmically. Such recession curves are used to calculate recharge volumes that occur between recessions.

The use of exponential regression models for derivation of baseflow recession curves (e.g., regression equations) uses automated calculations and processing of large amounts of data with reduced effort and time involved. Similar programs for automated estimation of groundwater recharge (e.g., RORA) were introduced in past years by Rutledge and Daniel (1994), Rutledge (1998), and Arnold and Allen (1999). The recharge algorithm they developed is an automated derivation of the Rorabaugh (1964) hydrograph recession curve displacement method that uses daily streamflow. The US Geological Society has a computer program called RECESS that is used to determine the recession index and to define the master recession curve from analysis of streamflow records. (For information on the RECESS and RORA models, see https://water.usgs.gov/ogw/recess). TR-55 is a method developed by the US Department of Agriculture (1986) that provides several techniques for baseflow separation used for small watersheds and works especially well with urbanized watersheds. The presented method predicts the peak rate of runoff as well as the total volume.

Other methods for baseflow separation are provided in the references and include Lyne and Hollick (1979), Pettyjohn and Henning (1979), Chapman (1991), Szilagyi and Parlange (1998), and Eckhardt (2005).

6.4.1. Graphical Separation of Baseflow Using Hydrographs

Three basic graphical methods of baseflow separation are the (a) *straight line method*, (b) *fixed base method*, and (c) *variable slope method.*

The *straight line method* (Figure 6.9a) involves drawing a horizontal line from where the rising limb of runoff begins (B) to where it intersects the recession limb. This method is applicable to ephemeral streams where baseflow is negligible. An improvement to this method is to connect the beginning point of the rising limb at an angle to the recession limb where normal baseflow begins at D (Chow, Maidment, and Mays 1988).

For the *fixed base method* (Figure 6.9b), the runoff is assumed to end at a fixed time N after the hydrograph peak. The baseflow before the surface runoff begins is projected ahead to the time of the peak. A straight line is used to connect this projection at the peak to the point on the recession at time N after the peak. Linsley, Johler, and Paulhus (1975) and Pettyjohn and Henning (1979) define N as $A^{0.2}$ where A is the area of the drainage basin and 0.2 is an empirical constant.

In the *variable slope method* (Figure 6.9c), the baseflow curve

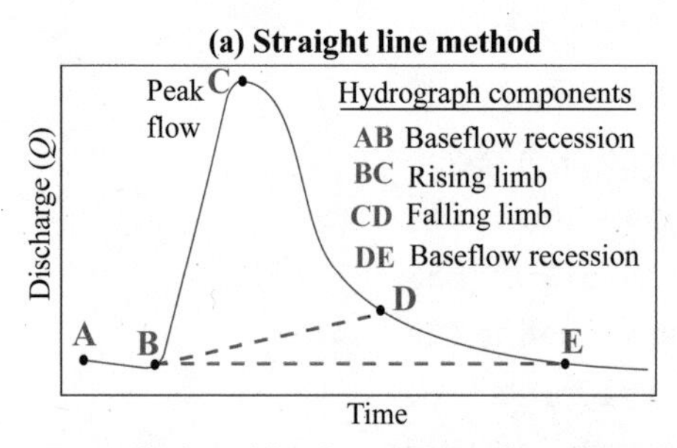

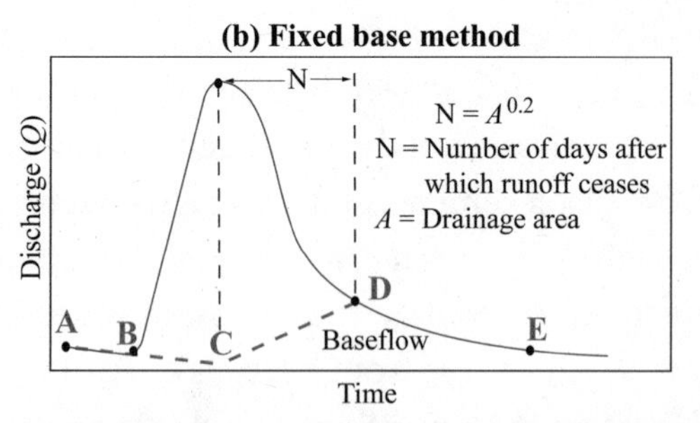

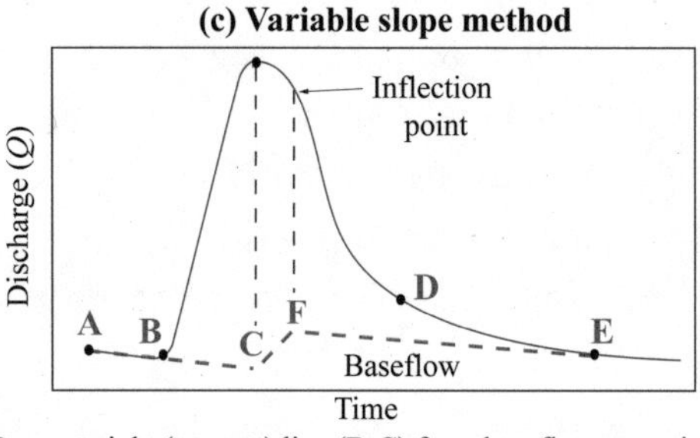

Figure 6.9. Graphical methods for separating baseflow from quickflow.

before the surface runoff begins is extrapolated forward to the time of peak discharge, and the baseflow curve after surface runoff ceases is extrapolated backward to the time of the point of inflection on the recession limb. A straight line is used to connect the endpoints of the extrapolated curves.

Case Study 6.2

Indirect Groundwater Discharge to the Great Lakes Using Hydrograph Separation

D.J. Holtschlag and J.R. Nicholas, 1998, "Indirect Ground Water Discharge to the Great Lakes," US Geological Survey Open-File Report 98-579, 25 pp.

Prior to 1998 estimates of basin water supply for the Great Lakes, direct groundwater discharge generally was assumed to be small, typically within the uncertainty limits associated with estimates of precipitation, surface runoff, diversions, and evaporation (Holtschlag and Nicholas 1998). Grannemann and Weaver (1998), summarizing studies of direct groundwater discharge to the Great Lakes, generally corroborates this assumption. For Lake Michigan, the only Great Lake for which enough information is available to make a lake-wide estimate, direct groundwater discharge is only about 5% of the basin water supply for the lake.

Direct groundwater discharge to the Great Lakes is difficult to estimate because it is determined by geologic and hydraulic properties that are difficult to measure locally and are highly variable from place to place. Therefore, indirect groundwater discharge has not been explicitly considered in estimates of basin water supply for the Great Lakes. Instead, it has been incorporated into the streamflow component of basin water supply. This perspective, however, does not account for the substantial amount of groundwater that flows indirectly into the Great Lakes as a component of streamflow.

Streamflow includes a surface runoff component and a groundwater component. The groundwater component of streamflow in the Great Lakes Basin constitutes indirect groundwater discharge to the Great Lakes. Indirect groundwater discharge can be readily estimated from streamflow records that are available from gauging stations throughout the Great Lakes Basin.

Holtschlag and Nicholas (1998) estimated the groundwater component of streamflow based on hydrograph separation by analyzing 5,735 days of daily streamflow values from 195 stream gauging stations, an average of 29.4 years per station. Among the selected gauging stations, the average groundwater component of streamflow was 67.3%. The indirect groundwater discharge to the Great Lakes ranges from 22% (Lake Erie) to 42% (Lake Ontario) of the basin water supply for the Great Lakes (Table 6.1). Thus, even without explicitly accounting for direct discharge, groundwater is a large component of the basin water supply of the Great Lakes (Holtschlag and Nicholas 1998).

In general, the highest percentages of groundwater discharge are associated with basins having the highest percentages of land area covered by coarse, textured soils and undisturbed vegetation. Lowest percentages of groundwater discharge are associated with basins having the highest percentage of impervious areas or

Table 6.1. Basin water supply for the Great Lakes

Lake[a]	Overlake precipitation (%)	Surface runoff (%)	Indirect groundwater discharge (%)
Superior	56.3	11.0	32.7
Huron	42.2	16.3	41.5
Michigan	56.2	9.3	34.5
Erie	53.5	24.3	22.2
Ontario	34.8	22.8	42.4

Source: Modified from Holtschlag and Nicholas (1998).
[a] Lakes listed from largest to smallest surface area.

basins having the highest percentage of thin, tight soils overlying less permeable geologic materials (Holtschlag and Nicholas 1998).

6.5. Losses

In stormwater and flood hydrology, *losses* refer to any rainfall that is not converted to quickflow. The amount of loss subtracted from storm rainfall is *rainfall excess*, or quickflow that is produced by the rainfall excess on the drainage basin. Losses are a result of evapotranspiration after being intercepted by vegetation or stored in surface depressions (puddles, ditches, stormwater ponds, etc.) or infiltrated into baseflow through lakes or streambeds (*transmission loss*). Estimation of losses can be calculated using historic data where there are data on runoff, catchment area and rainfall depth; losses can also be estimated as the difference between rainfall volume and volume of the quickflow hydrograph (stormwater hydrograph with the baseflow removed) (Hill, Mein, and Siriwardena 1998).

The loss models used in flood modeling are often simple, based on two parameters, one to characterize the *initial loss* (the water required to wet up the catchment) and one to characterize the *continuing loss*. The output of these models is the rainfall excess that is then used to generate a direct flow hydrograph. Loss models can derive detached individual models (e.g., the rainfall excess can be calculated separately) or be integrated within a watershed modeling system (Ball et al. 2016).

6.5.1. Basic Processes of Infiltration

In areas where hydrograph data are not available, losses must be determined by calculating infiltration as well as separately estimating *detention*, or *depression storage*, and *interception*. Typically, the simple empirical loss models account for infiltration losses only and depression, interception, and evapotranspiration losses are usually not directly defined.

Hillel (1982, 211) defines infiltration as "the term applied to the processes of water entry into the soil by downward flow through all or part of the soil surface. The rate of this process, relative to the rate of water supply, determines how much will enter the root zone and how much, if any, will runoff." He further defines the volume flux of water flowing into the profile per unit of soil area as the *infiltration rate* or *infiltration velocity* as flux has units of velocity [Lt^{-1}] (212).

6.5.1a. Capillary Potential (Suction) and Adsorption

The only two forces acting on saturated flow in an unconfined porous medium are gravity and friction (see section 9.7). However, for *unsaturated flow*, a third force, *suction* (also termed *tension*), is involved. The porous medium, or lithology, consists of a matrix of particles where the void spaces are only partially filled with water during unsaturated conditions (Figure 6.10). The water is attracted to the particles because of electrostatic or cohesive forces among the water molecule polar bonds of the liquids and adhesive forces between the liquid and particle surfaces (see sections

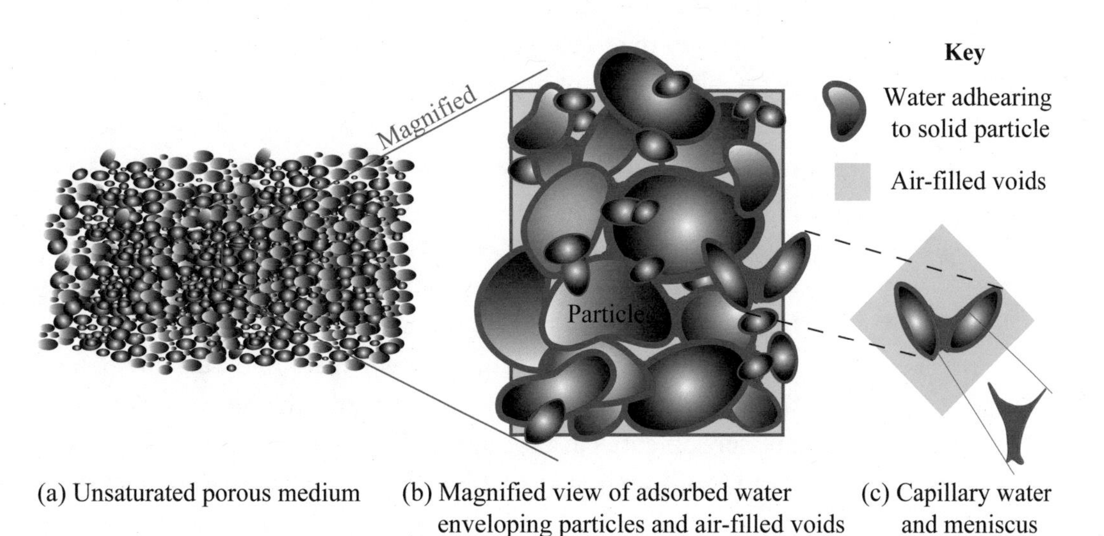

Figure 6.10. Process of capillary formation and adsorption in an unsaturated porous three-phase (gas, liquid, and solid) medium.

2.1 and 4.2.2). The water adhesion to the particle surfaces draws the water up around the particles, leaving air in the void centers between particles and causing a capillary effect (Figure 6.10b and 6.10c). (For further discussion on capillary flow as it relates to pressure potentials, see section 4.2.2 and Figure 4.5.)

Capillary potential or *negative pressure potential* in the soil matrix is due to the cohesive and adhesive bonds that cause soil particles to attract and bind water thus lowering its potential energy to that below saturated conditions (superatmospheric, or *positive* pressure potential) and atmospheric conditions (neutral). In an unsaturated three-phase (gas, solid, and liquid) soil system, capillary conditions result from surface tension of water and its contact angle with the solid particles (Figure 6.10c). In this unsaturated soil matrix, curved menisci form and the equation of capillary potential (Hillel 1971) can be applied such that

$$P_0 - P_c = \Delta P = \gamma\left(\frac{1}{R_1} + \frac{1}{R_2}\right) \tag{6.13}$$

where P_0 is the atmospheric pressure, conventionally taken as zero; P_c is the pressure of soil water, which can be less than atmospheric; ΔP is pressure deficit, or *subpressure*, of soil water; γ the surface tension of water; and R_1 and R_2 are the principle radii of curvature of a point on the meniscus. From equations (6.13) and (4.4) it is evident that the smaller the radii of the meniscus or capillary tube, the greater the suction or subpressure. This inverse relationship of negative potential and capillary tube size is analogous to soil systems where soils with smaller pores (clays) hold water at a more negative *pressure head* than soils with larger pores (coarse sands).

A second mechanism, *adsorption*, in addition to capillary action, also impacts negative pressure potentials in soils. Adsorption results in forming a hydration envelope or film around the particles (Figure 6.10b and 6.10c). Thus, equations of capillary potential are not enough to adequately describe the suction or negative pressure potential in a soil system. Adsorption is especially important in soils with high suctions, such as clays, that have an electric negative and positive double layer and exchangeable cations that increase adsorption dramatically (Hillel 1982). In sandy soils, adsorption is relatively minor and capillary forces predominate.

However, negative pressure potential, in general, results from the combined effect of both mechanisms, which cannot be separated as they are in a state of internal equilibrium. For example, if the adsorptive films change, their corresponding capillary wedges will also change (Figures 6.10b and 6.10c). Therefore, soil scientists are now using the more descriptive term of *matric potential*, as it expresses the total effect as a result of water's bipolar affinity for the whole soil matrix, which includes pores and sediment surfaces.

The effect of soil suction can be observed by placing a column of dry soil, with its bottom placed vertically, in a container of water. Moisture will be drawn up into the sand column above the water surface at a point where the effects of soil suction and gravity forces are equal and where the height is determined by the lithology. For example, the height ranges from a few millimeters in coarse sand to several meters in a clay soil due to the clay's smaller capillaries and higher adsorption capability.

6.5.1b. Vertical Movement of Water in an Unsaturated Porous Medium

The principles discussed in sections 9.3–9.8 on the development of groundwater flow illustrate how *Darcy's law* describes flow through a porous medium such that

$$q = -K\frac{dh}{dl} \tag{6.14}$$

where hydraulic conductivity K depends on the properties of the fluid and porous lithologic material. The negative sign indicates that head is decreasing in the direction of flow due to friction (see section 9.8). Darcy's law demonstrates that the volumetric flow rate is a function of the flow area, elevation, fluid pressure, and a proportionality constant K. It may be stated in several different forms depending on the flow conditions. Since its innovation, it has been found valid for any Newtonian fluid, and while it was established under saturated flow conditions, it may be adjusted to account for unsaturated flow such as infiltration.

To accommodate for infiltration, the third force of suction must be incorporated into the Darcy's law equation. In a saturated porous medium, the head h is measured in dimensions of height that represent the energy per unit of weight of the fluid (see sec-

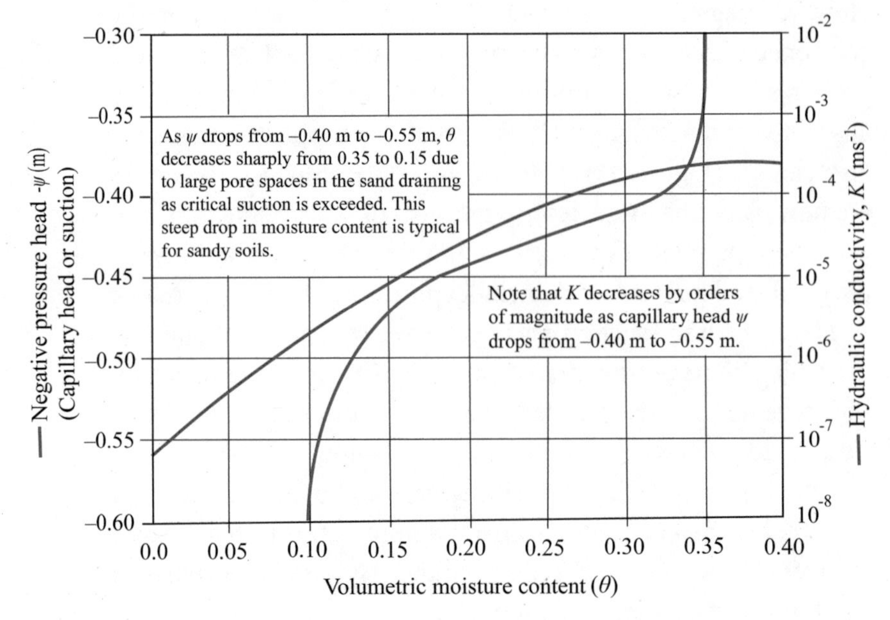

Figure 6.11. Comparing the change in capillary head and hydraulic conductivity with variation in moisture content for a fine sand. Modified with permission from Hornberger et al. (2014).

tions 9.3–9.6). In an unsaturated porous medium, the total energy within the fluid due to soil suction is also referred to as *capillary pressure head* ψ. The suction head, as well as the hydraulic conductivity, will vary with the soil moisture content θ of the soil matrix (Figure 6.11). The total head h is the sum of the suction and gravity heads

$$h = \varphi + z \tag{6.15}$$

where z is the elevation head. As demonstrated in Figure 5.29, *capillary pressure head* (suction) is a function of the moisture content θ, therefore Darcy's equation can be expressed as

$$q = -K(\theta)\frac{dh}{dl}. \tag{6.16}$$

Considering movement of water in the unsaturated zone as one-dimensional flow in the vertical (z) direction, referring to Hubbert's Treatise on Fluid Potential (see section 9.7) that states the velocity head in a porous medium is so small as to be negligible, and substituting for h, then equation (6.16) can be written as

$$q_z = -K(\theta)\frac{d(\varphi + z)}{dz}$$

or

$$-K(\theta)\frac{d\varphi}{dz} + 1 \tag{6.17}$$

The continuity equation (see section 9.11 and Figure 9.21) is required due to the transient nature of unsaturated flow. For flow in the vertical direction, the equation of continuity is

$$\frac{\partial\theta}{\partial t} = \frac{\partial q_z}{\partial z} \tag{6.18}$$

where the left side of the equation is the rate of mass θ change (volumetric wetness) in a control volume (see Figure 4.21) and the right side is the difference between rate of flow into and out of the control volume in the vertical direction. By applying the continuity equation to Darcy's equation

$$\frac{\partial\theta}{\partial t} = \frac{\partial}{\partial z} - \left[K\theta\left(\frac{\partial\varphi}{\partial z} + 1\right)\right] \tag{6.19}$$

$$q = -\left(K\frac{d\varphi\,\partial\theta}{\partial\theta\,\partial z} + K\right) \tag{6.20}$$

$$= -\left(D\frac{\partial\varphi}{\partial\theta} + K\right) \tag{6.21}$$

where D is the soil water diffusivity [Lt^{-1}] (see Philip 1957), or $K(\partial\varphi/\partial\theta)$.

Then, reapplying the continuity equation,

$$\frac{\partial\theta}{\partial t} = \frac{\partial}{\partial z}\left(D\frac{\partial\theta}{\partial z} + K\right) \tag{6.22}$$

which is the basic equation governing flow for one-dimensional transient or unsteady flow in an unsaturated porous medium and the foundation for a number of models used to estimate infiltration rates and cumulative infiltration. This equation was first described by Richards (1931) and termed the *Richard's equation.*

6.5.2. Infiltration Rate and Cumulative Infiltration

For the purposes of mass balance and investigating the effects of rainfall on watersheds and flow to lakes, the concern is with the infiltration rate (f) and cumulative infiltration (F) as it applies to losses. The processes of soil infiltration or the rate at which a soil can absorb water supplied to its surface (Hillel 1971) are affected by a variety of factors, such as:

a. Time from the onset of the rain: dependent on the soil moisture, the infiltration rate is usually high at the beginning, then decreases and eventually approaches a constant rate dependent on the characteristic of the soil;
b. Initial water content: the wetter the soil, the lower initial infiltrability (owing to smaller suction gradients) and the quicker it approaches its constant rate;
c. Hydraulic conductivity: the higher the hydraulic conductivity, the higher the infiltrability;
d. Soil surface conditions: highly porous and "open" structured soils tend to have a higher initial infiltrability than that of uniform soils, but the final infiltrability remains unaffected as it is limited by the lower conductivity transmission zone beneath. However, if the soil surface is compacted or covered by a lower conductive crust than the infiltration rate will be lower than the underlying layer;
e. Layered soil profiles: impeding layers within the profile may retard water movement during infiltration; and
f. Other such external factors as slope of the landscape, vegetation cover, rainfall intensity and land use.

If water is ponded on the surface, then infiltration $[Lt^{-1}]$ is at its maximum, or *potential* infiltration rate. In contrast, the water supply from a light rainfall is less than the potential infiltration rate, thus the actual infiltration will be less than the potential. Because most rainfall runoff investigations are concerned with flooding and runoff management, most infiltration equations and models describe or include the potential infiltration rate (Chow, Maidment, and Mays 1988).

The accumulated depth of infiltration over a time period is the *cumulative infiltration* (F) and is equal to the integration of the infiltration rate over the same time period:

$$F(t) = \int_0^t f(\tau)\,d\tau \tag{6.23}$$

where τ is a dummy variable of time in the integration. The infiltration rate is the time derivative of the cumulative infiltration (Chow, Maidment, and Mays 1988):

$$f(t) = \frac{dF(t)}{dt} \tag{6.24}$$

6.5.3. Hydrologic Soil Group

Pedogenesis, the process of soil formation, can be attributed to the combined influences of climatic, biotic, and topographic factors. Soils can be formed in situ from parent rock or deposited as *allochtonous* material by glaciers or rivers. The fundamental significance of soil in watershed dynamics is that it acts to either contain or discharge water (Gregory and Walling 1973). The soil infiltration capacity has a similar effect to the dominant rock type in a drainage basin. Unconsolidated soils allow water to infiltrate increasing watershed storage to the baseflow aquifer or reducing travel time as throughflow. This reduces the peak discharge while increasing the lag time of a stream or wetland. Conversely, soils with low infiltration capacities, such as fine clay soils, increase runoff, thus reducing the lag time of a stream or wetland system. The way water is stored and transmitted through soils influences the production of sediment and waterborne solutes that, in turn, ultimately influence the hydrologic and morphologic framework of the streams and wetlands (Gregory and Walling 1973).

When investigating runoff processes, the area soil characteristics and their spatial distribution are incorporated into the majority of the models and equations developed to estimate runoff. Hydrologic soil groups are based on estimates of runoff potential. Soils in the United States are classified into four hydrologic groups (A, B, C, D) and three dual classes (A/D, B/D, and C/D) (US Department of Agriculture 2005). These groups are based on their infiltration rate when the soils are not protected by vegetation, are thoroughly wet, and receive precipitation from long-duration

Table 6.2. Hydrologic soil group ratings

Soil group	Infiltration rate (cm/hr)	Runoff potential
A	>0.762	Low
B	0.382–0.762	Moderate
C	0.128–0.381	High
D	0.0–0.127	Very high

storms. The group's general infiltration rates are given in Table 6.2 and are defined as follows:

Group A soils have a high infiltration rate (low runoff potential) when thoroughly wet. These consist mainly of deep, well-drained to excessively drained sands or gravelly sands. These soils have a high rate of water transmission.

Group B soils have a moderate infiltration rate when thoroughly wet. These consist chiefly of moderately deep or deep, moderately well-drained or well-drained soils that have moderately fine texture to moderately coarse texture. These soils have a moderate rate of water transmission.

Group C soils have a slow infiltration rate when thoroughly wet. These consist chiefly of soils having a layer that impedes the downward movement of water or soils of moderately fine texture or very fine texture. These soils have a slow rate of water transmission.

Group D soils having a very slow infiltration rate (high runoff potential) when thoroughly wet. These consist chiefly of clays that have a high shrink-swell potential, soils that have a high water table, soils that have a claypan or clay layer at or near the surface, and soils that are shallow over nearly impervious material. These soils have a very slow rate of water transmission.

For a soil assigned to a dual hydrologic group (A/D, B/D, or C/D) the first letter is for drained areas and the second is for undrained areas. Only the soils that in their natural condition are in group D are assigned to dual classes.

A unit on a soil map represents an area dominated by one or more major classifications of soils based on precisely defined limits and soil properties (Figure 6.12). A general soils map and data for

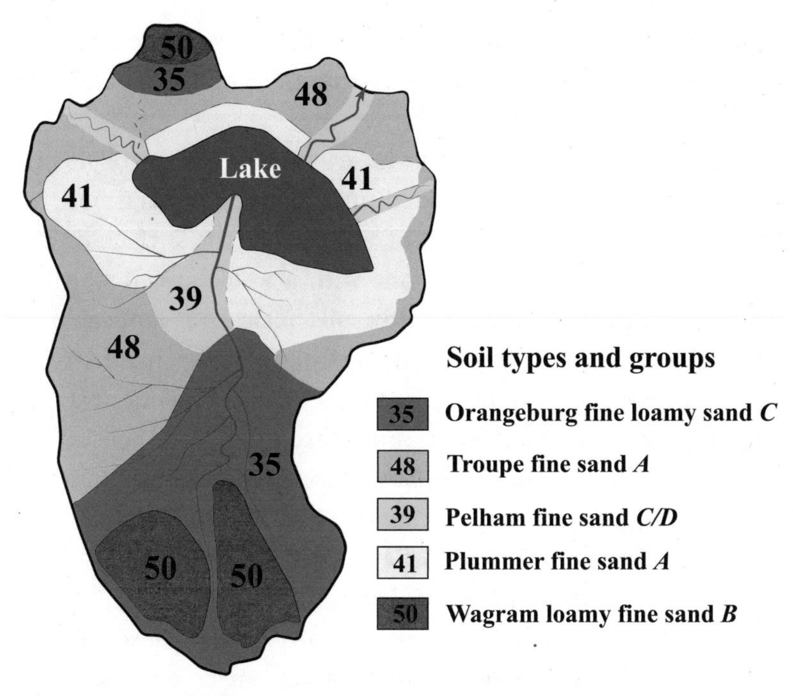

Figure 6.12. Example of a soil map for a hypothetical drainage basin.

any specific area in the United States can be generated from the Natural Resources Conservation Service website (https://websoilsurvey.nrcs.usda.gov/app/WebSoilSurvey.aspx).

Soils are a natural phenomenon and thus can vary over time and space. Thus, the range of some of the observed properties (i.e., infiltration rate) can vary and include some of the characteristics of the surrounding soils due to erosion and deposition, sedimentation, and other geologic processes. Therefore, to better understand the infiltration characteristics of the soil, it is necessary to do site-specific infiltration testing. One common method is to use a double-ring infiltrometer analysis. The general concepts of ring infiltrometers are discussed in section 10.3.2 and Case Study 7.1. (For details on ring infiltrometer methodology, see Bouwer 1986.)

6.5.4. Approximate and Empirical Infiltration Models

Various complex equations and approaches for modeling movement of water in soil in both unsaturated and saturated conditions have been developed to represent the reduction of infiltration capacity with time. These models attempt to simplify the processes involved in infiltration, such as applying the physical principles governing infiltration for simplified boundary and initial conditions.

Approximate models, such as those by Green and Ampt (1911) and Philip (1957), assume ponded surface conditions at time zero as well as uniform movement of water from the surface down through deep homogeneous soils with a well-defined wetting front, which may be valid for sandy soils but are not applicable to clay soils (Hillel 1989; Haverkamp, Rendon, and Vachaud 1987). These simplifying assumptions reduce the amount of physical soil data required from that of a numerical solution, but their validity under changing initial and boundary conditions are limiting (Haverkamp, Rendon, and Vachaud 1987).

Empirical models such as Kostiakov (1932), Horton (1940), and Holtan (1961) commonly relate infiltration rate or volume over a time period modified by certain soil properties. These models incorporate parameters usually estimated from a measured infiltration rate for a given soil condition. Models such as the rational method and Soil Conservation Service method assign runoff coefficients derived from such factors as land use and soil type to estimate losses from infiltration.

The problem with *simplifying models* is in accurately specifying soil parameters related to the area soil type and initial conditions that are applicable for use on a certain catchment (Maidment 1993). For example, Siriwardene, Cheung, and Perera (2003) did infiltrometer testing at twenty-one sites in eight urban catchments to estimate the infiltration parameters used for Horton's model and acknowledged the difficulty in selecting representative values for the infiltration parameters. They attributed this to "the significant variability with respect to soil type and land use in the catchment." Thus, in practice, the uncertainty of soil conditions and areal variability of soil properties do not justify the use of anything more but the simplest of models (Mein and Goyen 1988).

The Green-Ampt method is one of the more commonly used infiltration models with simplifying assumptions of flow resulting in an analytical solution utilizing a Darcy-type of equation. Skukla, Lal, and Unkefer (2003) and Williams (1996) compared Green-Ampt with other models and found that although successful in its application the results were not on average superior to models using simple empirical models. Skukla, Lal, and Unkefer (2003) used double-ring infiltrometers to analyze ten infiltration models, including Green-Ampt, and stated that Horton's model resulted in the best estimates for most land use conditions.

The Green-Ampt and Horton models are examples of approximate and empirical methods still widely in use. A summary of their governing equations is presented next. For more detailed discussion on these and other infiltration models mentioned, see Hillel (1971, 1982, and 1989); Mein and Larson (1973); Chu (1978); Chow, Maidment, and Mays (1988); Haverkamp, Rendon, and Vachaud (1987); Maidment (1993); Lee and Lim (1995); King (2000); and Xu (2003).

6.5.4a. Green-Ampt Method

When water is applied to a soil system, such as by rainfall or irrigation, there are four basic moisture zones associated with infiltration (Chow, Maidment, and Mays 1988) (Figure 6.13a):

1. *Saturation zone* near the surface;
2. *Transition zone* of unsaturated flow and relatively constant moisture content θ_t;
3. *Wetting zone* in which the moisture content θ_i decrease with depth; and
4. *Wetting front* where the change of moisture content between the wet soil above and dry soil below appears as a sharp plane of discontinuity.

Green-Ampt is a simplified approach that works especially well for infiltration into initially dry soils that are coarse in texture and exhibit a sharp wetting front. The main assumption is that there is a well-defined wetting front where the suction remains constant regardless of time and position. The next assumption is that behind the wetting front the soil is uniformly wet and of constant hydraulic conductivity (Hillel 1971). This implies an infiltrated uniform wetted zone separated by a sharp plane from a totally unfiltered zone where the K value changes abruptly at a suction value ψ and prevails at the wetting front with specific moisture content θ (Figure 6.13b).

These simplifying assumptions allow for an analytical solution using a Darcy-type equation

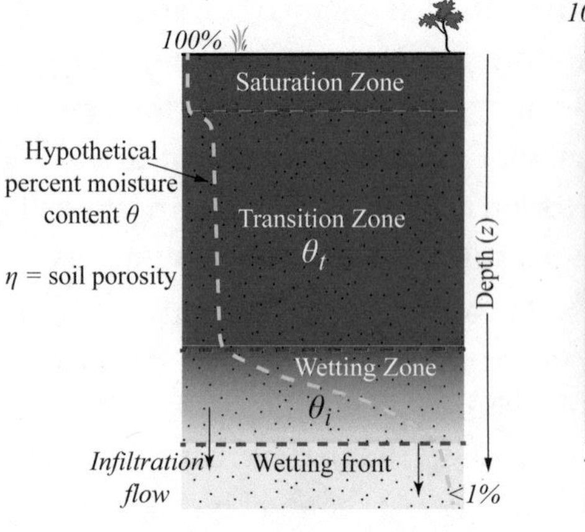

(a) Moisture zones during infiltration
$\Delta\theta = (\theta_t - \theta_i)$

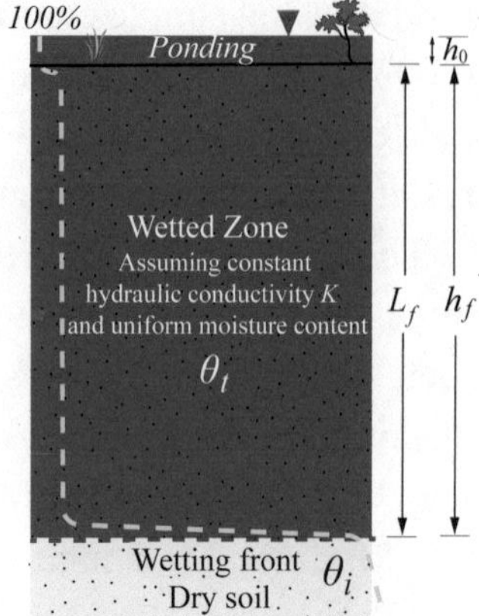

(b) Green-Ampt model
Ponding indicates that the potential infiltration rate is met.

Figure 6.13. Comparison of four moisture zones created during infiltration (a) with the Green-Ampt model assumptions (b).

$$f = \frac{dI}{dt} = K\frac{h_o - h_f}{L_f} \tag{6.25}$$

where f is the infiltration flux, or rate into the soil and through the transmission zone; F is the cumulative infiltration; K is the hydraulic conductivity of the transmission zone; h_0 is the pressure head at the surface entry (assuming a potential infiltration rate with ponding); and L_f is the distance of the surface from the wetting front. Assuming that ponding is negligible, then

$$\frac{dF}{dt} = K\frac{\Delta h_p}{L_f} \tag{6.26}$$

where Δh_p is the pressure decrease from the surface to the wetting front and L_f is the length of the wetting zone. This indicates that the pressure drop, or infiltration rate, linearly varies inversely to the length of the wetting front, or the infiltration rate decreases with depth.

One of the assumptions of Green-Ampt is a constant or uniform wetted zone from the surface to the wetted front, consequently the cumulative infiltration I should be equal to the product of the depth L_f and the wetness increment $\Delta\theta$ where

$$\Delta\theta = \theta_t - \theta_i \tag{6.27}$$

and θ_t is the transmission or wetted zone wetness during infiltration and θ_i is the initial soil wetness that occurs beyond the wetting front (Figure 6.13a and 6.13b). If θ_t is at saturation, then it is equal to the porosity (η_p) of a medium, or $\Delta\theta = \eta_p - \theta_i$. The moisture content θ is the ratio of the volume of water to the total volume within the control surface; hence, the increase in storage within the control volume as a result of infiltration is $L_f(\eta_p - \theta_i)$ for a unit cross section (Chow, Maidment, and Mays 1988). Therefore, this quantity is equal to F, the cumulative depth of water infiltrated into the soil, or

$$F(t) = L_f\,\Delta\theta \tag{6.28}$$

6.5.4b. Horton's Model

Horton's infiltration model (Horton 1933, 1939, and 1940) is one of the earliest and best-known models still widely used today. Horton recognized that during a wetting event (e.g., rainstorm) infiltration capacity, or potential infiltration rate (f_p), decreased with time until it reached a minimum constant rate (f_c). Horton attributed this decrease in infiltration primarily to factors operating at the soil surface rather than to flow processes within the soil (Xu 2003).

Horton's equation, which describes a decrease in infiltration rate as a function of exponential decay (Turner 2006), is derived from the basic relationship

$$\frac{-df_p}{dt} = k\left(f_p - f_c\right) \text{ or } \frac{-df_p}{f_p - f_c} = kdt \tag{6.29}$$

where he termed k a proportionality factor (Horton 1940). Then he integrated

$$\ln\left(f_p - f_c\right) = -kt + C \tag{6.30}$$

where C is the constant of integration. When $t = 0$, $f_p = f_0$ then $C = \ln(f_0 - f_c)$ and

$$ln\frac{f_p - f_c}{f_0 - f_c} = kt \tag{6.31}$$

or

$$\frac{f_p - f_c}{f_0 - f_c} = e^{-kt} \tag{6.32}$$

The final form is derived by multiplying both sides by the denominator from the left side followed by the addition of f_c to both sides, resulting in

$$f_p = f_c + (f_0 - f_c)e^{-kt} \tag{6.33}$$

where

f_p = the infiltration capacity or potential infiltration rate [Lt^{-1}];
f_0 = the initial infiltration rate at t = 0 [Lt^{-1}];
f_c = the final infiltration rate [Lt^{-1}];
k = soil parameters (t^{-1}) that control the rate of decrease of infiltration and depend on initial water content θ_i and an application rate R [Lt^{-1}]; and
t = time after start of infiltration.

The resulting model is in the form of an exponential decay function. Recall that the cumulative infiltration F is equal to the integration of the infiltration rate over the same time period. Conversely, the infiltration rate f_p is the time derivative of the cumulative infiltration as demonstrated in Figure 6.14.

The Horton model consists of three characterizing parameters

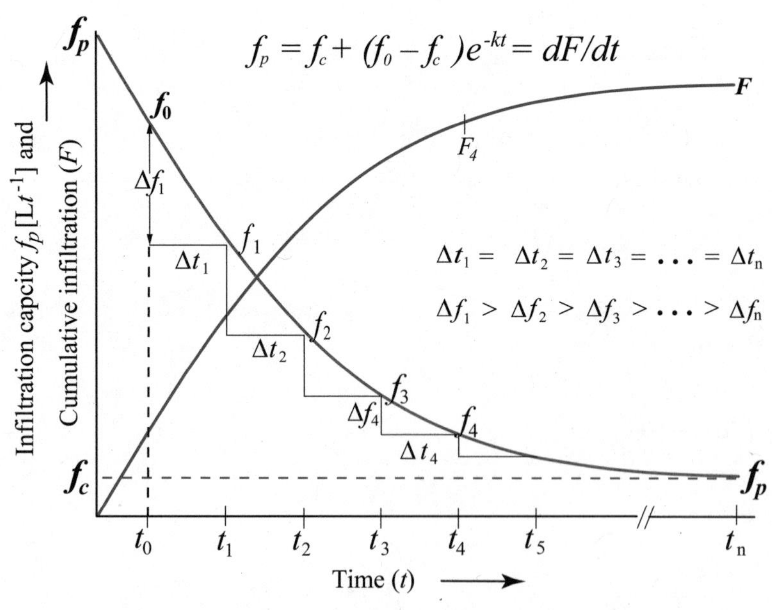

Figure 6.14. Demonstration of Horton's infiltration rate and cumulative infiltration.

(f_0, f_c, and k) that must be determined empirically from field measurements. An equation of a straight line can be derived by subtracting f_c from both sides of equation (6.33) and then taking the natural log (ln) of each side, so that

$$\ln(f_p - f_c) = \ln(f_0 - f_c) - kt \tag{6.34}$$

When the value of f_c is subtracted from the values for i and the natural log of the resulting values are plotted as a function of time, k can be determined from the slope and f_0 can be found from the intercept (Turner 2006). Other methods for finding parameters include the least squares method (Blake 1968).

Horton's equation is empirical such that k, f_c, and f_0 must be calculated from experimental data versus measured in a laboratory, which makes this model cumbersome to practice (Hillel 1989. Further criticism includes that it is only applicable when rainfall intensity exceeds f_c, which limits its use (Rawls et al. 1993), and that it neglects the role of capillary potential gradients in the decline of infiltration capacity over time and assumes that control is due almost entirely to surface conditions (Beven 2004).

Despite its limitations, Horton's equation has been widely used because it generally provides a good fit to the data and because it reflects the laws and basic equations of soil physics (Chow, Maidment, and Mays 1988).

6.6. Urban Runoff and Consumptive Use

6.6.1. Urban Runoff

One of the most important factors influencing runoff is the physical characteristics of the surface or landscape the stormwater flows over. For example, a sandy lithology will have a higher permeability or infiltration capacity than clay sediments. Impervious surfaces such as streets, parking lots, and roofs allow for almost no infiltration. Thus, areas with a high percentage of impervious surface or *impervious cover* from urbanization will have a large amount of stormwater runoff as well a variety of other impacts to the watershed. These impacts are generally classified according to one of four broad categories: hydrologic, physical, water quality, or biological indicators. More than 225 research studies have documented the adverse impact of urbanization on one or more of these key indicators (Center for Watershed Protection 2003).

For example, several investigations have shown that urbanization causes an increase in peak flows by a factor of 10 to floods in the range of one to four exceedances per year with diminishing impacts on larger floods (Tholin and Keifer 1959; American Society of Civil Engineers 1975; Espey and Winslow 1974; Hollis 1975; Cordery 1976; Ferguson and Suckling 1990).

Native flora and fauna species, those that occur in the region in which they evolved, are an important consideration in an urban environment. Plants especially evolve over geologic time in response to physical and biotic processes characteristic of a region. This includes climate, soils, rainfall cycles and seasonality, drought, frost, and interactions with the other species inhabiting the local community. Hence native plants have certain traits that make them uniquely adapted to local conditions. One strategy to mitigate the effects of urbanization on runoff is to maximize such water-use efficiencies as the development of root systems of native flora, which improves soil and creates more pathways for infiltration and root uptake of water. In a forested watershed, most of the precipitation infiltrates the soil and subsequently percolates deeper into groundwater or is evapotranspired back to the atmosphere (Figure 6.15a).

Land use is especially important as it applies to the development, destruction, and disturbance of natural ecosystems and subsequent native vegetation, soils, shallow depressions, and overall changes in topography and drainage patterns. Land development (including consumptive use of water resources) typically introduces impervious surfaces, removes or reduces native vegetation, compacts or removes soils, and in general results in decreased infiltration and increased runoff (Figure 6.15b). As urbanization occurs and the percentage of impervious surface increases, an increasing amount of precipitation runs off the landscape and eventually is discharged to receiving waters.

In a natural forest ecosystem about one-third of annual rainfall is recycled to the atmosphere through evapotranspiration while another third recharges the groundwater through infiltration. Overland flow is minimal. Urbanization impacts can increase runoff to 30% of annual precipitation depending on the extent and connectedness of impervious surfaces. Runoff can exceed 70% of annual rainfall in areas of ultraurban development (Hinman 2005).

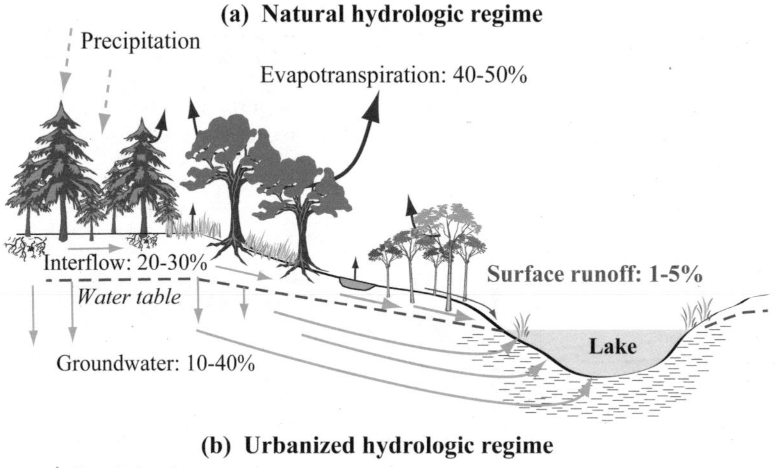

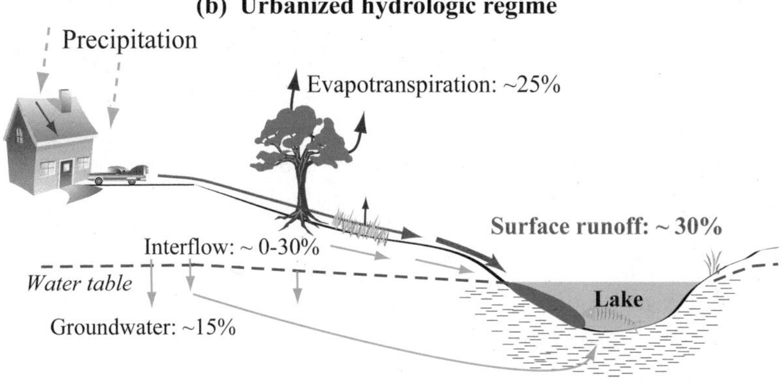

Figure 6.15. Impact of urbanization on the hydrologic cycle. Note the significant differences in percentages of evapotranspiration and water taken to the lake via each type of flow path. There is much less natural infiltration and evapotranspiration and much more surface runoff. Data from Hinman (2005).

With less storage capacity for water in urban basins and more rapid runoff, urban streams rise more quickly during storms and have higher peak discharge rates than do rural streams. In addition, the total volume of water discharged during a flood tends to be larger for urban streams than for rural streams (Konrad 2003). For example, during a storm on February 1, 2000, streamflow in Mercer Creek, an urban stream in western Washington, increased earlier and more rapidly, had a higher peak discharge and volume, and decreased more rapidly than in Newaukum Creek, a nearby rural stream (Figure 6.16a). It should be noted that the differences between the two streams may not be a result solely of land use but also such factors as differences in geology, topography, basin size and shape, and storm patterns. In general, however, the annual maximum discharge in a stream will increase as urban development happens, although the increase is sometimes masked by substantial year-to-year variation in storms. Regression analysis of the data showed a clear trend of increase in the an-

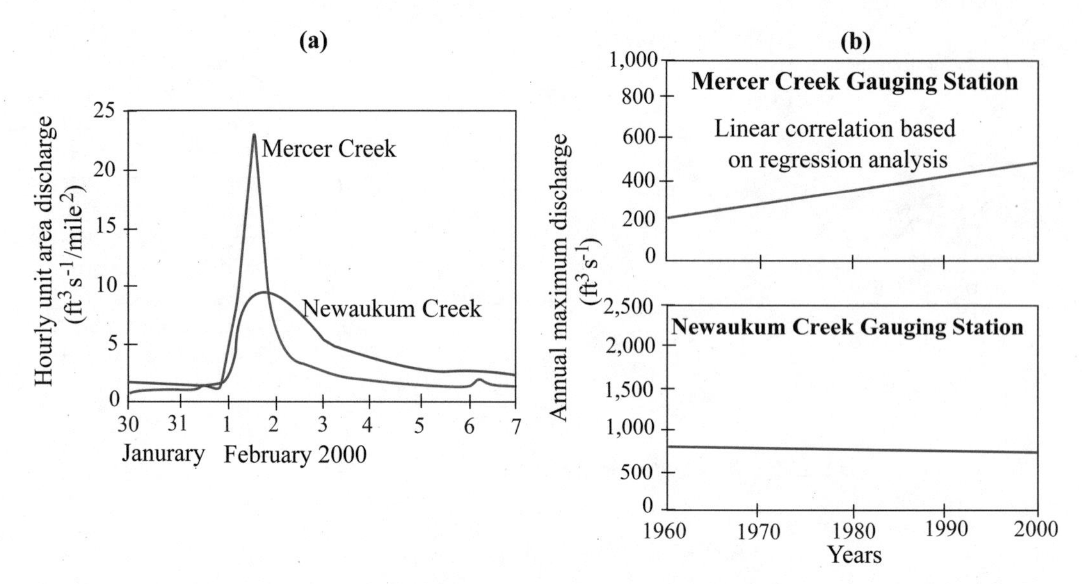

Figure 6.16. (a) Streamflow in Mercer Creek and nearby Newaukum Creek during a one-day storm on February 1, 2000. (b) Annual maximum discharge increased with urban development in Mercer Creek from 1960 to 2000 but remained essentially the same in nearby rural Newaukum Creek during that period. Modified from Konrad (2003).

nual maximum discharge for Mercer Creek from 1960 to 2000. In comparison, the annual maximum discharge for rural Newaukum Creek varied during the period but showed no clear trend (Figure 6.16b).

The transition from native ecosystems to agricultural or open space to urbanized land use can have an extreme effect on water movement and water quality within a watershed due to the changes in landscape and increased runoff volume (Q), velocity, and peak discharges (Q_p) during a storm event. Installation of urban stormwater drainage systems increases the efficiency with which runoff is delivered to the stream (e.g., curbs and gutters, storm drainpipes). Consequently, a greater fraction of annual rainfall is converted to surface runoff, runoff occurs more quickly, and peak flows become larger (Center for Watershed Protection 2003). Figure 6.17 illustrates the change in stream hydrology due to increased urban runoff as compared to predevelopment conditions.

Schueler (1987) demonstrated that runoff values derived from forty-four small watersheds across the country were directly related to subwatershed impervious cover (Figure 6.18b). Runoff data was derived from forty-four small catchment areas across the

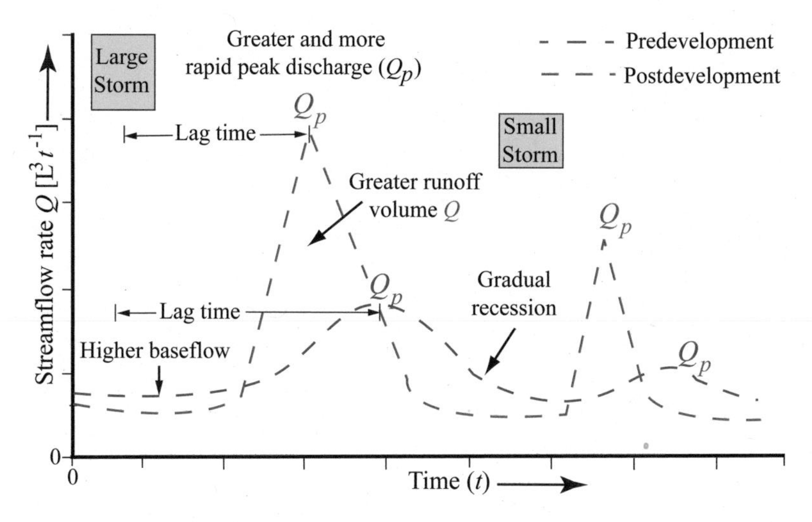

Figure 6.17. Generalized change of stream hydrograph in response to urbanization. Urbanization results in a decrease in lag time and baseflow and an increase in peak discharge of a stream. Modified from Schueler (1987) with permission from the Metropolitan Washington Council of Governments.

country for the Environmental Protection Agency's Nationwide Urban Runoff Program. Conversely, a decline in groundwater recharge is common due to urbanization and decreases in infiltration (Foster, Morris, and Lawrence 1994). Changes in streamflow and related decreases in groundwater recharge were also quantified by Bhaduri et al. (1997) when investigating the impacts of urbanization on the hydrology of a watershed in the Midwest.

Klein (1979) measured baseflow in twenty-seven small watersheds in the Maryland Piedmont and reported an inverse relationship between impervious cover and baseflow (Figure 6.18a). Spinello and Simmons (1992) demonstrated that baseflow in two urban Long Island streams declined seasonally as a result of urbanization. Saravanapavan, Anbumozhi, and Yamaji (2004) also found that percentage of baseflow decreased in direct proportion to percentage of impervious cover for thirteen subwatersheds of the Shawsheen River watershed in Massachusetts.

Dry weather flow in streams may decrease because less groundwater recharge is available. Over time, watershed development can alter or eliminate a significant percentage of the perennial stream network. In general, the loss of stream network becomes quite extensive when a watershed's impervious cover exceeds 50%. This loss is striking when pre- and postdevelopment stream networks and wetlands are compared. Figure 6.19 illustrates how the drainage network of Rock Creek in the Four Mile Run, Virginia, watershed changed in response to watershed development including

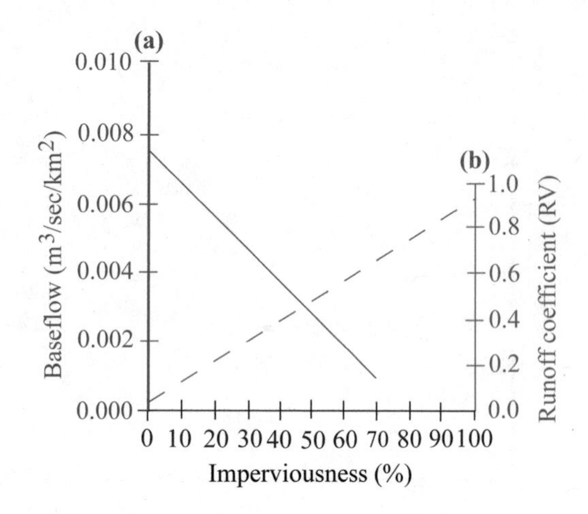

Figure 6.18. Decreases in baseflow (a) and increases in runoff (b) in relation to percentage of impervious cover in Maryland's Piedmont area. Modified from the Center for Watershed Protection (2003) with baseflow data from Klein (1979).

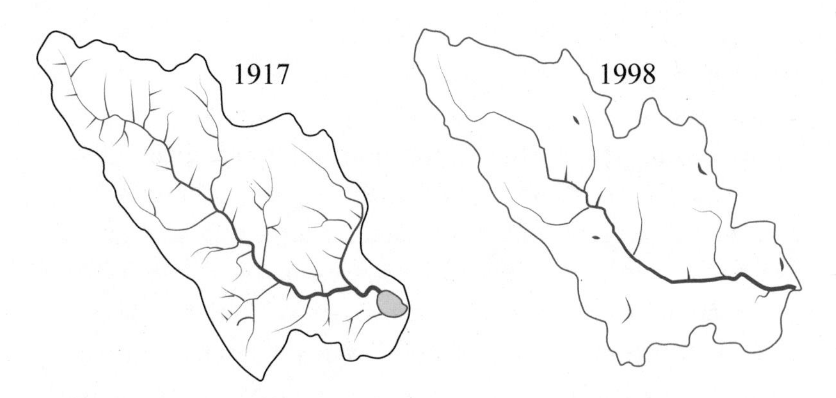

Figure 6.19. Reduction of Rock Creek drainage network due to urbanization between 1917 and 1998 in Four Mile Run watershed, Virginia. Modified from Northern Virginia Regional Commission (2001).

the loss of the lake within the southern portion of the watershed (Center for Watershed Protection 2003).

6.6.2. Consumptive Use

Urbanization includes consumptive use of water resources by humans (e.g., agriculture, personal hydration, industry). The benefit of water consumption for these applications is it increases economic productivity and stability; however, as populations increase, long-term use and increased withdrawal rates often result in sustained reduction in lake inflows and levels adversely affecting lake systems and related wetland ecosystems, causing environmental concerns. For example, many of the world's saline lakes are shrinking at disturbing rates, reducing water bird habitat and economic benefits while threatening human health (Wurtsbaugh et al. 2016).

Large saline lakes represent a diverse array of lake systems and are mostly located in arid endorheic basins. These lakes represent 44% of the volume and 23% of the area of all lakes on earth (Messager et al. 2016). Saline lakes are diverse in habitat and ecosystems as well as viable economic uses, such as mineral extraction and fishing. For example, the Caspian Sea comprises 41% of global saline lake volume and supports thriving fishing, shipping, and mineral industries.

Water use, especially for agricultural irrigation, is increasing and resulting in lake desiccation to the point that saline lakes are shrinking on a global scale (Williams 1996; Messager et al. 2016; Wurtsbaugh et al. 2016). For example, agricultural water development in the Aral Sea watershed has reduced lake area by 74% and volume by 90% (Micklin 2007). Owens Lake in eastern California was completely desiccated by 1940 after the city of Los Angeles diverted streams for agricultural and urban use and California's Salton Sea has suffered a recent and precipitous decline of over seven meters since 2000; a result of management decisions that decreased water flowing into the lake (Wurtsbaugh et al. 2016; Case et al. 2013). The near-complete desiccation of Lake Urmia increased salinity, eradicated brine shrimp, and resulted in the subsequent loss of flamingos and other birds (Lotfi 2012). Similarly, water diversions from the Aral Sea increased salinity above levels tolerated by fish, leading to a collapse of the commercial fishery that had once harvested 40,000 metric tons annually and provided 60,000 jobs (Micklin and Aladin 2008).

Although these examples are of large saline systems, freshwater lakes and wetlands of all types and sizes are negatively impacted due to human actions. To monitor and preserve these invaluable lentic systems, mass balance investigations are needed to quantify the relative contributions of natural processes of inputs and outputs in comparison to our impacts on lake inflows. With a reliable water balance, causes of lake decline from water diversions or climate variability can be identified, the inflow needed to maintain lake health can be maintained, and other best management practices can be defined.

Case Study 6.3 is an example of the importance of a water budget analysis after government managers initially blamed the declining lake levels of Utah's Great Salt Lake on natural rainfall cycles without direct analysis of the cause. Completion of the water budget analysis clearly revealed that the cause of the lake's desiccation was due to water diversions for agriculture and other human uses.

Case Study 6.3

Impacts of Water Development on Great Salt Lake and the Wasatch Front

W. Wurtsbaugh,. Miller, S. Null, P. Wilcock, M. Hahneneberger, and F. Howe, 2016, "Impacts of Water Development on Great Salt Lake and the Wasatch Front" (*Watershed Sciences Faculty Publications*, Paper 875).

Utah's Great Salt Lake is immeasurably valuable as an environmental, cultural, and economic resource with an economic value at approximately $1.32 billion per year for mineral extraction, brine shrimp cyst production, and recreation. The abundant food and wetlands of the lake attract 3 million shorebirds, during spring and fall migrations, designating it as a Western Hemisphere Shorebird Reserve Network Site (Wurtsbaugh et al. 2016).

Great Salt Lake is an *endorheic* (closed) lake system that lies in a terminal basin with water input from precipitation, groundwater, and inflows from the Bear, Weber, and Jordan Rivers (Figure 6.20). Water loss from the lake is mainly a result of evaporation (Wurtsbaugh et al. 2016). As this is an endorheic lake system, its stage, area, and volume naturally rise during wet periods and fall during droughts as those components are dependent on river flow, precipitation, and groundwater inputs.

Water supply to the lake has decreased over time from agricultural, industrial, and urban uses, causing an overall reduction in lake size and resource. Since American settlers arrived in 1847, the lake's elevation has decreased by 3.4 meters and its volume by 48%, exposing much of the lake bed (Wurtsbaugh et al. 2016). Table 6.3 shows the types and impacts of human water consumption on the Great Salt Lake.

There has been no significant long-term change in precipita-

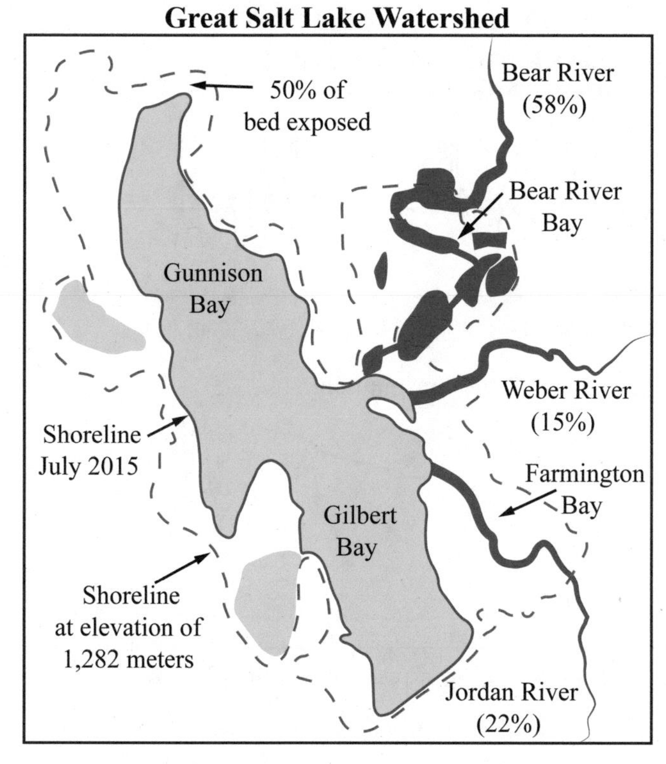

Figure 6.20. Map of Great Salt Lake showing the "natural" shoreline elevation at 1,282 meters above mean sea level (NGVD) and the major rivers and their relative contribution (%) along with the shoreline from a July 2015 NASA satellite photo at near record low lake stage exposing approximately 50% of the lake bed. Modified with permission from Wurtsbaugh et al. (2016).

Table 6.3. Types of human water consumption and impact on decreasing Great Salt Lake stage

Source and percentage of water use	Median estimated decrease in lake stage[a]
Agricultural (63%)	2.10 meters
Mineral extraction; salt pounds (13%)	0.43 meters
Municipal and industrial (11%)	0.40 meters
Impounded wetlands (10%)	0.34 meters
Reservoir evaporation (3%)	0.09 meters

Source: Modified with permission from Wurtsbaugh et al. (2016).
[a] Total lake stage decrease = 3.36 meters.

tion and water supply from mountain tributaries since the pioneers arrived in 1847 (Figure 6.21a) (Wurtsbaugh et al. 2016). Consumptive water use, however, increased dramatically over the subsequent 170 years, reducing the net river inflow to the lake by 39% (Figure 6.21b). Although wet periods like those in the mid-

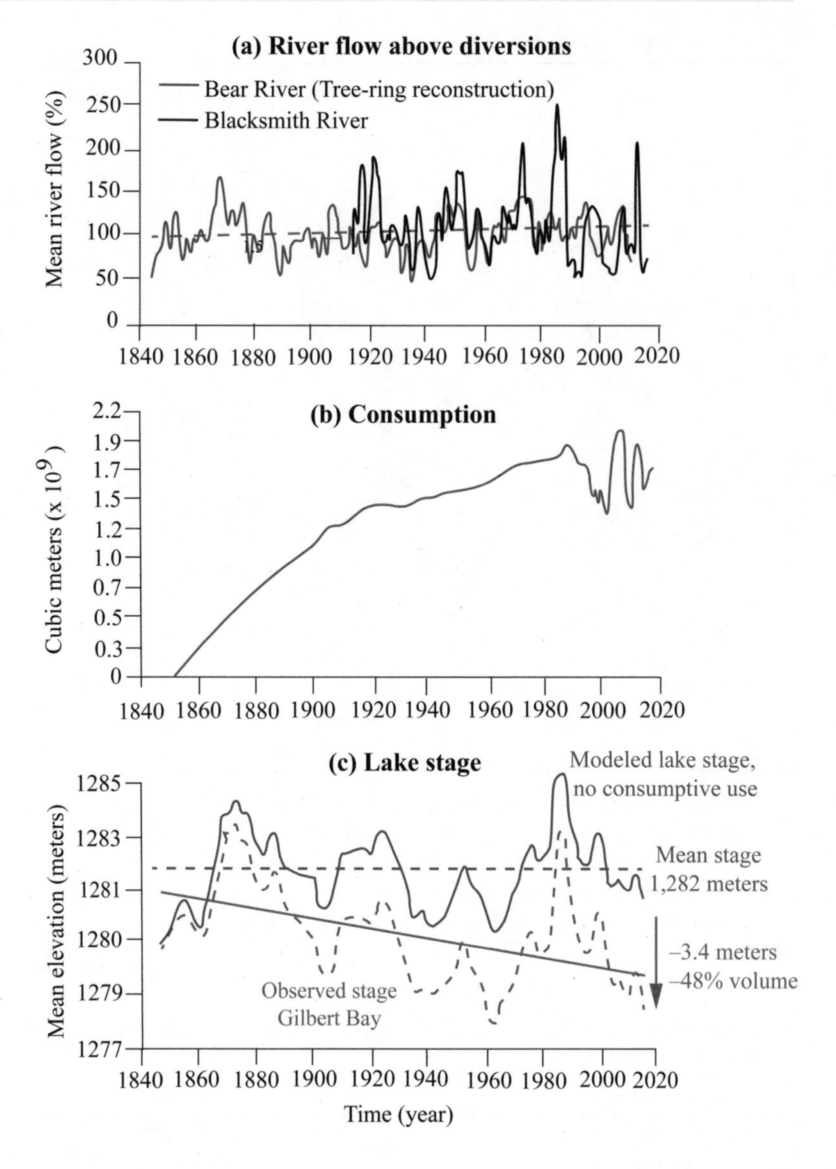

Figure 6.21. (a) Water flow in headwater streams (Blacksmith River Gauge; Bear River based on tree ring reconstructions); (b) Estimated consumptive use from Table 6.3; (c) Great Salt Lake stage data (dashed red line). Blue line is from modeled stage data with the consumptive use removed. Red solid line indicates declining values. Modified with permission from Wurtsbaugh et al. (2016).

1980s and the current drought cause water supply and lake levels to fluctuate, the lake level has persistently declined (Figure 6.21c, red line) whereas the flow in the upper river watersheds above the diversion area has remained relatively constant (Figure 6.21a).

The blue line in Figure 6.21c represents modeled lake stage where consumptive use was removed. This analysis reveals relatively little variation in overall lake stage with a mean elevation of 1,282 meters (above mean sea level) over the last 170 years. This contrasts sharply with the observed stage data illustrated in the red dashed line. Linear regression analysis on these data (red line) indicates a strong correlation for the overall decline of lake stage over this time period (Wurtsbaugh et al. 2016).

Utah has taken steps to increase water conservation and has reduced urban per capita use by 18%; however, overall municipal water use has increased by 5% because of a growing population. To significantly reduce water use, a balanced conservation ethic needs to consider all uses, including agriculture, which consumes 63% of the water in the Great Salt Lake basin (Wurtsbaugh et al. 2016).

Increased awareness of how water use is lowering Great Salt Lake will help avoid the fate of other salt lakes, such as the Aral Sea in Central Asia or California's Owens Lake, both of which have been desiccated and are now causing severe environmental problems.

Mass balance studies are applicable to all types and sizes of lake systems, both fresh and saline. With increasing population and potential effects of climate change, water budget analyses are necessary for providing a scientific basis for implementing solutions for best management practices and for discussions on the tradeoff between consumptive use and maintaining wetland systems at a sustainable level.

6.6.3. Implications

Research has demonstrated that the effect of watershed urbanization on peak discharge is more marked for smaller storm events. In particular, the *bank-full flow*, or channel-forming flow, is increased in magnitude, frequency, and duration. Increased bank-full flows have strong ramifications for sediment transport and channel enlargement. All these changes in the natural water balance have impacts on the physical structure of streams, lakes, and wetlands and ultimately affect water quality and biological diversity (Center for Watershed Protection 2003).

Over time, the hydrology of watersheds can be altered, and the natural water cycle disrupted. Increase of streamflow and velocity can result in the scouring of streambeds and streambanks, changing the stream channel geometry and direction. As streambanks are eroded, the vegetation that protects the banks are exposed and are more likely to be uprooted during storms, increasing the vulnerability of the bank to scouring. Sediment from channel erosion and other sources upstream can change stream channel and lake morphology due to shifting deposition of muds, silts, and sands, such as sandbars and other features covering the historical channel bed.

The quality of rainwater that flows over urban landscapes is impacted as well. Runoff picks up and carries a wide variety of pollutants, including fertilizers, pesticides, sediments from construction sites or other disturbed areas, automobile oils, metals and grease, and a host of many more contaminants that can adversely impact receiving waters.

The 2016 National Land Cover Database (www.mrlc.gov) can be used to estimate the percent of imperviousness of each land use (e.g., commercial, residential) within each catchment. Use of generic runoff coefficients based on land use type, is generally insufficient to adequately estimate stormwater runoff volumes. In addition to impervious coverage by land use, incorporation of the relative permeability, and native soil retention coefficients should be used to adjust runoff generation from each land use type.

6.7. Summary

At any point in a stream or lake at a specific time during a rainfall and subsequent runoff event, the flowing water will be contributed to by an assortment of complex pathways and processes that all come together to make up the flow at that instant. Each drop of water within the flow would have originated as rainfall traveling on a variety of journeys, depths, and flow velocities throughout the catchment: one drop of streamflow may have started as rainfall on the water surface a short distance upstream, another may have come from rain falling on saturated soil beside a lakeshore, while another fell on unsaturated soil at the far reaches of watershed. Yet another may have originated from a previous storm event and traveled to the lake or stream via groundwater.

Streamflow and quickflow to a lake derived from rainfall pass through various storages. Groundwater represents long-term storage. Temporary storage, lasting the duration of the runoff event, consists of water in transit in each element of the drainage system, including water in the rivers, streams and tributaries, hill slopes, and overland flow. Floodplains and wetlands can temporarily store water. There is also riverbank and lakeshore storage, water wetting up the bank profile at the start of an event and later flowing back into the stream or lake as the water level drops and hydraulic head reverses. Hence, understanding the overall catchment characteristics is essential when trying to derive stormflow impacts as well as lake mass balance or water budgets both during and after a storm event.

The most prominent topographic characteristics of watersheds influencing hydrograph behavior are drainage density, relief and slope, shape and pattern, and area (Gregory and Walling 1973). Watershed dynamics are also affected by geology, soils, vegetation coverage, climate, and anthropomorphic (e.g., urban and agriculture) land and consumptive use.

The time distribution of runoff, or shape of the hydrograph, at the basin outlet is influenced mainly by storm events (climatic), topographic, and geologic factors. The rising limb portion of the hydrograph is mainly a result of the storm event and topographic factors whereas the recession limb is mainly controlled by geologic factors. The pedology and geology of the catchment influence primarily the groundwater component and the losses. High infiltration rates reduce the surface runoff; high permeabilities combined with high transmissivities substantially enhance the baseflow component.

The geology of a drainage basin expressed as structural control and lithologic composition exerts a major dynamic effect on the basin's physical characteristics, such as relief, slope, shape, drainage density, stream and wetland patterns, erosion, and sedimentation. Variation in rock types and soils influence stream morphology due to differences in lithologic porosity, permeability, and erodibility (sedimentary vs. crystalline rocks). Porous and permeable lithology increases infiltration, facilitating storage and baseflow, while impermeable lithology increases runoff discharge. The type of rock affects weathering rates, sediments yields, and chemical solute amounts supplied to streams and lakes (Schumm 1977).

The interaction between climate, soil, and vegetation influence watershed dynamics in a variety of ways. Rainfall is a function of climate, which determines the quantity of water input as a key factor of such flow regimes as drainage density and formation of lakes and wetlands. Precipitation is essential to chemical and physical weathering of lithologic material contributing to the formation of different soils. A winter storm (i.e., snow) will result in an increase in runoff discharge when the snow melts, creating a significant lag time dependent upon the climatic temperature.

Typically, if the climate's been hot and dry or freezing cold, the ground will be hard, resulting in reduced infiltration, increased overland flow, a reduction in the lag time, and subsequent increase in the peak discharge of runoff to a lake. Climate is also important in influencing the vegetational cover of the watershed, as seen in climate-controlled rainforest versus desert or arctic ecosystems.

Vegetation affects the amount and movement of water and sediment through the watershed system (Gregory and Walling 1973). Vegetation cover reduces precipitation input to the basin by interception and evapotranspiration but stores water within the catchment in the soil and plant mass. Because vegetation impedes runoff, maximum infiltration occurs in fully established forest; minimum infiltration and maximum runoff is seen in areas where vegetation has been removed.

Land use is understood to be an alteration of the composition of the watershed landscape due to human activity (Trimble 1997). The processes of agriculture and, especially, urbanization change the vegetational, pedogenic, and morphologic composition of the watershed. Hence, they influence water runoff, infiltration, evapotranspiration, and erosional dynamics of the basin. These changes, in turn, affect the amount and composition of water and sediment delivered to streams and lakes.

Lastly, land use can significantly influence the runoff coefficient. Urbanized areas may have a runoff coefficient of almost 100%, whereas natural vegetation may have low runoff. Plowing, drainage, cropping intensity, deforestation, and so on also have considerable effects on runoff, as does consumptive water use. Water balance studies have revealed a definite decline of lake level, volume, and area from persistent use of water for agriculture and other forms of consumptive use for most of the large saline lakes throughout the world (Wurtsbaugh et al. 2016).

CHAPTER 7

Methods for Estimating Storm Runoff

7.1. Introduction

Looking back at Figure 6.5, runoff results from rainfall not lost to infiltration, interception, depression storage, or evapotranspiration. This chapter discusses the fundamentals of estimating stormwater runoff rates and volumes from rainfall storms using mathematical models. These models are designed to simulate the rainfall-runoff processes based on the hydrologic cycle of rainfall input to basin and subsequent water exchange with the various hydrologic interactions illustrated in Figures 6.1 and 6.2.

It is important to remember that the drainage basin is the fundamental unit of investigation. The processes that convert rainfall to runoff at any location in a basin are complex and dependent on atmospheric interactions, geology, geomorphology, vegetation, and soils and strongly influenced by human activities such as urbanization and agriculture. These processes vary on a temporal basis that is generally dependent on the duration, intensity, and amount of precipitation as well as on evapotranspiration since the last rainfall event. Thus, for a certain drainage basin, the rainfall-runoff relationship is very dependent on basin wetness and is sensitive to surface-soil water content, type and state of the vegetation, and dominant runoff processes for a specific basin (e.g., urban versus natural land use).

Assessing the quantity of runoff from rainfall measurements is dependent on the time scale or storm duration, as well as the size of the basin being examined. It is easier to determine runoff from longer duration storms then storms of short duration (measured in hours). During longer storms, rainfall losses become maximized and better estimated, allowing for more accurate linear correlations between rainfall and runoff. The rainfall-runoff correlations are more accurately and easily derived in smaller basins with more evenly distributed amounts of rainfall and less diverse landscapes. Basin shape can also influence runoff characteristics (see next section). Drainage basins come in a wide assortment of shapes. Circular shapes are common as are more elongated and narrow shapes. For a circular drainage basin, the river's hydrograph can often be described as "flashy" because it will have a relatively steep rising limb and a high peak discharge. This is because all points in the drainage basin are (somewhat) equidistant from the river so all the precipitation reaches the river at the same time.

The size of the drainage basin obviously has an impact on the hydrograph. Large basins will have high peak discharges because they catch more precipitation, but at the same time they'll have longer lag times than small basins because the water takes longer to reach the rivers.

In general, all runoff computation methods are mathematical expressions of the hydrological cycle, treating rainfall as an input and producing runoff as the output. Each method used for this transformation incorporates mathematical approximations that are dependent on model complexity, data availability, and accuracy and produce a range of results of runoff volume or rate estimates.

As such, conversion from rainfall to runoff cannot be determined mathematically with exact certainty. The theoretical or mathematical models and equations using simplifying assumptions and empirical data can be developed to simulate these processes and predict generalized estimates of resultant runoff volumes with acceptable accuracy. To develop and calibrate rainfall-runoff models, local data are essential for valid application to an actual catchment and cannot be overemphasized. For example, a specific storm may produce a large flood in one catchment but in a similar nearby sub-basin produce little or no runoff. The primary criterion for the selection of a certain runoff estimation method is that the techniques should be based on observed and correlated flood/runoff data in the region of interest and have been peer reviewed by the profession (Ball et al. 2016).

A greater understanding of climate, climatic influences, and land-use effects, and increases in hydrologic databases coupled with technological innovation, has resulted in updated and more complex models and techniques to investigate a full spectrum of rainfall events and stormwater flow. These more advanced models use a variety of deterministic, probabilistic, or statistical techniques that include numerous more computationally intensive procedures and continuous simulation approaches, which are introduced in chapter 11.

The more advanced methodology is beyond the scope of this text, yet references such as Ball et al. (2016) are provided for further investigation. The focus of this chapter are the fundamentals of computing rainfall-produced stormwater runoff rates and volumes through the use of various simpler yet fundamental mathematical methods. Many of these models estimate runoff by incorporating rainfall losses through infiltration, interception, and depression storage. Although evapotranspiration is a dominant influence in the hydrologic cycle, its impact decreases significantly during a storm. Many of these methods and models discussed here have been derived for immediate and practical use and are available to download for free from such government agencies as the US Geological Society, US Department of Agriculture, US Environmental Protection Agency, and Geoscience Australia (see *Australian Rainfall and Runoff* by Ball et al.).

The various methodologies for estimating rainfall and runoff have been developed to predict and manage extreme storm events that result in flooding. Extreme rainfall events can cause devastating floods and result in loss of life, significant property damage, and financial injury for many communities as well as have a negative impact on lakes and wetlands. Storm runoff flooding results from short-duration highly intense rainfall or long-duration low intensity rainfall (or snowmelt). Modeling runoff impacts from past extreme rainfall events are essential for predicting and managing flooding runoff through such tactics as sewer systems, holding ponds, and other flood reducing structures.

A simple design event is an invented storm defined by a time-intensity relationship and a spatial precipitation pattern of a past storm with a known annual exceedance probability and duration. The simple design event method represents common practice for flood investigation and traditionally includes the use of the rational and modified-rational methods, unit hydrographs, Soil Conservation Service method, hydrographs, and runoff-routing procedures (Haan and Schulze 1987; Cordery and Pilgrim 2000; McKerchar and Macky 2001; Smithers 2012; Ball et al. 2016).

7.2. Characterizing Rainfall Events

In all runoff models, rainfall is the fundamental input and is characterized by its size, intensity, and frequency of occurrence. Thus, the first step in stormwater analysis is the estimation of the rainfall that will fall on the area of interest for a given period of time. As discussed in section 5.3, rainfall can be quantified by the following characteristics:

Duration: length of time [t] of the storm event (hours)
Depth: total amount [L] of rainfall occurring during the rainfall duration
Intensity: rate [Lt^{-1}] of rainfall or depth divided by the duration

The size of a storm is the total precipitation over a specific duration. How often storms of a certain size range are likely to reoccur is called the *recurrence interval.* For example, a rainfall of a specific size and duration that occurs on average every fifty years is called a fifty-year storm. These types of storms, however, can occur randomly over a given interval and are shown to be mathematically random events. Their frequency of occurrence can also be expressed as an *annual probability* where

Annual probability (%) = 100 / recurrence interval (in years).

The amount of rainfall is rarely constant for a storm's duration. Thus, measuring depth and time is rain *intensity*. Intensity is the amount of rain that has fallen over a unit of time. The average storm intensity is calculated by dividing a rainfall depth for a storm event by the duration, which is the time over which the rainfall accumulated. The rate of runoff [L^3t^{-1}] is related to the rainfall intensity where the peak intensity produces the largest runoff rate.

7.2.1. Annual Exceedance Probability and Average Recurrence Interval

Annual exceedance probabilities (AEPs) and average recurrence intervals (ARIs) are both a measure of the rarity of a rainfall event requiring ten or more years of data to perform a frequency analysis for the determination of these values. The larger the historical time series data, the more significant the statistical analysis.

Table 7.1. Storm event recurrence intervals and probabilities of occurrences

Average recurrence interval (years)	Probability of occurrence in any given year	Percent chance of occurrence in any given year (%)	Rain volume (mm/day) in Leon County, Florida
100	1 in 100	1	280
50	1 in 50	2	254
25	1 in 25	4	230
10	1 in 10	10	200
5	1 in 5	20	165
2	1 in 2	50	114

Source: Rainfall data from Hershfield (1963).

Laurenson (1987) defined AEP as "the probability that a given rainfall total accumulated over a given duration will be exceeded in any one year" and ARI as "the average, or expected, value of the periods between exceedances of a given rainfall total accumulated over a given duration" (Teegavarapu 2012, 213).

For example, a municipality may want to protect an urban lake area from flooding by designing a stormwater retention system for the volume of stormwater runoff produced from an actual extreme storm event, such as a 100-year 24-hour storm. AEP and ARI are generated by looking at statistical patterns of rainfall duration and amount of rain in a specific area over a number of years, as seen in Table 7.1 for Leon County, Florida, from data published by Hershfield (1963).

The chance or probability that a two-year twenty-four-hour storm (which is about 114 mm/day of rain for Leon County, Florida) will occur in any given year is 50%. For a twenty-five-year twenty-four-hour storm it is 4%, and so on, to 1% for a 100-year storm. In terms of average recurrence, a rainfall total of 254 mm in a consecutive twenty-four-hour period is said to have a fifty-year recurrence interval.

The use of ARI can lead to confusion in the minds of some decision-makers and members of the public such that it is implied that the associated magnitude (e.g., a 100-year storm) will be exceeded only within the elapsed time frame before the next occurrence. Hence, to avoid misunderstanding, it is preferable to express the rarity of a rainfall event in terms of the AEP (Pilgrim 1987). For example, a rainfall total of 280 mm falling in ten hours at the Tallahassee Airport (Leon County, Florida) has a 0.010 (i.e., 1%) probability of being equaled or exceeded in any one year, which can be easier to understand than the equivalent statement of a rainfall total of 280 mm in ten hours has an average recurrence interval of 100 years. The latter can be misconstrued to mean this storm occurs every 100 years since the last occurrence.

7.2.2. Design Storms

Design storms are simulated storms generated from AEP data and based on statistical assumptions of depth, intensity, and duration over a specified time interval (e.g., twenty-four hours) for a certain area or region. Many of the same computational processes may be used to estimate rainfall and runoff estimates for both actual and design storms. Thus, the data generated from a design storm can be used to approximate rainfall and runoff from a similar actual storm.

Even though the design storm is used to estimate actual storm events, it would be very unlikely that an actual storm event would replicate all the design storm assumptions. Hence valid application of design storms requires recognizing the differences between the design and actual storm. For example, runoff from an actual storm replicates the specific storm pattern and hydrologic conditions prevailing at the time of the storm. For estimating runoff in the design storm then, the factors used must be represented as closely as possible to the actual storm, resulting in a determinist historical approach to modeling.

In contrast, the design storm does not represent the actual characteristics of an historical storm but a probabilistic estimate of a number of similar historical storms all with varying characteristics and whose magnitude will be equaled or exceeded with a given frequency (e.g., 0.1). Even though the mathematical model equations for estimating runoff may be the same as the actual storm, the model is probabilistic with the intention of estimating runoff with a selected exceedance frequency.

With this approach, a rainfall event with preselected AEP and duration is transformed into a flood hydrograph by a simple hydrologic model (or *transfer function*). The approach is termed *deterministic* in the sense that the single resulting flood output is uniquely derived from a set of inputs that are explicitly selected. The transformation often involves the application of two modeling steps, namely:

1. runoff production modeling, to convert the storm rainfall input at any point in the catchment into rainfall excess (or runoff) at that location; and
2. hydrograph formation modeling, to simulate the conversion of rainfall excess into a flood hydrograph at the point of interest (Ball et al. 2016).

7.2.3. Intensity-Duration-Frequency Curves

If one is interested in predicting estimates of runoff into a specific lake system during different potential storm conditions, the most common approach is to use a design storm event or events that involve the interactions between rainfall intensity (or depth), storm duration, and the return period, or *frequency*, based on the historical storm data for the area of investigation. For most large cities in the United States, these types of predictive rainfall data are available as *intensity-duration-frequency* (IDF) *curves* (Figure 7.1) that allow calculations of the average design-rainfall intensity

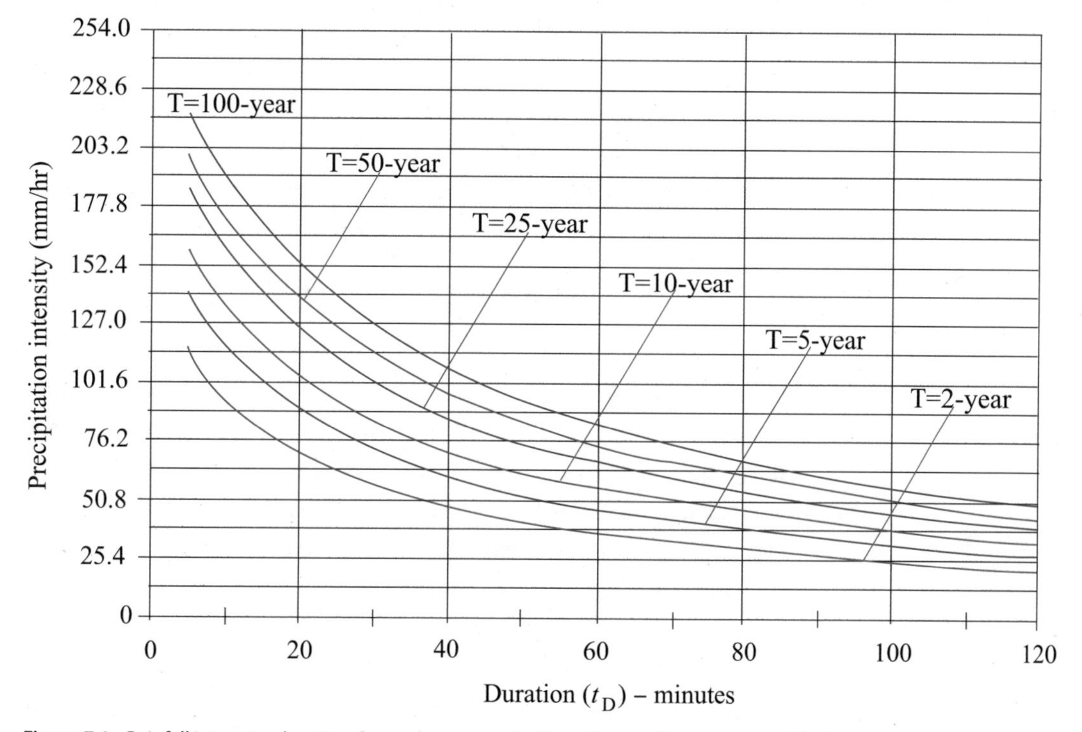

Figure 7.1. Rainfall intensity-duration-frequency curves for Knox County, Tennessee. Modified from 2008 International Association for Measurement and Evaluation of Communication (AMEC) data from Weather Bureau Technical Papers TP-25 and TP-40 (Hershfield 1963) and National Weather Service Publication HYDRO-35 (Fredrick, Myers, and Auciello 1977).

for a given exceedance probability over a range of durations. IDF curves can be described mathematically as

$$i = \frac{P}{t_D} \tag{7.1}$$

to facilitate calculations where i is the average intensity [Lt^{-1}], P is the rainfall depth, and t_D is the duration, usually in hours (Chow, Maidment, and Mays 1988).

For example, using the IDF curve in Figure 7.1 we can see that the design precipitation intensity i for a 20-minute (0.333-hour) duration storm with a 25-year return period in Knox County, Tennessee, is 127 mm. Then, by modifying equation (7.1), we find the estimated rainfall depth: $i \times t_D = P$, or 127 mm × 0.333 hr = 42.33 mm.

7.3. Runoff Models for Small- to Medium-Sized Catchments

Estimates of runoff peak flows on small- to medium-sized drainage basins are one of the most common applications for runoff and flood estimations and management. The exact size for defining a small to medium basin is variable, but the upper limits of 25 km^2 (10 mi^2) and 500 km^2 (200 mi^2), respectively, can be considered a general guide (Maidment 1993). Two widely used methods for estimating flood runoff estimates are the rational method and US Soil Conservation Service method.

7.3.1. Rational Method

The rational method is an empirical relation between rainfall intensity and peak flow and can be traced back to the mid-nineteenth century. It is still probably the most used method for design of urban drainage (Pilgrim 1986; Linsley 1986) and is characterized as an approximate deterministic model representing the runoff peak discharge from a given or design rainfall event. The method was derived from a simplified analysis of runoff with the *runoff coefficient* (C) being the ratio of peak runoff to the rainfall intensity (i) (Linsley 1986).

Although this method has its critics, especially as it is applied to large multibasin areas, it continues to be used because of its simplicity (Chow, Maidment, and Mays 1988). This method is appropriate for estimating peak discharges for small drainage areas of up to about 200 acres (0.809 km^2) with no significant flood storage (Maidment 1993). The method provides the designer with a peak discharge value but does not provide a time series of flow or flow volume.

7.3.1a. Flow/Peak Discharge Q *Calculation*

The concept for the rational method is that rainfall of intensity i begins instantaneously and continues indefinitely; the rate of runoff will increase until the *time of concentration* (t_c) when the entire watershed is contributing to flow at a common outlet (e.g., lake, stormwater pond, or sewer). The product of rainfall intensity i and watershed area A is the inflow rate (iA) for the system (or property/area under analysis), and the ratio of this rate of peak discharge Q_p (which occurs at t_c) is the runoff coefficient C ($0 \leq C \leq 1$). This is expressed in the rational formula

$$Q_p = CiAF \tag{7.2}$$

where

Q_p = maximum rate of runoff or peak discharge (cfs or m^3/s),
C = runoff coefficient (dimensionless),
i = average rainfall intensity (in/hr or mm/hr),
A = drainage area (acre or hectare), and
F = unit conversion factor, 1 for English, 0.278 (1/3.6) for SI units.

The rational equation was originally derived using units of cubic feet per second, inches per hour and acres. Hence, F (= 0.278) is a unit conversion factor for converting Q to SI units of m^3s^{-1} (Maidment 1993).

In urban areas, the drainage area usually consists of subareas or subcatchments of different surface characteristics (e.g., forest, lawn, asphalt). As such, a composite analysis is required that accounts for the different surface features. The area of the subarea is denoted A_j and the runoff coefficient of each subarea is denoted as C_j. The peak runoff is then computed using the following form of the rational equation:

$$Q_p = iQ_p = i\sum_{j=1}^{m} C_j A_j \tag{7.3}$$

where m is the number of subareas drained by the larger catchment.

7.3.1b. Runoff Coefficient

The runoff coefficient C represents the integrated effects of infiltration, evaporation, retention, and interception, all of which affect the volume of runoff. The determination of C requires judgment based on experience and understanding on the part of the investigator.

The runoff coefficient C is the least precise variable in the rational method due to the variability of such surface characteristics as percent imperviousness, slope of the area, type of surface (e.g., a forest versus a meadow), changes in infiltration rates over time, proximity to the water table, and initial moisture condition of the soil. Other factors include soil compaction, porosity of the subsoil or underlying sediments, depression storage, and vegetation type. However, these factors influence all runoff calculations not just the rational method.

Thus, a reasonable coefficient must be chosen to represent the integrated effects of all these factors. To do this, as much data for an area as possible must be investigated, including surface type or land use, topography, flow directions and slope analysis, depth of local water table, permeability of soils and underlying geology, and whatever further data is available in terms of flood zones, rainfall, evaporation, and other similar hydrologic data. Table 7.2 is an example of C values used by Knox County.

The more typical rational coefficients in Table 7.2 are applicable for storms of five- to ten-year frequencies. For higher intensity, less frequent storms, infiltration and other losses have a proportionally smaller effect on runoff (Wright-McLaughlin Engineers 1969). For major storms with a recurrence interval of twenty-five (AEP less than 4%) or more years, Table 7.3 provides a recommended adjustment factor C_f for the rational method coefficient during major storms events, modifying equation (7.2) to

$$Q_p = C_f C i_A. \qquad (7.4)$$

The general procedure for rational method calculations for a single catchment is as follows:

Table 7.2. Rational method, recommended runoff coefficient values

	Runoff coefficient (*C*) by hydrologic soil group (A–D) and topography slope (%)											
Land use	A			B			C			D		
Slope (%)	<2	2–6	6	<2	2–6	>6	<2	2–6	>6	<2	2–6	>6
Forest	0.08	0.11	0.14	0.10	0.14	0.18	0.12	0.16	0.20	0.15	0.20	0.25
Meadow	0.14	0.22	0.30	0.20	0.28	0.37	0.26	0.35	0.44	0.30	0.40	0.50
Pasture	0.15	0.25	0.37	0.23	0.34	0.45	0.30	0.42	0.52	0.37	0.50	0.62
Farmland	0.14	0.18	0.22	0.16	0.21	0.28	0.20	0.25	0.34	0.24	0.29	0.36
Residential												
1 acre	0.22	0.26	0.29	0.24	0.28	0.34	0.28	0.32	0.40	0.31	0.35	0.46
½ acre	0.25	0.29	0.32	0.28	0.32	0.36	0.31	0.35	0.42	0.34	0.38	0.46
⅓ acre	0.28	0.32	0.35	0.30	0.35	0.39	0.33	0.38	0.45	0.36	0.40	0.50
¼ acre	0.30	0.34	0.37	0.33	0.37	0.42	0.36	0.40	0.47	0.38	0.42	0.52
⅛ acre	0.33	0.37	0.40	0.35	0.39	0.44	0.38	0.42	0.49	0.41	0.45	0.54
Industrial	0.85	0.85	0.86	0.85	0.86	0.86	0.86	0.86	0.87	0.86	0.86	0.88
Commercial	0.88	0.88	0.89	0.89	0.89	0.89	0.89	0.89	0.90	0.89	0.80	0.90
Street ROW[a]	0.76	0.77	0.79	0.80	0.82	0.84	0.84	0.85	0.89	0.89	0.91	0.95
Parking	0.95	0.96	0.97	0.95	0.96	0.97	0.95	0.96	0.97	0.95	0.96	0.97
Disturbed	0.65	0.67	0.69	0.66	0.68	0.70	0.68	0.70	0.72	0.69	0.72	0.75

Source: Modified with permission from standards used by Knox County, Tennessee (AMEC Earth & Environmental Inc. 2008).
[a] Right of way.

Table 7.3. Frequency adjustment factor (C_f) for annual exceedance probability (AEP) of major storm events

Recurrence interval in years (% AEP)	C_f
10 or less (10%–50% AEP)	1.0
25 (4% AEP)	1.1
50 (2% AEP)	1.2
100 (1% AEP)	1.25

1. *Define boundary conditions.* Delineate the catchment boundary and determine its area.
2. *Define the flow path from the uppermost portion of the catchment to the design point.* Divide the flow path into reaches of similar flow type or land use (e.g., overland flow, shallow swale flow, forest, etc.). Determine the length and slope of each reach.
3. *Determine the time of concentration (t_c) for the selected waterway.*
4. *Find the rainfall intensity* (i) for the design storm using the calculated t_c and the rainfall intensity-duration-frequency curve (see section 7.3.2).
5. *Determine the runoff coefficient* (C).

6. *Calculate the peak flow rate* (Qp) from the catchment using equation (7.3).

Case Study 7.1

Stormwater Runoff Assessment Using Rational Method

W.L. Evans, 2018, *Stormwater Management Report: Analysis of Stormwater Runoff at CASA 12 Property 2410 Monday Road, Tallahassee, Florida* (Tallahassee: EIII Environmental Consulting Inc.), 15–24.

This is a simple case study demonstrating the basic rational methodology estimations applicable to small area requiring a basic stormwater runoff assessment for stormwater retention system. Step 4 in section 7.3.1b was eliminated due to the small area of investigation. Although this example does not demonstrate peak flow to a lake or catchment outlet, it does use the basic rational method applicable to any small catchment system as well as to designing a flood urban runoff system to protect a lake from the effects of urban runoff.

Background

The investigation was for an underfunded, volunteer, not-for-profit organization with limited resources as part of their facility expansion. The city was requiring (without a stormwater assessment) an elaborate and expensive state-of-the-art infiltration pond system that was cost prohibitive. This pro bono investigation was to determine stormwater requirements as part of a request to build a more basic yet superior runoff retention system at 80% of the cost of the proposed elaborate swale infiltration system.

Area of Investigation

The area of investigation was a small 2.62-acre (0.01 km^2) tract with relatively straightforward runoff flow directions based on one-foot topographic contours (Figure 7.2).

The infiltration rate for the property was determined by two tests using the standards set forth for infiltrometer testing by American Society for Test Methods for infiltration rate of soils in field using double-ring infiltrometer (ASTMD 3385) (see Bouwer [1986] for details) as well as the requirements for the City of

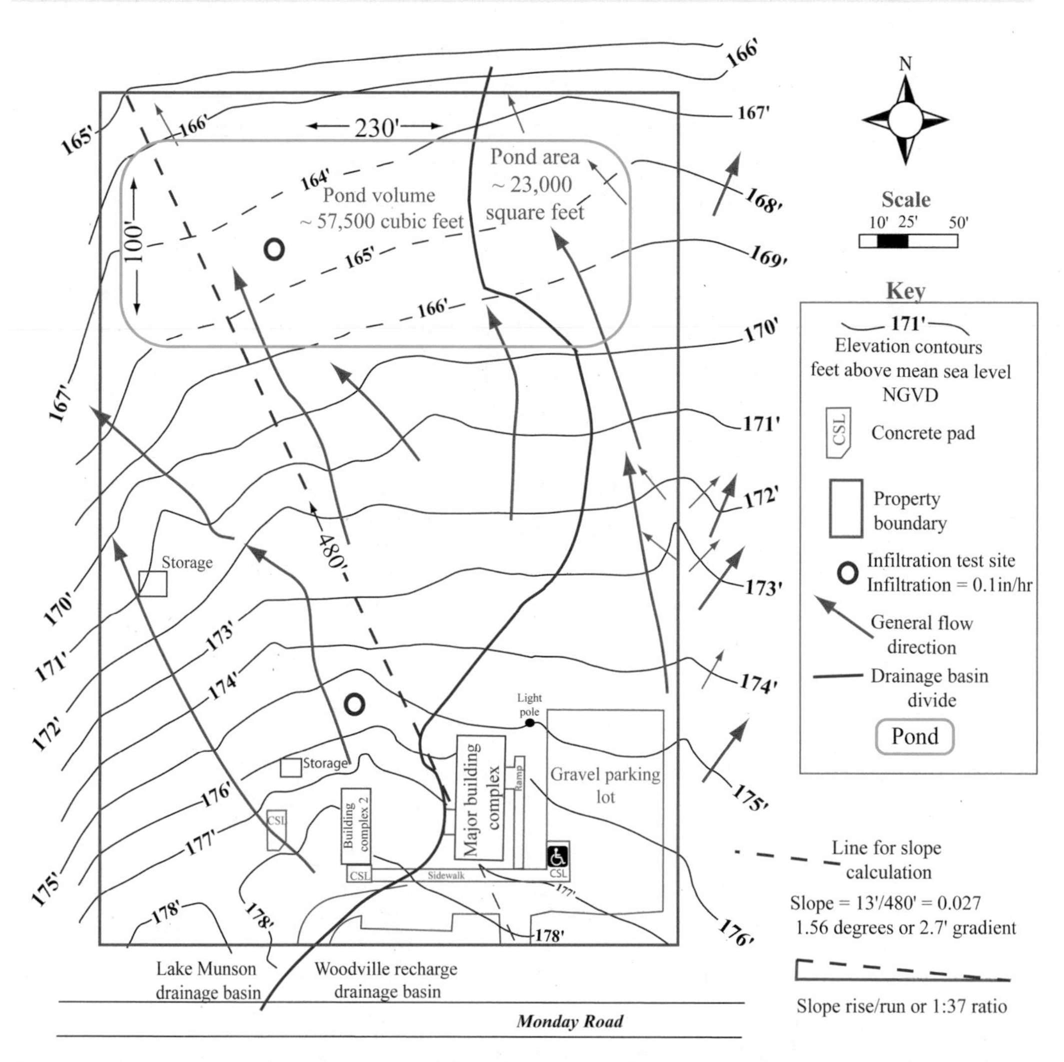

Figure 7.2. Stormwater runoff, flow direction, and slope determination for a commercial property, including one-foot contours, flow patterns, retention pond, and other stormwater data. Modified from Evans (2018) and EIII Environmental Consulting Inc. with permission.

Tallahassee. The infiltration rate for both sites were calculated at 0.1 in/hr (0.254 cm/hr), which correlated with the rates typical of the Orangeburg fine loamy sand that was located on and around the study property. This soil is classified as a well-drained and medium runoff–type soil with a hydrologic rating of C (see Table 6.2 and Figure 6.12).

Flow direction is based on the topographic patterns of one-foot contours. Approximately 16% of the property flows away from the proposed retention pond area. This flow is included in the peak volume estimates to the stormwater pond to demonstrate maximum stormwater impacts for retention pond design and in case future construction innovations require diversion of runoff to the existing stormwater management system.

Overall patterns of flow are used to determine the placement of the stormwater system and gradient direction for best of slope delineation. The slope was estimated at 0.027, which equates to a 2.7% gradient (2.7-foot drop for every 100 feet) or 1.41-degree angle of slope. The slope results, in correlation with the storm events and land use, were used in the rational runoff coefficient selection from Table 7.2.

Most stormwater retention structures and calculations are designed to handle a 25-year 24-hour storm (9 inches/day). However, a 100-year storm event (11 inches/day) and Tropical Storm Faye (11.44 inches/day) were included to cover a worst-case scenario and observe maximum volume capabilities for natural drainage on the property.

Results

Based on experimental data analysis from various researchers, a number of tables have been developed to help in selecting a sensible coefficient *C*. The coefficients used in this analysis are from Chow, Maidment, and Mays (1988), which incorporated many of the factors described herein. The rational basic equation

Table 7.4a Land use and runoff volumes for entire commercial property

Land use	Subarea (ft^2)	Subarea (acres)	Percent area (%)	Runoff coefficients (*C*) 25-year storms	100-year storms	Runoff for 25-year storm (ft^3/day)	Runoff for 100-year storm (ft^3/day)	Tropical Storm Faye11.44 in/day at 100-yr *C* (ft^3/day)
Open lawn/grass	70,791	1.65	64.28	0.42	0.49	22,115	31,534	33,297
Urban forest	13,334	0.316	12.11	0.40	0.47	3,967	5,697	6,117
Landscape	5,719	0.135	5.19	0.44	0.51	1,872	2,652	2,836
Gravel parking	7,969[a]	0.183	7.24	0.86	0.95	2,549	3,442	2,842
lots (2)	3,985	0.092	3.62					
Concrete	3,043	0.070	2.76	0.88	0.97	1,991	2,683	2,796
Existing buildings	2,249	0.052	2.04	0.88	0.97	1,472	1,983	2,077
Proposed building	2,000	0.046	1.82	0.88	0.97	1,863	2,430	1,838
Miscellaneous	1,042	0.024	0.94	0.42	0.49	325	416	484
Property Totals	110,132	2.57	100	N/A	N/A	36,154	50,837	52,287

Source: Land use data from Evans (2018). Runoff coefficients from Chow, Maidment, and Mays (1988, table 15.11, slope = 2.7, or average range of 2%–7%).

[a] Gravel deemed 50% impervious by Leon County Storm Water Management.

Table 7.4b. Land use and runoff volumes for west drainage area

Land use	Area (ft²)	Area (acres)	Percent area (%)	Runoff coefficients (C) 25-year storms	100-year storms	Runoff for 25-year storm (ft³/day)	Runoff for 100-year storm (ft³/day)	Tropical Storm Faye 11.44 in/day at 100-yr C (ft³/day)
Open lawn/grass	10,890	0.25	59	0.42	0.49	3,402	4,851	5,049
Urban forest	7,405	0.17	41	0.40	0.47	2,203	3,164	3,290
Drainage totals	18,295	0.42	100	N/A	N/A	5,605	8,015	8,339
Flow to pond						30,549	42,822	43,948
Percentage of total volume (%)						18.3	18.8	18.9

(7.4) results for all three design storms are given in Table 7.4a, including flow to the southwest away from the pond in Table 7.4b.

7.3.2. Time-Area Method Rainfall Intensity *i*

The time-area method for runoff estimation can be considered an extension of and improvement on the rational method because the unrealistic assumption made in the rational method of uniform rainfall over the entire catchment and during the whole of t_c is avoided in the time-area method, where the catchment contributions are subdivided in time. The peak discharge Q_p is the sum of the flow contributions from subdivisions of the catchment, which are defined by equal flow time to the stream system where Q_p is required, such as flow into a lake or reservoir, or at catchment outlet. These areas are delineated by equal time contours or isochrones (Figure 7.3). Determining isochrones requires extensive knowledge of the drainage basin, usually by observations and measurements during critical times of wet or saturated periods or times of floods.

The runoff flow from each area bounded by two isochrones $(t-\Delta t,t)$ is estimated from the product of the average intensity of effective rainfall (i) from time $t-\Delta t$ to time t and the difference in area ΔA (Figure 7.3a). Therefore, the flow at Q_4 at point b into the lake at time 4 hours (Shaw 1988) is

$$Q_4 = i_3\Delta A_1 + i_2\Delta A_2 + i_1 A_3 + i_0 A_4$$

or

$$Q_t = \sum_{k=1}^{t} i_{(t-k)}\Delta A_{(k)} \tag{7.5}$$

The average rainfall intensity i has a duration equal to the given or design storm duration equal to the *time of concentration* (t_c). Time of concentration is an idealized concept and is defined as the time taken for a drop of water falling on the most remote location (point a) in the drainage basin to travel to the basin lake (point b), where remoteness relates to the time of travel rather than distance (Maidment 1993). This is the time after the beginning of a storm when all portions of the catchment are contributing simultaneously to flow at the outlet and is analogous to Figure 6.4c in the *variable source area* illustration. Modifying equation (7.3), the peak flow at point b is

$$Q_t = \sum_{k=1}^{n} i_{(n-k)} \Delta A_{(k)}$$

where n is the number of incremental areas between successive isochrones given by $t_c/\Delta t$, and k is a counter.

From the hyetograph (Figure 7.3b) and peak flow equation (Figure 7.3c), a hydrograph can be constructed for peak flow into the lake (Figure 7.3d). By making the discrete intervals Δt very small and considering the contributing subcatchments to be continuously adding to the lake with increasing time, the time-area curve can be generated with limits of the total area of the catchment at t_c (Figure 7.3e). Thus, for any time t, the corresponding area A gives the maximum flow to the river outfall as a result of a rainfall duration t. The derivative of the time-area curve (dA/dt) gives the rate of increase in a contributing area with time. The length of the time axis is equal to the time of concentration of the catchment and is termed *the time-area-concentration curve* (Shaw 1988).

The variation in rainfall intensities (i) within a storm event are averaged over discrete periods depending upon the time-area isochrone selected. Thus, the areas that contribute the most rapid runoff (e.g., urban parking lots) can be examined by their isochrone's time-area location and incorporating a critical sequence of design storm intensities to determine their maximum peak flow. When investigating such differences within a catchment, however, determination of t_c can be problematic.

There are many similar methods for estimating t_c. The equations for estimating t_c are empirically derived based on the analysis of one or more data sets. Two examples representative of different

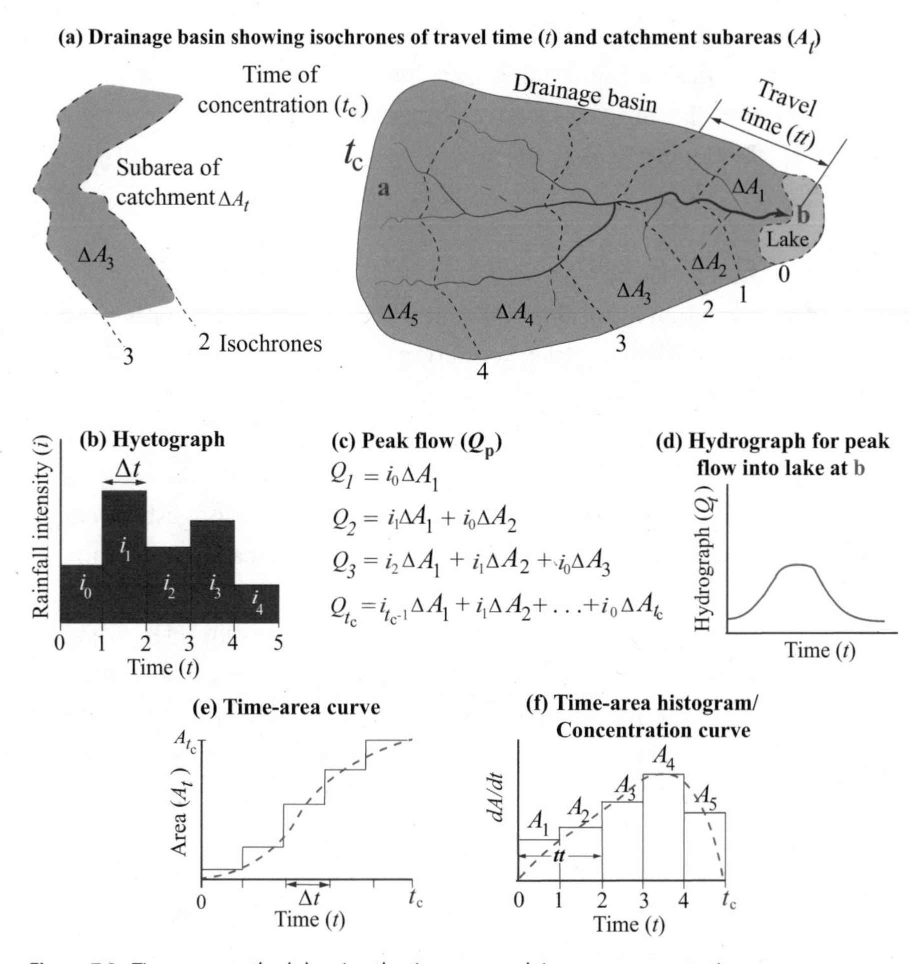

Figure 7.3. Time-area method showing the time-area and time-area concentration curves.

types of land use, such as urban versus natural or agricultural, are the Morgali and Linsley (1965) method and the Kirpich (1940) method.

7.3.2a. Morgali and Linsley Method

For small urban areas with drainage areas less than ten or twenty acres, and for which the drainage is basically planar, the method developed by Morgali and Linsley (1965) is useful. It is expressed as

$$t_c = \frac{0.94(nL)^{0.6}}{i^{0.4}S^{0.3}} \tag{7.6}$$

where

t_c = time of concentration (min),
i = design rainfall intensity (in/hr),
n = Manning surface roughness (dimensionless),
L = length of flow (ft), and
S = slope of flow (dimensionless).

7.3.2b. Kirpich Method

For small natural or agricultural drainage basins with streamflow, the Kirpich (1940) equation can be used:

$$t_c = 0.0025\left(\frac{L}{\sqrt{S}}\right)^{0.80} \tag{7.7}$$

where L is the length of the catchment along the longest stream or river channel and S is the slope of the overall catchment (mm^{-1}).

The time-area method has been revised and extended since its inception in the 1920s. Many of the modifications are specific to catchments and conditions and can be found throughout the literature and advanced hydrological textbooks referenced throughout this section.

7.3.3. Soil Conservation Service Method

The US Department of Agriculture's Soil Conservation Service (SCS) was renamed as the Natural Resources Conservation Service (NRCS) in 1994 to reflect its broader mission. However, throughout the literature this method is still termed the SCS method, as it will be referred to in this text.

The Soil Conservation Service (1972) developed a method for computing losses from the storm rainfall that is widely used for estimating runoff on small- or medium-sized ungauged drainage basins (see US Department of Agriculture 2004, *National Engineering Handbook*, chapter 10, for details of the following derivations). In the United States, it has replaced the rational method due to a larger database and methodology that better exemplify the physical characteristics for stormwater flow (Maidment 1993).

The assumptions of the SCS method is that for a storm event, the direct or cumulative runoff (Q) is always less than or equal to the depth of precipitation (P) due to abstractions or losses where a certain amount of rainfall is retained within the catchment as storage. Similarly, after runoff begins, the additional depth of water retained in the catchment (F) is less than or equal to some potential maximum retention (S), where F is the abstraction after runoff begins.

During the early part of the storm, there is some amount of precipitation where no runoff will occur due to initial abstraction (I_a). Initial abstraction includes rainfall retained in surface depressions, intercepted by vegetation, and absorbed by infiltration. Thus, the potential runoff is the precipitation minus the initial abstraction $Q = P - I_a$ (Figure 7.4). The premise of the SCS method is that the ratios of the two quantities are equal such that

$$\frac{F}{S} = \frac{Q}{P - I_a}. \tag{7.8}$$

From the conservation of mass or principle of continuity

$$P = Q + I_a + F \tag{7.9}$$

and F, the water retained as loss to the catchment after runoff begins, is

$$F = P - (Q + I_a).$$

Combining equations (7.8) and (7.9) and solving for runoff gives

$$Q = \frac{(P - i_a)^2}{P - i_a + S} \tag{7.10}$$

Through studies of numerous watersheds, the Soil Conservation Service discovered an empirical relationship from observed data that gave an approximation of the initial abstraction where I_a = $0.2S$ (US Department of Agriculture 2004) thus

$$Q = P - 0.2S$$

and

$$Q = \frac{(P - 0.2S)^2}{P + 0.8S} \tag{7.11}$$

When the P and Q data from numerous drainage basins were plotted, the SCS generated curves of the type seen in Figure 7.4. For convenience these curves were standardized with a *dimensionless curve number* (CN) defined such that $0 \leq CN \leq 100$. For impervi-

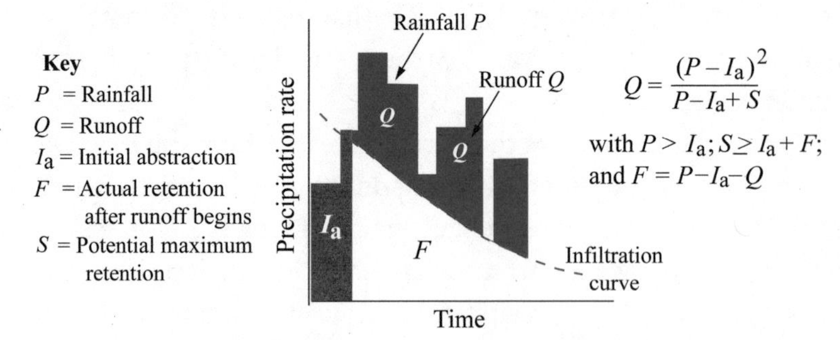

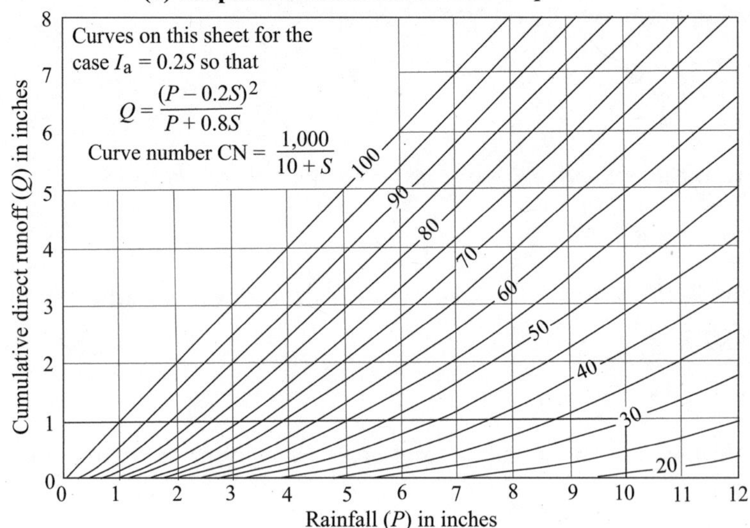

Figure 7.4. Relationship between runoff and rainfall as defined by the Soil Conservation Service (SCS). Graphical solution (b) for the SCS runoff equation (a) is shown here for rainfall less than or equal to 8 inches, estimating the volume of storm runoff for various hydrologic soil-cover complexes indicated by curve numbers. Modified from US Department of Agriculture (2004, Figure 10-2).

ous and water surfaces, $CN = 100$. For natural surfaces, $CN < 100$. S is related to the soil and cover conditions of the watershed and is related to CN by

$$S = \frac{1000}{CN} - 10 \tag{7.12}$$

where S is in inches.

7.3.3a. *Retention Parameters*

The initial abstraction I_a can be considered the boundary between storm size that produces runoff and the storm event that produces no runoff. The *maximum retention* of a catchment S is excess to I_a and dependent upon the soil cover complex, and in principle it should not vary from storm to storm. As precipitation becomes large as defined by the storm volume being greater than the maximum retention S, then

$$\textit{Maximum possible loss} = I_a + S. \tag{7.13}$$

The parameter F is the actual retention for a storm event greater than the initial abstraction such that

$$\textit{Actual abstraction} \text{ (loss)} = I_a + F. \tag{7.14}$$

The variability in the curve number results from rainfall intensity and duration, total precipitation, soil moisture conditions, cover density, stage of vegetation growth, and temperature. These are collectively called the *antecedent runoff condition* (ARC). ARC is divided into three classes: I for dry conditions, II for average conditions, and III for wetter conditions. The curve numbers shown in Figure 7.4 apply for normal antecedent moisture conditions (ARC II). For dry conditions, (ARC I) or wet conditions (ARC III), equivalent curve numbers can be generated by

$$\mathrm{CN(I)} = \frac{4.2\mathrm{CN(II)}}{10 - 0.058\mathrm{CN(II)}} \tag{7.15}$$

and

$$\mathrm{CN(III)} = \frac{23\mathrm{CN(II)}}{10 + 0.13\mathrm{CN(II)}}. \tag{7.16}$$

7.3.3b. Hydrologic Soil Groups

As discussed in section 6.4, infiltration rates of soils vary widely and are affected by subsurface permeability as well as surface intake rates. Soils are classified into four hydrologic soil groups (A, B, C, and D) according to their minimum infiltration rate (Table 6.2), which is obtained for bare soil after prolonged wetting.

7.3.3c. Hydrologic Soil-Cover Complex

A combination of a hydrologic soil group (soil) and a land use and treatment class (cover) defines a *hydrologic soil-cover complex*. Chapter 9 of the *National Engineering Handbook* gives tables and graphs of runoff curve numbers assigned to such complexes (see US Department of Agriculture 2004, table 9-5). These tables were determined empirically with the four soil groups and a wide range of land use, including (a) agricultural, (b) national and commercial forest, and (c) urban and residential. From the example above to compare the SCS and rational method, the selected curve numbers are given in Table 7.5 and are selected from the *National Engineering Handbook* urban and residential tables 9-5 and agricultural land use table 9-1 (property urban forest land use).

Table 7.5. Runoff curve numbers selected for urban areas[a] and agricultural lands, table 9-5 and table 9-1, respectively, from the *National Engineering Handbook* (ARC condition II, $I_a = 0.2S$)

Cover description, cover type, and hydrologic condition	Average percentage impervious area (%)[b]	Curve number by hydrologic soil group			
		A	B	C	D
Open space (lawns, parks, cemeteries, etc.)[c]					
Poor condition (grass cover < 50%)		68	79	86*	80
Fair condition (grass cover 50% to 75%)		49	69	79	84
Good condition (grass cover 75% to 100%)		39	61	74*	80
Impervious areas					
Paved parking lots, roofs, driveways, etc., excluding right of way		98	98	98*	98
Streets and roads		98	98	98	98
Paved curbs and storm sewers		83	89	92	93
Paved open ditches		76	86	89	91
Gravel, including right of way		76	85	89*	91
Dirt, including right of way		72	82	87	89
Urban districts					
Commercial and business	86	89	92	94	96
Industrial	72	81	88	91	93
Residential areas by average lot size					
⅛ acre, townhouses	65	77	85	90	92
¼ acre	38	61	75	83	87
½ acre	25	54	70	80	86
1 acre	12	51	68	79	84
2 acres	12	46	65	77	82
Wood or forest area					
Poor condition (litter, small trees, brush absent from fire/grazing)		45	66	77	83
Fair condition (grazed not burned, some forest litter, cover soil		36	60	73	79
Good condition (protected from grazing with litter and brush cover)		30	55	70*	77

Source: Modified from US Department of Agriculture (2004).
* Indicates curve numbers applicable to Case Study 7.2.
[a] Average runoff condition for urban areas and $I_a = 0.2S$.
[b] The average percentage of impervious area shown was used to develop the composite curve numbers. Other assumptions are as follows: impervious areas are directly connected to the drainage system, impervious areas have a curve number of 98, and pervious areas are considered equivalent to open space in good hydrologic condition.
[c] Curve numbers shown are equivalent to those of pasture. Composite curve numbers may be computed for other combinations of open space type.

The general procedure for SCS method calculations for a single catchment is as follows:

1. *Define boundary conditions.* Delineate the catchment boundary and determine its area.
2. *Define the flow path* from the uppermost portion of the catchment to the design point.
3. *Divide the areas* into similar hydrologic soil groups (A–D) and land-use hydrologic soil complexes (urban, forest, etc.), and determine the antecedent runoff condition.
4. From the three defined areas, derive the curve number from the SCS tables in chapter 9 of the *National Engineering Handbook* (summarized here in Table 7.5).
5. *Define the rainfall precipitation P* for a design storm or from gauge data.
6. *Find runoff Q* using graphical solution (Figure 7.4) or equation (7.10).

Case Study 7.2

Stormwater Runoff Assessment Using Soil Conservation Service Method

W.L. Evans, 2018, *Stormwater Management Report: Analysis of Stormwater Runoff at CASA 12 Property 2410 Monday Road, Tallahassee, Florida* (Tallahassee: EIII Environmental Consulting Inc.), 15–24.

For a comparison of the SCS method against the rational method, runoff Q in this case study was calculated using the SCS graphical solution in Figure 7.5 with the land-use data from Table 7.6. To simplify the results only the 25-year 24-hour storm and 100-year 24-hour design storm were used in the calculation.

Overall, the SCS estimations were approximately 30.5% higher than the rational method estimations, with 53,534 ft^3 versus 36,154 ft^3 (32% higher) and 70,928 ft^3 versus 50,837 ft^3 (28% higher) for the two design storms. One reason for this variability is that the rational method assumes a constant runoff ratio for a storm event whereas the SCS ratio of runoff to rainfall increases as duration of the storm increases, which is a more accurate representation of a storm event.

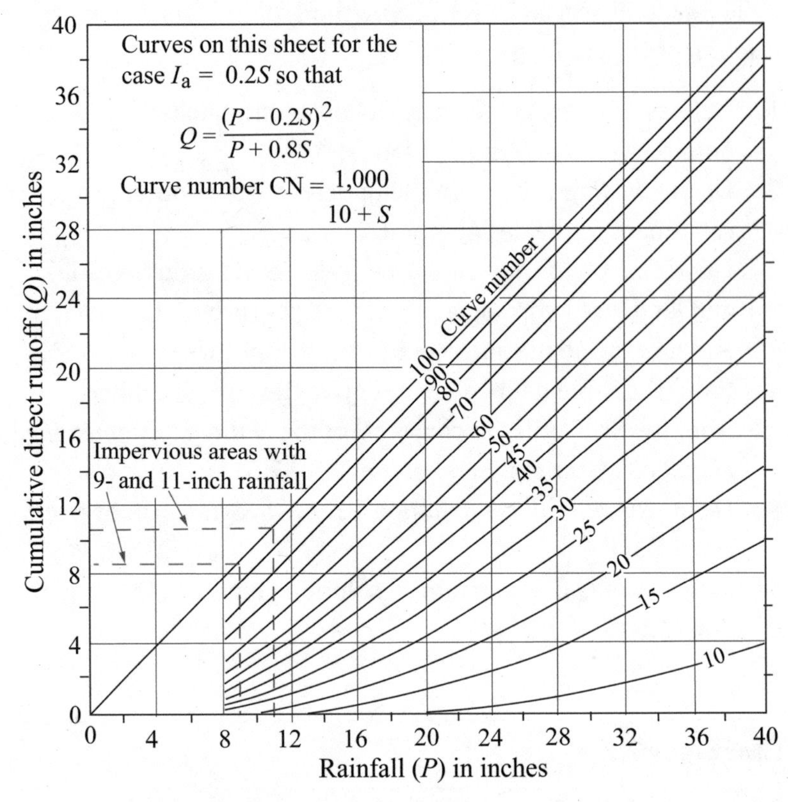

Figure 7.5. Soil Conservation Service chart for estimating the volume of storm runoff for various hydrologic soil-cover complexes indicated by curve numbers for rainfall greater than 8 inches. Modified from US Department of Agriculture (2004).

Table 7.6. Land use and runoff volumes for commercial property using Soil Conservation Service method under average runoff condition II where $I_a = 0.2S$.

Land use	Subarea (ft²)	Subarea (acres)	Percent area (%)	25-yr storm P = 9 in. Curve number (CN)	Rate of discharge Q (in.)	Runoff for 25-year storm (ft³/day)	100-yr storm P = 11 in. CN	Q	Runoff for 100-year storm (ft³/day)
Open lawn/grass	70,791	1.650	64.28	74	5.8	34,216	74	7.7	45,424
Urban forest	13,334	0.316	12.11	70	5.2	5,778	70	7.1	7,889
Landscape	5,719	0.135	5.19	80	6.6	3,122	80	8.4	4,003
Gravel parking	7,969	0.183	7.24	89	7.1	4,749	89	9.7	6,442
	3,985	0.092	3.62						
Concrete	3,043	0.070	2.76	98	8.5*	2,155	98	10.7	2,713
Existing buildings	2,249	0.052	2.04	98	8.5*	1,593	98	10.7	2,005
Proposed building	2,000	0.046	1.82	98	8.5*	1,417	98	10.7	1,783
Miscellaneous	1,042	0.024	0.94	74	5.8	504	74	7.7	669
Property Totals	110,132	2.568	100	N/A	N/A	53,534			70,928

* Use Soil Conservation Service chart in Figure 7.5 for runoff Q if precipitation P is greater than 8 inches.

Studies by Mousavi, Nekoei-Meher, and Mahdavi (1998) and Khosroshahi (1991) revealed that *synthetic unit hydrographs* used in the SCS method (see section 7.4.3) generated watershed flood estimates with less error than the rational method and could be used for peak flow estimation in the similar condition watersheds.

The current Natural Resources Conservation Service has a computer model called TR-55 Urban Hydrology for Small Watersheds that uses the SCS methodology to estimate runoff and peak discharge for small catchments (US Department of Agriculture 1986). From this computer program the unit hydrograph for a basin can be generated and used for the discussion on the use of hydrographs for stormwater analysis. The emphasis of the TR-55 model is on urban and urbanizing areas, but the model is applicable to a variety of small drainage basins.

7.4. Hydrographs

Hydrographs can be used to estimate quickflow Q as well as baseflow and a host of other storm characteristics, and they are essential in flood calculations and modeling. They can also be used to demonstrate the hydrologic effects of existing or proposed land use changes to a watershed and how that impacts lake or wetlands.

There are three types of hydrographs that can be used to describe catchment systems and runoff from storm events (Chow, Maidment, and Mays 1988):

Natural hydrographs: wherein the data is collected directly from flow records of a gauged stream or lake such as seen in the Peachtree hydrograph for a 51 mm storm and discussed in section 5.4.3 (see Figure 5.18).

Unit hydrographs: "the hydrograph of one-inch (or one cm or mm if using SI units) storm runoff generated by a rainstorm of fairly uniform intensity occurring within a specific period of time" (Dunne and Leopold 1978, 329).

Synthetic hydrograph: generated by a design storm with catchment characteristics to simulate a natural hydrograph.

7.4.1. Natural Hydrographs

Hydrographs are graphs that display the change of a hydrologic variable over time and can be used to define certain hydrological characteristics of a drainage basin. For example, long-term hydrographs, such an *annual or seasonal hydrograph*, which plots lake stage or streamflow versus time over a longer temporal span can, be used to ascertain certain correlations between rainfall, runoff, evapotranspiration, and lake level or streamflow in a drainage basin. Figure 7.6 is a hydrograph of Lake Fuqua—a dammed reservoir lake near Duncan, Oklahoma—showing summer lake levels and cumulative precipitation from April through July 2017. Note the rainfall data is cumulative over this period, and the spikes in precipitation indicate a storm event with a correlating rise in lake stage. As this is a closed (*endorheic*) lake, stage decline is associated with evapotranspiration and seepage.

Storm hydrographs are used to determine baseflow and runoff in a drainage basin. A storm hydrograph measures the time distribution and runoff volume generated during a storm event mainly to determine correlations that could predict and plan for flooding and for best management practices in protecting natural resources. The shape of the hydrograph is dependent on many of the drainage basin characteristics, such as topography, size and

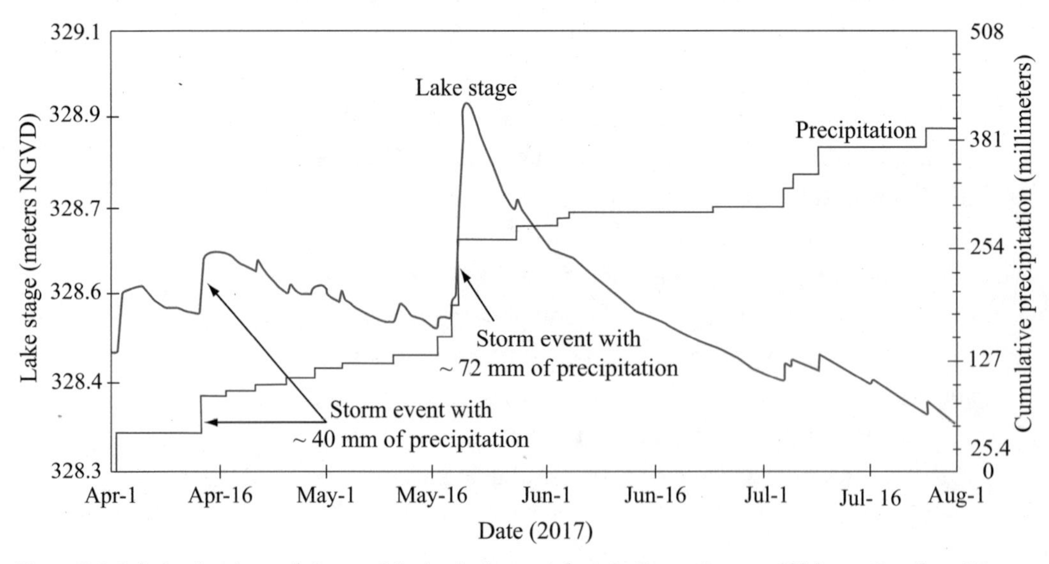

Figure 7.6. Lake level and cumulative precipitation hydrograph for Lake Fuqua, Duncan, Oklahoma. Data from US Geological Society (2017).

shape of the catchment area, patterns and distribution of soil types, underlying geology, land use, stream patterns, stream density, and climate. Chow (1959) describes the hydrograph as an "external expression of the physiographic and climatic characteristics that govern the relations between rainfall and runoff of a particular drainage basin" (Chow, Maidment, and Mays 1988, 132).

7.4.2. Unit Hydrograph

One of the major advances in hydrological analysis has been the theoretical concept of the *unit hydrograph*, introduced by Sherman (1932, 1940). He defined the unit hydrograph as one of surface runoff resulting from effective rainfall falling in a unit of time, such as one hour or one day, and producing uniformly in space and time over the total catchment (Sherman 1942).

The unit quantity of effective rainfall is normally taken as one inch or, in SI units, one millimeter or one centimeter, and the outflow hydrograph is expressed by discharge in m^3/s. The unit duration may be of one hour or more, depending upon the size of the catchment, storm characteristics, and type of data measurement. The unit duration, however, cannot be more than the time of concentration (t_c) or basin lag or period of rise.

The unit hydrograph, by definition, specifies that the input rainfall excess is assumed to be uniform over the basin and thus defined by a single hyetograph. In other words, the unit hydrograph is the basin response as a lumped system to a standard input of rainfall. This deterministic analogy was discussed in section 5.4.

The unit hydrograph methodology has undergone numerous enhancements from various researchers; however, the basic ideas presented by Sherman remain valid. Simply stated, for a given drainage basin, all hydrographs resulting from rains of the same period of excess (unit duration) have equal time bases (t_b). The ordinates of each hydrograph are proportional to the volume of direct runoff (quickflow) if the time and areal distribution of the precipitation are similar. Thus, a unit hydrograph is a hydrograph for a specific time period of rainfall excess (quickflow or runoff) and uniform distribution and whose volume of quickflow is equal to one-inch of water over the entire watershed.

The basic premise is that watershed-controlling features such as basin size and shape, geology, patterns of soil and vegetation

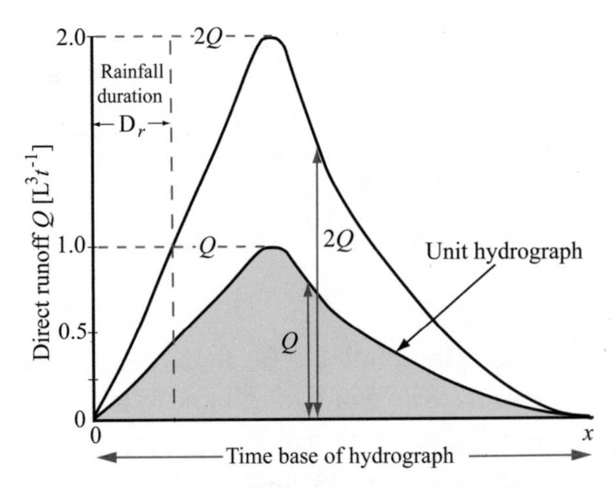

Figure 7.7. A two-unit hydrograph generated by doubling the ordinates of the unit hydrograph (shaded) illustrating the principle of superposition.

distribution, and land use as well as stream characteristics such as patterns, channel size, gradients, and densities are assumed constant from storm to storm. As such, given a storm of a specific duration, the expectation is that the shape of the hydrograph (showing temporal behavior of runoff) would be the same from storm to storm.

For example, if the time base X for a hydrograph is fixed for storms of a specific duration D_r and the temporal behavior of runoff is a function of the watershed features specified earlier, then the direct runoff ordinate of the hydrograph should be proportional to the total volume of storm runoff produced (Figure 7.7). This unit hydrograph is by definition a one-unit storm of duration D_r. Thus, all storms of duration D_r will have the same time base X. If the storm volume is two units, then the ordinates of the hydrograph will be twice that of the unit hydrograph. This is the *principle of proportionality* (or *superposition*), and with it the volume of direct runoff produced by storms of similar duration but different intensity can be predicted by altering the ordinates of the unit hydrograph.

This is an approximate methodology with problems of a theoretical nature; however, the results are sufficiently accurate for most runoff estimations and planning purposes, usually predicting runoff peaks within plus or minus 25% of their true value (Dunne and Leopold 1978). Unit hydrographs are generally derived from streamflow data and estimates of temporal distribution of rainfall excess, or by synthesis using drainage basin physiographic data.

The method can also be used for predictions in areas with relatively short records of rainfall and runoff and is a suitable technique in areas of urbanization. Hence, the unit hydrograph is one of the most common techniques for runoff prediction.

The unit hydrograph is a simple linear model that can be used to derive any hydrograph resulting from any amount of excess rainfall. Inherent in this model are the following assumptions:

1. The excess rainfall has a constant or uniform intensity or rate within the effective duration. For example, if the average rainfall over a certain basin during a five-hour storm is 100 mm, a unit hydrograph of five-hours duration can be derived only if the intensity of rainfall is more or less 20 mm/hour over five hours. Such a storm would be termed a *unit storm*. If the same amount of rainfall is distributed with varied intensity, the unit hydrograph cannot be precisely estimated by this simple method.
2. The excess rainfall is uniformly distributed throughout the entire catchment area.
3. The excess rainfall of equal (unit) duration will produce *direct runoff hydrographs* (DRH) having the same or constant time base time.
4. The ordinates of all DRHs of a common time base (i.e., hydrographs due to effective rainfalls of different intensity but equal duration) are directly proportional to the total amount of direct runoff represented by each hydrograph. Therefore, the direct runoff response to rainfall excess is assumed to be linear. A linear response means that if an input $x_1(t)$ causes an output $y_1(t)$ and an input $x_2(t)$ causes an output $y_2(t)$, then an input $x_1(t) + x_2(t)$ gives an output of $y_1(t) + y_2(t)$. Thus, if $x_2(t) = rx_1(t)$, then $y_2(t) = ry_1(t)$.

The assumption of linear response enables the use of the method of superposition to derive direct runoff hydrographs. One illustration of this principle would be if for a one-hour duration storm with discharge or direct runoff (ordinate axis) units

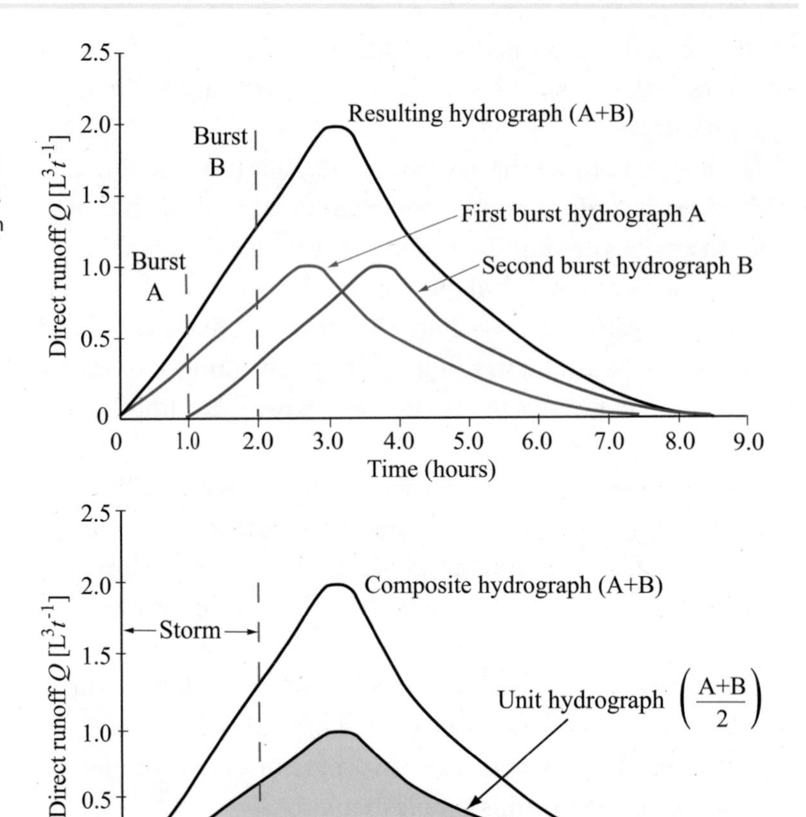

Figure 7.8. Principle of Superposition: (a) A two-hour storm with two consecutive bursts lasting one hour each and generating hydrographs A and B. The two-hour storm hydrograph is produced by adding hydrographs A and B; and (b) a unit hydrograph is produced by halving the ordinates of the storm hydrograph (A+B/2) for a two-burst two-hour duration storm.

of 0, 0.25, 0.5, 1.0, 0.5, 0.25, 0 respectively, the excess rainfall of two units of precipitation for a one-hour duration storm would result in doubling the ordinates to 0, 0.5, 1.0, 2, 1.0, 0.5, 0 (Figure 7.8a). A second example would be if there was a one-unit rainfall for multiple yet equal durations, such as two hours of one unit per hour producing a two-unit hydrograph (i.e., two one-hour bursts or two consecutive one-hour storms), then the direct runoff hydrograph discharge would be calculated by summing the corresponding ordinates of the two-unit hydrographs. The unit hydrograph for this certain storm would be generated by dividing the storm hydrograph by two (Figure 7.8b).

For a specific catchment, the hydrograph from a given excess rainfall reflects all of the combined unchanging physical characteristics of the watershed. This implies that the direct runoff hydrograph from a given drainage basin due to a given pattern of effective rainfall will always be the same irrespective of the time of occurrence (e.g., even if the basin characteristics change with season, etc., the unit hydrograph remains the same). This assumption is called *principle of time invariance.*

7.4.2a. Limitations of the Unit Hydrograph

The unit hydrograph is arguably the most direct way of describing the response of a catchment to storm rainfall, as it takes into account of all factors that influence the flood response to actual excess rainfall input. The basic concept underlying the approach is simple to understand and easy to apply, being suited to both hand and spreadsheet calculations. The unit hydrograph approach can be applied with some confidence where its main assumptions are at least approximately satisfied: spatial uniformity of rainfall excess and linearity of catchment response (Ball et al. 2016).

In theory the unit hydrograph should be applicable to catchments of all sizes. However, like most natural systems, general assumptions are never perfectly met. Field data and laboratory tests have shown that the assumption of a linear relationship is not exactingly true. As discussed in section 5.2, a uniform storm distribution over the entire catchment is not a realistic assumption for the majority of storm events, and the larger the drainage basin area, the less chance there is for the validity of unit hydrograph assumptions one and two discussed earlier. A more realistic approach is to refer only to a storm whose spatial distribution reflects basin characteristics, such as topography or a typical storm pattern for the area.

The principle of linearity is also not entirely valid due to variability in proportion of surface, subsurface, and groundwater runoff components during smaller and larger storms of the same duration; the maximum peak of the unit hydrograph derived from a smaller storm will be less than the one derived from a larger storm. As such, the nature and duration of the recession limb that is a function of the peak flow will also vary. When appreciable nonlinearity is observed, it is essential to use derived unit hydrographs only for reconstructing events of similar magnitude.

When the area of the drainage basin surpasses a few thousand square kilometers, the reliability of the unit hydrograph method diminishes. When this area is exceeded, the catchment must be divided into sub-basins and unit hydrographs developed for each sub-basin (Chow, Maidment, and Mays 1988).

No two rainstorms have the same pattern in space and time. Although it is not feasible to derive separate unit hydrographs for each possible time-intensity pattern, it is necessary to construct several unit hydrographs (grouping storms into categories according to their length) to cover different durations of storms. Investigations have demonstrated that a tolerance of ± 25% in unit hydrograph duration is acceptable (Dunne and Leopold 1978). Thus, a two-hour unit hydrograph can be applied to storms of 1.5 to 2.5 hours in duration.

When snow conditions prevail either as snowfall precipitation or rainfall onto an existing snowpack, the unit hydrograph method cannot be applied. During snowfall, the snowmelt runoff is governed mainly by temperature changes. In addition, if rainfall gets mixed in with an existing snowpack, the delayed runoff produced will vary depending upon the different conditions of the snowpack.

As seasons and land use change, so do conditions of flow and other physical characteristics of the drainage basin. Thus, the *principal of time invariance* is only valid when the time and conditions of the catchment are specified.

7.4.2b. Unit Hydrograph Applications

Despite the unrealistic assumptions, the unit hydrograph is still a very valuable and flexible tool for developing synthetic hydrographs from design storms due to the fundamental principles of *invariance* and *superposition*. By understanding and remaining within the various ranges and restrictions discussed so far, the unit hydrograph has a number of applications:

a. Creating a runoff or flood hydrograph with a limited record of data
b. Estimating total runoff volume and its distribution over time
c. Validating the reliability of flows obtained by statistical methods
d. Easy calculating of peak flows and runoff in simple computer models

7.4.2c. Unit Hydrograph Construction and Application for Simple Storms

A unit hydrograph describes the characteristics of both the drainage basin and the rainstorm. As discussed, no two rainstorms are exactly alike; thus, to minimize the effects of individual rainstorms, data from several events with similar excess rainfall and duration should be used. After deriving the unit hydrograph for each storm, the arithmetic mean of these hydrographs should be calculated to produce a representative unit hydrograph that reduces the influence of the individual storm biases. Unit hydrographs for single peak storms are constructed using the following steps:

1. Select four or more hydrographs from storms of intense, moderately uniform rain quantities and similar duration.
2. Plot each hydrograph and separate the quickflow from the baseflow, creating a direct flow hydrograph using one of the methods discussed in section 6.4.
3. Calculate the total amount of storm or quickflow [L^3] using stream gauge, computer model, or graphical techniques.
4. Reduce the ordinates of the quickflow graph to their equivalent values for one unit (e.g., 1 mm or 1 inch) of quickflow by dividing each ordinate by the ratio of the total amount of quickflow to one unit. It should be noted that no two hydrographs are identical though they will have the same general shape.
5. Plot the reduced hydrographs together and superimpose them all beginning at the same time (Figure 7.9).
6. Set the peak of the unit hydrograph by calculating the arithmetic mean discharge [L^3t^{-1}] of all peaks flows (U_p), and their average time (t_p) of occurrences (see data in Figure 7.9).
7. Draw the unit hydrograph to conform to the average shape of the reduced hydrographs, passing through the calculated peak and having a volume of 1.0 unit (see red line in Figure 7.9).

In Figure 7.10, a unit hydrograph is derived with $Q = 1$ unit (e.g., cubic millimeters or cubic inches) using the steps just outlined. Three other hydrographs for $Q = 0.5$, 2, and 3 units were drawn using the principle of superposition (see Figures 7.7 and 7.8). The

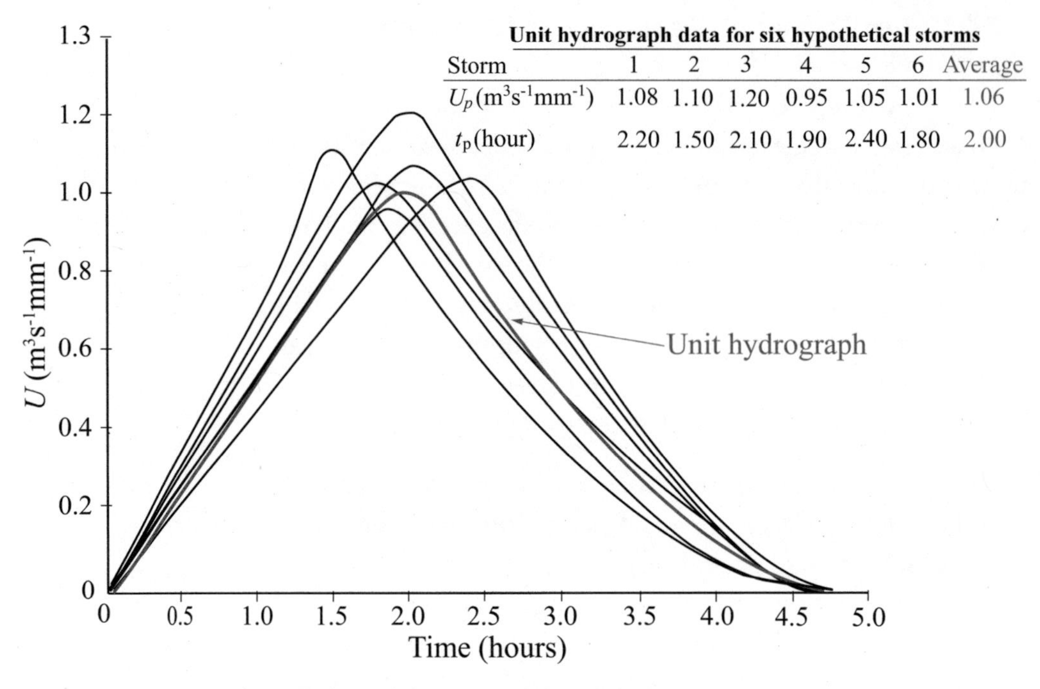

Figure 7.9. Construction of a unit hydrograph from six simple hypothetical storms.

values of Q were chosen for ease of derivation and visual example. The principle of proportionality would work for any value of Q (e.g., 1.33, 2.78, 3.19, etc.). The ratios for the other three hydrographs are calculated as

$$\frac{\textit{measured quickflow } Q_{measured}}{\textit{unit quickflow } Q_{unit}} = \frac{\textit{measured discharge } Q_{measured}}{\textit{unit discharge } Q_{unit}} \quad (7.17)$$

At time = 2 hours and for a measured quickflow (Q) of 1 unit, the discharge is 200 [L^3t^{-1}]. Using equation (7.17), the discharge for 3 units ($3Q$) of quickflow is

$$\frac{3}{1.0} = \frac{x}{200} =$$

$$(1.0)(x) = (3.0)(200)$$

$$x = 600 \ [L^3t^{-1}]$$

These calculations can be applied for all coordinates on the $Q = 3$ hydrograph. For example, at time = 1.5 hours the unit discharge is 150 [L^3t^{-1}] and the discharge for the $3Q$ is 450 [L^3t^{-1}].

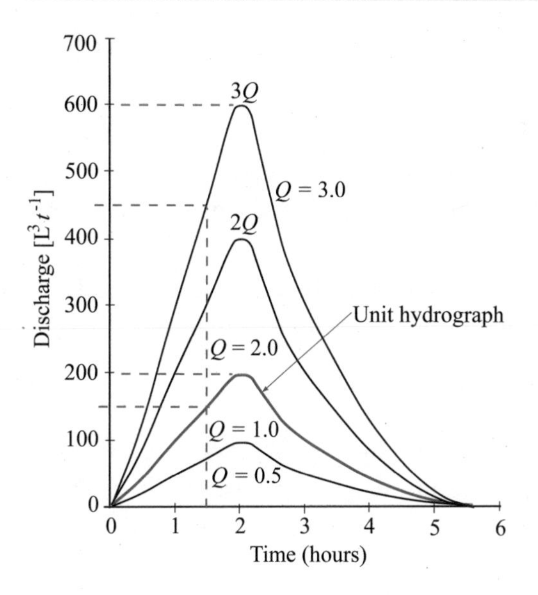

Figure 7.10. Demonstration of the principle of proportionality (superposition) for computing storm hydrographs (equation 7.17 of the same duration with different discharges *Q* of 0.5, 2.0, and 3.0 units (e.g., mm or in) from a unit hydrograph.

Similarly, for a $0.5Q$ hydrograph at time = 2 hours and 1.5 hours, the discharges would be 100 and 75 $[L^3t^{-1}]$, respectively.

Derivation of unit hydrographs from composite and multiperiod storms and hyetographs displaying complex patterns that require more sophisticated techniques are not in the scope of this introductory text. For further information, see Maidment (1993), Shaw (1988), and chapter 16 of the *National Engineering Handbook* (US Department of Agriculture 2004).

7.4.3. Synthetic Unit Hydrograph

Unit hydrographs derived from a specific drainage basin from rainfall and gauged stream data are only applicable to that watershed and the gauging point of the stream. *Synthetic unit hydrograph* methods are used to develop unit hydrographs for other ungauged areas, such as a sub-basin in the same watershed or nearby drainage basin with similar geomorphic characteristics (e.g., drainage area, channel gradient, climate, and drainage density). These relationships may be derived by considering a number of catchments in a reasonably homogeneous area for which unit hydrographs have been derived from recorded rainfall and streamflow data. These relationships are empirical and as such cannot be expected to be universally applicable. In general, their application should be restricted to the region in which the relationships were derived.

There are three types of synthetic unit hydrographs (Chow, Maidment, and Mays 1988):

1. Those base on models of watershed storage (Clark 1945).
2. Those based on a dimensionless unit hydrograph (Soil Conservation Service 1972).
3. Those related to such hydrograph characteristics as peak flow rate as well as base time to drainage basin characteristics (Snyder 1938; Gray 1961).

Of the various synthetic unit hydrograph approaches available, the only ones to have found widespread use are those based on the model of Clark and the dimensionless unit hydrograph from the SCS.

Clark's model is commonly referred to as the *Clark-Johnstone model*. The Clark-Johnstone method has been simplified by Cordery and Webb (1974) to produce a model that is suitable for some limited applications; it involves the translation of excess rainfall to the outlet and routing this translated flow through a *lumped* concentrated storage at that location. It has been used quite widely for synthetic unit hydrograph derivation and has been shown to be applicable to most of the east coast of Australia (Cordery, Pilgrim, and Baron 1981). The basic assumption is that the shape of the unit hydrograph may be determined from two parameters, namely the base length of the time-area curve (C) and the catchment storage factor (K).

The SCS model originated from Snyder's model and is discussed in the next section. For more detailed description of these synthetic unit hydrograph approaches from Clark and Snyder and examples of their application, see Cordery (1987); Chow, Maidment, and Mays (1988); Maidment (1993); Dunne and Leopold (1978); and the original authors of Snyder, Gray (1961) and Clark (1945).

7.4.3a. Dimensionless Hydrograph

The SCS synthetic *dimensionless unit hydrograph* (DUH) method originated from Snyder (1938) where lag time was used for calculating incremental rainfall duration. Mockus (1957) used this approach to develop the DUH from the analysis of numerous small-gauged watersheds. Mockus expressed the unit hydrograph

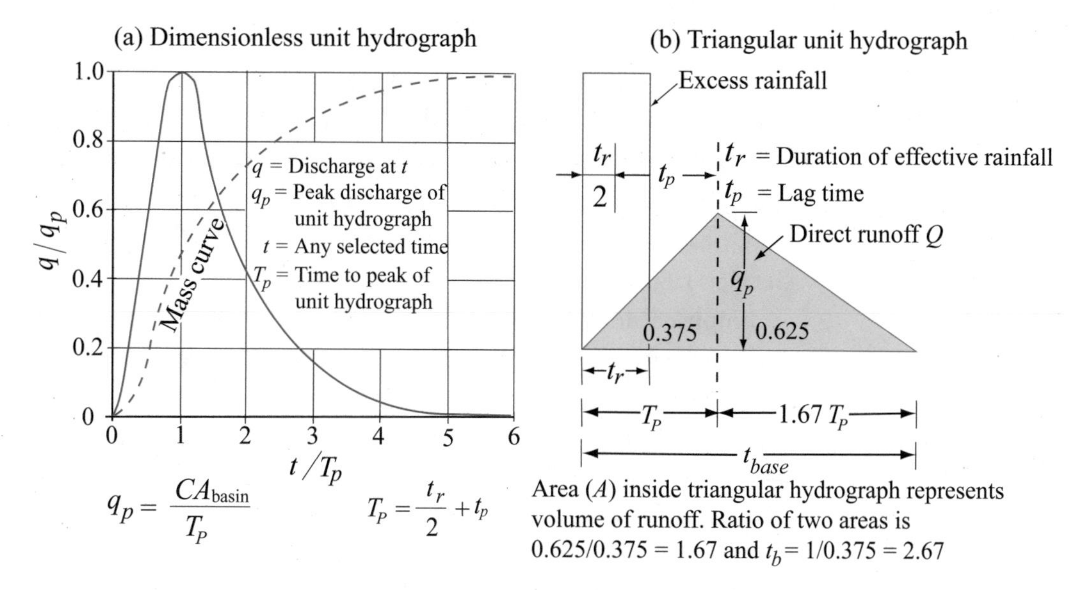

Figure 7.11. Examples of Soil Conservation Service synthetic unit hydrographs: (a) dimensionless hydrograph and (b) triangular unit hydrograph. Modified from the Soil Conservation Service (1972).

values in terms of ratios of discharge q to peak discharge q_P (i.e., q/q_P) and corresponding time to time to peak discharge (t/Tp) resulting in a dimensionless unit hydrograph.

Utilizing Mockus's concepts, the Soil Conservation Service (1972) developed a synthetic DUH based on the analysis of a large number of watersheds with widely varying locations, sizes, and physiography where streams and rainfall were gauged (Dunne and Leopold 1978). In the SCS dimensionless unit hydrograph method, all the hydrograph ordinates are given by ratios between instantaneous discharge (q) and peak discharge (q_p) (or total volume Q and accumulated volume Q_a) and between time (t) and time to peak (T_p) (Figure 7.11a). From measurements of peak discharge and the lag time for the duration of the excess rainfall, the unit hydrograph can be derived from the synthetic dimensionless hydrograph for a given basin. The mass curve is a plot of cumulative volume of water that can be stored from a streamflow versus time.

7.4.3b. Triangular Unit Hydrograph

The equation for calculating the peak rate for the SCS unit hydrograph is contingent on the shape and percent volume occurring before the peak compared with the percent volume after the peak and the curvature of the rising and receding limbs. Investigations have shown that the curvilinear shape of a dimensionless unit hydrograph can be approximated by a triangular-shaped unit with nearly the same results. This is done for the convenience of the manual calculations of q_p and T_p (Figure 7.11b). Time is in hours and the discharge is in $m^3/s \times cm$ or $ft^3/s \times in$ (Soil Conservation Service 1972).

Empirical analysis by SCS of various hydrographs indicated that for most watershed conditions the DUH consisted of 37.5% of the total runoff volume before the peak discharge with the remaining 62.5% occurring after the peak discharge. Thus, the time of recession (t_r) is estimated at 1.67 T_p. As the area under the unit hydrograph should be equal to a direct runoff Q of 1 cm (or 1 inch), then

$$Q = \frac{q_p T_p}{2} + \frac{q_p t_r}{2} \tag{7.18}$$

$$q_p(cm/hr) = \frac{2Q}{T_p + t_r} \text{ or } \frac{2Q}{T_p + 1.67T_p} = \frac{0.75Q}{T_p} \tag{7.19}$$

$$q_p = \frac{CA}{T_p} \tag{7.20}$$

where C is a user-definable variable termed the *peak rate factor* (PRF) and A is the drainage area in square kilometers (or square miles) of the watershed. The default value of 2.08 set by SCS is the metric conversion from 1 $cm/hr/km^2$ to $m/s/m^2 \times 0.75$ (the corresponding PRF of 484 comes from the English conversion of 1 in/hr/mi to $ft^3/s/mi \times 0.75$), which creates a hydrograph wherein three-eights (0.375) of its area is under the rising limb (Figure 7.11b).

The *National Engineering Handbook* (US Department of Agriculture 2007) recognized several studies after 1972 that depicted the variation in peak rate factors from below 0.43 (100 in English units) in flat watersheds to more than 2.58 (600 in English units) in mountainous regions. Hence, it is recommended that the PRF be modified accordingly depending upon the geographic conditions. In general, higher PRFs should be used for steeper watersheds, and lower PRFs should be used for watersheds with milder slopes (e.g., coastal watersheds).

Investigations of many large and small rural watersheds indicate that the basin lag t_p approximates $0.6T_c$ where T_c is the time of concentration of the watershed. As illustrated in Figure 7.11b, time of rise T_p can be evaluated in terms of lag time t_p and the duration of the effective rainfall t_r such that

$$T_p = \frac{t_r}{2} + t_p \tag{7.21}$$

Example: Derive a 20-minute (t_r) SCS unit hydrograph for a basin of area (A) of 6 km² and a time concentration (T_c) of 2.50 hours.

Duration of effective rainfall: $t_r = 20$ minutes, or 0.33 hours

Lag time: $t_p = 0.6T_c = 0.6 \times 2.50 = 1.5$ hours

Time of peak rise: $T_p = \frac{t_r}{2} + t_p = \frac{0.33}{2} = 1.5 = 1.67\,\text{hours} = \frac{0.33}{2} + 1.5 =$ 1.67 hours

Peak discharge: $q_p = \frac{CA}{Tp} = \frac{2.08\frac{\text{m}^3}{s} \times 6\,\text{km}^2}{1.67} = \frac{12.48}{1.67} = 7.47\ \text{m}^3/\text{s} \times \text{cm}$

Starting with the dimensionless unit hydrograph in Figure 7.11a, a converted hydrograph can be drawn by multiplying the x-axis (t/T_p) by q_p and y-axis (q/q_p) by T_p (Figure 7.12a). Or a triangular unit hydrograph can be drawn as time base $t_b = 2.67T_p$, or 4.46 hours (Figure 7.12b).

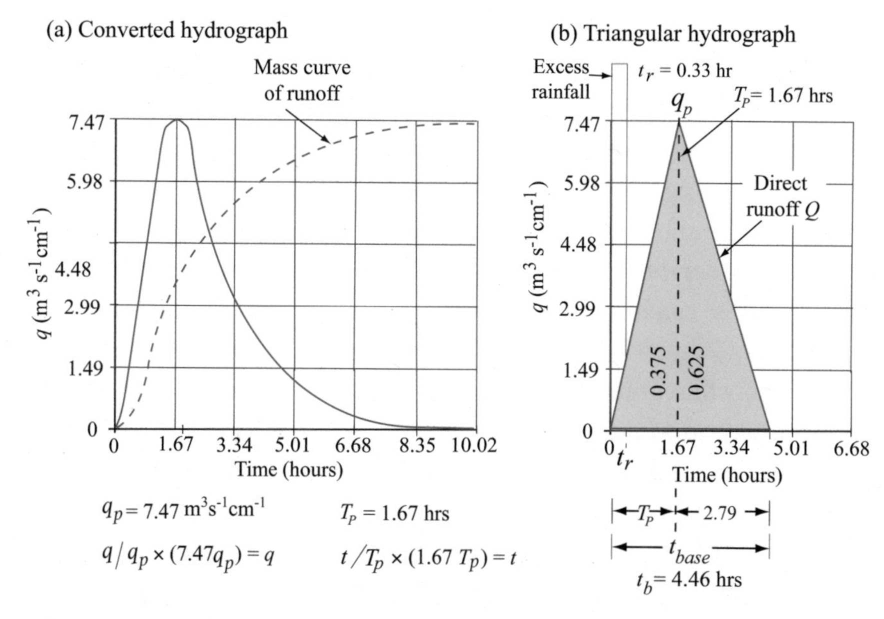

Figure 7.12. Construction of a twenty-minute Soil Conservation Service hydrograph for a basin of area 6.0 km² and time concentration of 2.5 hours: (a) hydrograph and (b) triangular unit hydrograph.

In short, the dimensionless unit hydrograph is useful for constructing a synthetic unit hydrograph for a wide variety of watersheds and design storms and is probably one of the most widely used in the United States for estimating floods on small urban basins in areas where there are no gauged data (Maidment 1993).

7.4.4. Influences Affecting Unit Hydrograph Shape

Drainage basin physical characteristics influence hydrograph behavior. Thus, unit hydrographs vary from watershed to watershed due to such factors as watershed size, geomorphic characteristics, hydrogeology, topography, soil, slope, shape (length and width ratios), amount of storage, stream network density and channelization, and land use (i.e., vegetation or urbanization). Larger basins generally lengthen the time axis, or base of the hydrograph, because of the longer response for the t_c to reach the basin outlet, the reduction of peak flow per unit area, and the decreasing influence of rainfall intensity for a given rainfall depth with increased basin area.

In a hypothetical geographic area, where all basin characteristics are homogeneous and rainfall input is uniform, then runoff would be dependent on area only. Thus, catchment area may be considered the most significant factor in runoff flow and has been defined by Morisawa (1962) as $Q = f(A)$. Despite the rarity of uniform homogeneous catchment areas, the discharge-to-area relationship continues to be used as a primary component for predicting runoff and flooding events.

The effect of catchment shape is illustrated in Figure 7.13 by considering the discharge hydrographs from two differently shaped catchments with the same surface area and slope and subject to rainfall of the same intensity. The isochrones to the outlet show that drainage basin B has the smaller time of concentration (2.3 hours) and reaches its peak discharge q_p much faster than drainage basin A. Basin A has a slower rising and lower flow peak than basin B because it takes longer for the area runoff to reach the basin outlet as the more elongated catchment needs four hours to reach the peak. Despite the dramatic differences in hydrograph behavior, calculation of total runoff (Q) with equation (7.18) for both basins gives similar volumes that flow into the basin lake outlets.

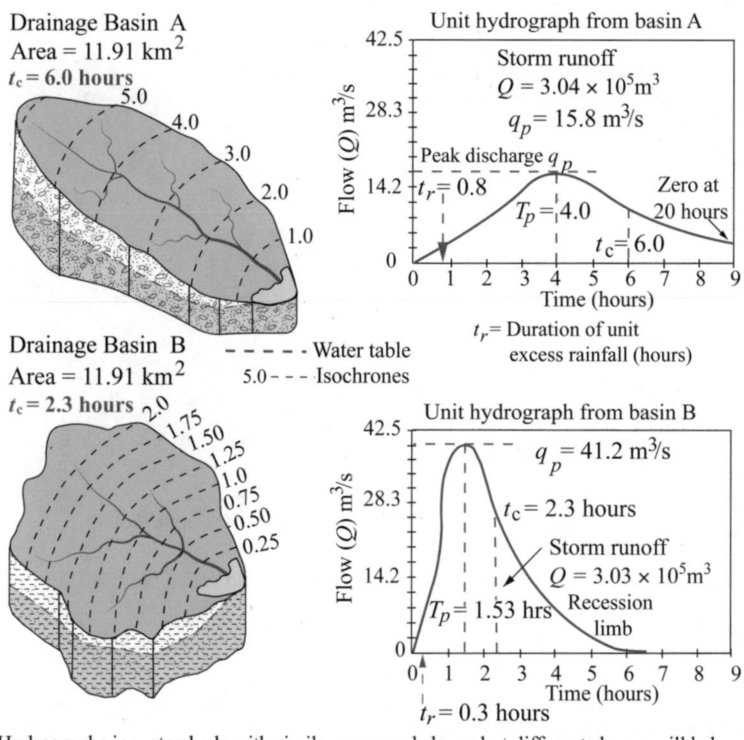

Hydrographs in watersheds with similar areas and slopes but different shapes will behave differently. For example, basin A has a gradualy rising and somewhat lower flow peak as well as a much longer time of concentration (t_c) than basin B even though the runoff volumes (Q) are similar.

Figure 7.13. Effect of basin shape on unit hydrographs. Modified from US Department of Agriculture (2004).

Figure 7.14 illustrates the effect of two sub-basins (C1 and C2) of different shapes flowing into the same lake or reservoir system that, in turn, flows into a third catchment (C3). A unit hydrograph is produced for flow into the lake outlet by combining the unit hydrographs from the two sub-basins. Total runoff Q to the lake is calculated with equation (7.18) by combing the volumes calculated from each sub-basin hydrograph.

When comparing watersheds with similar physical characteristics, such as stream types and density, basin area, shape, and other attributes, but different long profile slopes, the hydrographs will show higher runoff discharges, shorter discharge increase times, and shorter lag times for the basin with the higher relief gradient or slope angle (Figure 7.15). The higher energy provided by steeper relief increases the rate of runoff travel time and tends to lower infiltration capacities, thus increasing runoff volume (and sediment loads) as slopes get steeper.

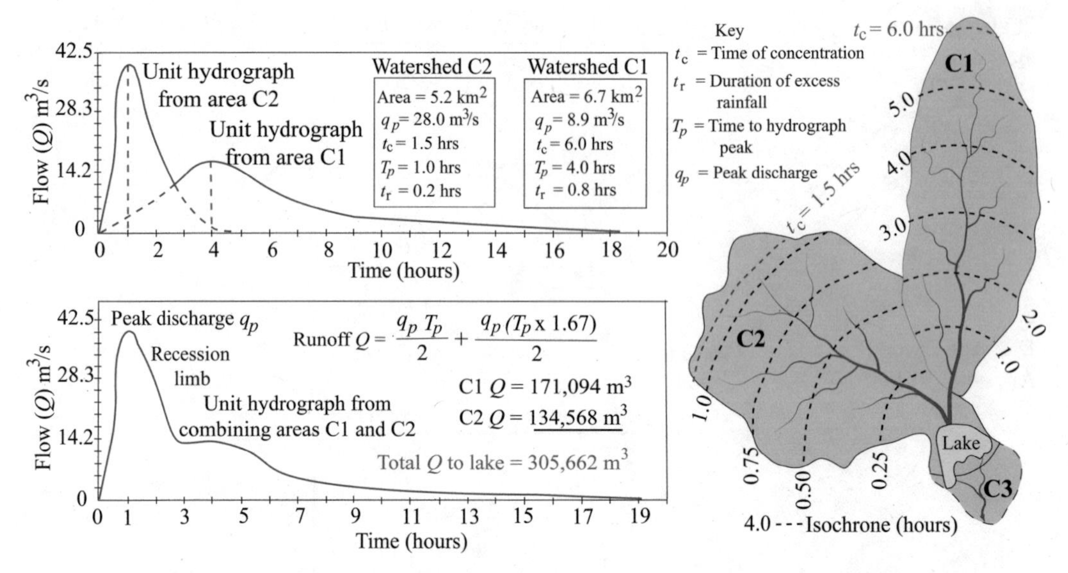

Figure 7.14. Effect on runoff unit hydrographs of two sub-basins flowing into a common lake outlet. Modified from US Department of Agriculture (2004).

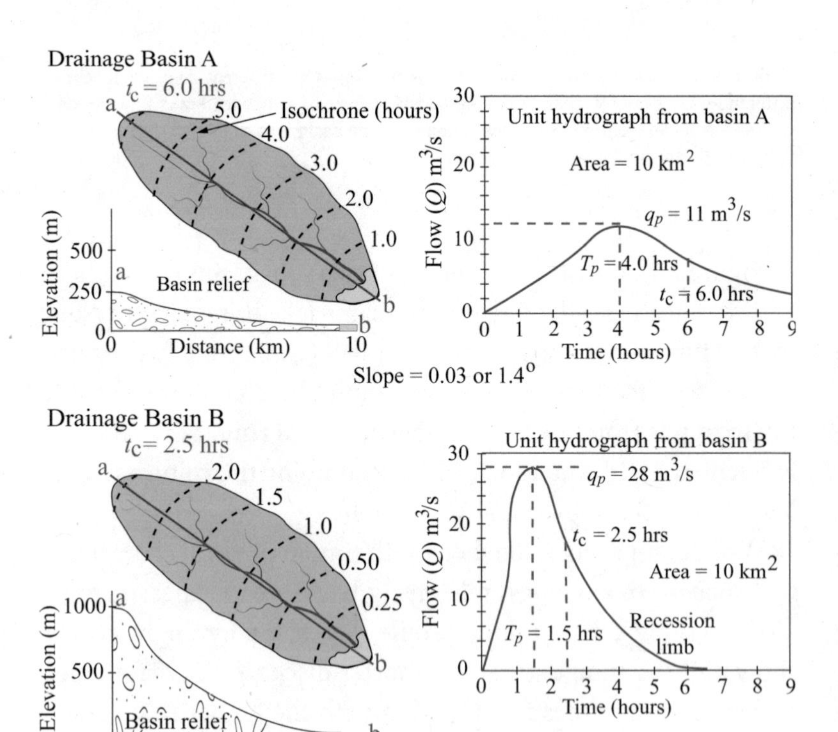

Figure 7.15. Effect of slope on hydrograph behavior illustrated by two identical hypothetical catchments with different slopes.

Watershed drainage patterns reflect the different influences of geology, vegetation, soils, land use, sedimentation, and other attributes. Drainage density, defined as the average length of streams within a basin per unit area, is a useful index that quantitatively describes drainage development (Horton 1945). Watershed stream patterns, density, location, and type (i.e., influent, effluent, or intermittent), all determine the efficiency of drainage within the catchment. The time for water to flow a given distance is directly proportional to the stream length and density. A high-density well-defined stream system will reduce the distance and time for runoff to reach a designated area, such as a lake or outlet. This is demonstrated in an outflow hydrograph with a short time to peak and higher peak flow rate than a catchment with lower stream drainage.

7.5. Runoff Hydrograph Estimation Models

Rainfall distribution over different parts of a watershed and corresponding runoff response may vary significantly depending on the details of the topography, drainage network, vegetation, land use, and other physical characteristics within the area of investigation. Hydrograph inputs from different parts of the watershed can be progressively combined on their way to the watershed outlet or to a lake or reservoir to determine the overall runoff response of the catchment at some point of interest.

This representation of the complexities of a watershed flow in a hydrograph estimation model is, by necessity, highly conceptualized as it attempts to represent the basin characteristics most responsible for estimating runoff responses. Thus, only some of the factors responsible for runoff are directly reflected in the combined hydrographs at downstream points of interest as others are excluded. Different catchment modeling approaches have therefore evolved to find an appropriate compromise between required model complexity and spatial resolution on the one hand, and desirable modeling efficiency on the other. These different modeling approaches can be applied with different degrees of complexity, and often more simple methods may be quite appropriate (Ball et al. 2016).

The different catchment representations in flood hydrograph

estimation models can therefore be classified based on how the different forms of temporary flood storage are conceptualized and in how much detail they are represented in the model. Other factors, such as losses, which determine the flood volume, and baseflow, which may modify the flood hydrograph shape and volume, must also be a part of the modeling of runoff hydrographs.

7.5.1. Runoff-Routing Approaches

The general term *runoff-routing* refers to runoff or flood hydrograph modeling approaches where a simplified conceptual representation is used to model the actual processes involved in the conversion of rainfall inputs to direct runoff. In a general sense, flow routing is an analysis to trace the flow through the watershed system given an input of precipitation (Chow, Maidment, and Mays 1988).

The purpose of flow routing in models is to provide a calculated estimate of the hydrograph at the downstream end of a reach given a hydrograph at the upstream end. Once such method already covered is the unit hydrograph technique, which is suitable for calculating stormwater flow into lake systems (discussed in section 6.5.4). However, as all lake systems are a component of drainage basin hydrology, this section briefly reviews catchment processes that are represented by routing methods in runoff/flood models. For further information on these methods, see Ball et al. (2016); Chow, Maidment, and Mays (1988); and Maidment (1993).

Direct runoff generated from storm rainfall in the upper parts of the catchment initially moves downhill as shallow overland flow and is changed by the various forms of detention storage as it moves over the watershed surface. It becomes gradually concentrated into minor drainage pathways and successively combined with baseflow and flows from other pathways. These flows eventually reach well-defined water courses, creeks, or rivers and move downstream, combining with other tributaries as it flows to the catchment outlet or as it moves through an interim lake or wetland system before reaching the outlet.

Thus, runoff is determined primarily by various forms of temporary storage as well as transmission losses (i.e., evapotranspiration) as it makes its way to the watershed outlet (Figure 7.16). The different types of temporary storage are distributed through-

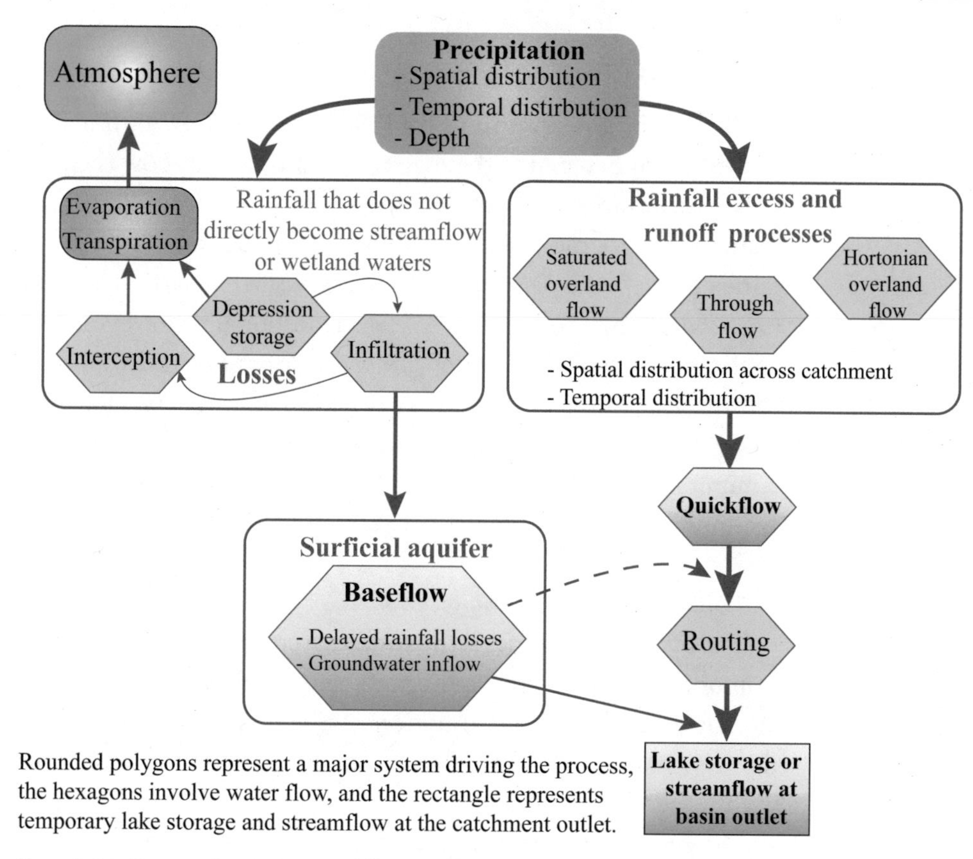

Figure 7.16. Process of converting rainfall to runoff and losses and routing flow to lakes and streams.

out the watershed and include catchment surfaces (quickflow), baseflow, stream channels and banks, floodplains, wetlands, and drainage channels (pipes). In runoff hydrograph estimation modeling, the different forms of storage do not need to be represented separately but can be modeled as combined (conceptual) storage components.

A significant degree of simplification in the representation of the actual processes in models is made possible by the fact that catchments act on rainfall inputs as systems with a high degree of damping. This means that the streamflow hydrograph output at the catchment outlet does not reflect the "high frequency" variations of the input in either the time or space dimensions. Similarly, small errors in modeling the various catchment processes may have little effect on the outflow hydrograph. This enables surprisingly accurate and useful results to be obtained from relatively simple models (Laurenson 1975).

7.5.2. Flow Routing Models

Chow, Maidment, and Mays (1988) define *flow routing* as a procedure to determine the magnitude and time of flow (i.e., the flow hydrograph) at a point in the watercourse from known or assumed hydrographs. Routing of flows in a catchment may be attained using *hydrologic* (*lumped* and *semidistributed*) or *hydraulic* (*distributed*) methods (Figure 7.17). Traditionally, however, flow routing is classified as either *lumped* or *distributed*. In lumped routing, the flow is calculated as a function of time at one location along the watercourse, whereas distributed or hydraulic rating is computed as a function of time simultaneously at several cross sections along the watercourse (Maidment 1993).

7.5.2a. Lumped Hydrologic Models

The hydrologic model is the simplest representation of routing in models that combines continuity with a relationship between storage and flow. The assumption of a hydrologic model is that there is limited spatial variation in rainfall and loss characteristics within the watershed such that it is acceptable to treat the drainage basin (or sub-basin) as a homogeneous unit (Figure 7.17a). Models that do not allow for spatial variation in runoff or routing characteristics within a catchment are referred to as a lumped

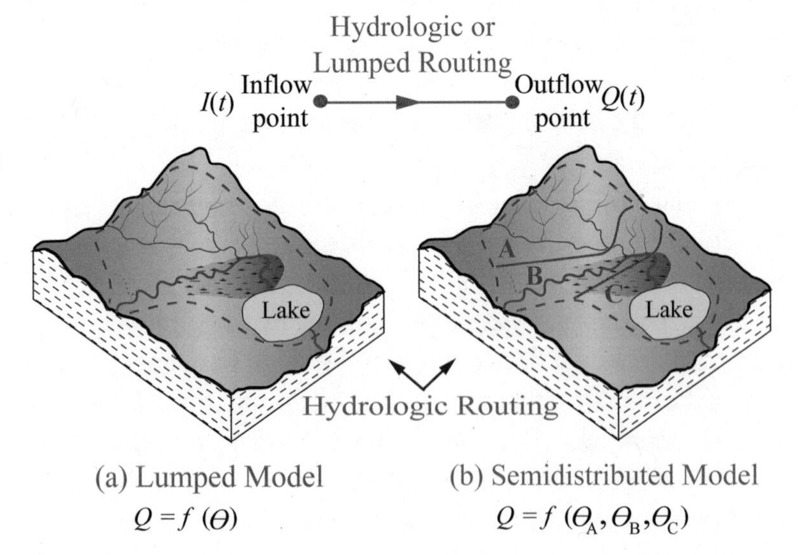

Figure 7.17. Spatial representation of hydrologic lumping and semidistribution storage/flow routing models. Modified from Ball et al. (2016).

approach as it considers only the inputs (inflows) and outputs (outflows) of the system without considering what is happening within the system (e.g., a black box; see Figure 11.2).

Lumped models treat a drainage area as a single unit and use averaged values of inputs and other catchment parameters. For example, spatially averaged rainfall is used as the main driver with single average values for initial and continuing loss. Simple routing approaches are used, perhaps based on the passage of a hydrograph through a single storage or separate storages for surface water and groundwater. Lumped models are less common in design flood estimation or flood forecasting but adequate for determining peak flow rates and runoff volume, such as seen in the unit hydrograph, which is an example of lumped methodology.

With this approach, flow paths in a catchment are divided into a series of components, where the volume of storage at any time is related to the flow (discharge) in each component. The result is that flow is calculated as a function of time at an actual location. Thus, *inflow* I at an upstream location and *outflow* Q at a downstream location are a function of time: $I(t)$ and $Q(t)$. Thus, the principle of *mass conservation* (*continuity*) requires that the difference between the two flows be equal to the time rate of change of the storage S within the stream reach (Maidment 1993) such that

$$I(t)-Q(t)=\frac{dS}{dt}. \tag{7.22}$$

The storage S is related to I and Q by a subjective empirical storage function. The simplest is a single-valued function of outflow Q (i.e., $S=f(Q)$) or of water surface elevation h (i.e., $S=f(h)$). This implies that water surface h is level throughout the watercourse, usually a lake or reservoir (Maidment 1993). If storage is a function of both inflow and outflow, such as open channels or long narrow reservoirs, then a more complex relationship exists. This lumped approach is also considered a *storage routing* procedure with different applications of storage routing principles focused on different types of systems with different forms of storage, such as level pool routing methods (concentrated storage as in reservoirs) and river routing methods (see Maidment 1993; Chow, Maidment, and Mays 1988; and Ball et al. 2016 for further information).

7.5.2b. Semidistributed Hydrologic Models

Semidistributed hydrologic models (Figure 7.17b) consider catchments as a number of reasonably large subareas. The spatial distribution of catchment rainfall is represented by the rainfall depth on each subcatchment, and losses and routing parameters can vary by subarea. This approach is commonly used in design flood estimation to represent areal variations in rainfall and losses, and the effects of varying flow distance to the catchment outlet (Ball et al. 2016). Semidistributed approaches can be used to create groups of hydrologic processes that are modeled in a consistent way. For example, the routing of flow downhill slopes can be modeled separately from flow routing in channels. Model setup then requires the explicit identification of hill slopes and channels that are to be modeled. The modeling equations, inputs, and parameters for these areas must be defined and integrated into the model.

7.5.2c. Semidistributed (Node-Link Type) Models

Semidistributed (*node-link*) models (Figure 7.18) allow the spatial variation of inputs and key processes to be modeled explicitly. This is particularly important in large catchments and subcatchments that have undergone extensive development and urbanization, resulting in modification of such natural runoff characteristics as construction of reservoirs, flood mitigation, and transport and drainage systems (Ball et al. 2016).

These models are flexible and relatively straightforward to generate and run for calculating runoff/flood hydrographs. As a result, the node-link models are currently the most widely used modeling approach for flood hydrograph estimation in Australia with a range of ready-to-use modeling systems available to set up models for watersheds of various sizes and complexity (Ball et al. 2016).

Semidistributed (node-link) models integrate the more significant features and flow characteristics of the watershed to be represented in the model system but do so in an extremely conceptualized process. Some degree of lumping is required when conceptualizing a complex system, such as spatial averaging of inputs, but the models can be modified to adapt to the different catchment flow elements most influential to that specific watershed under investigation. A simple conceptual representation of the runoff-routing process shown in Figure 7.18 illustrates how

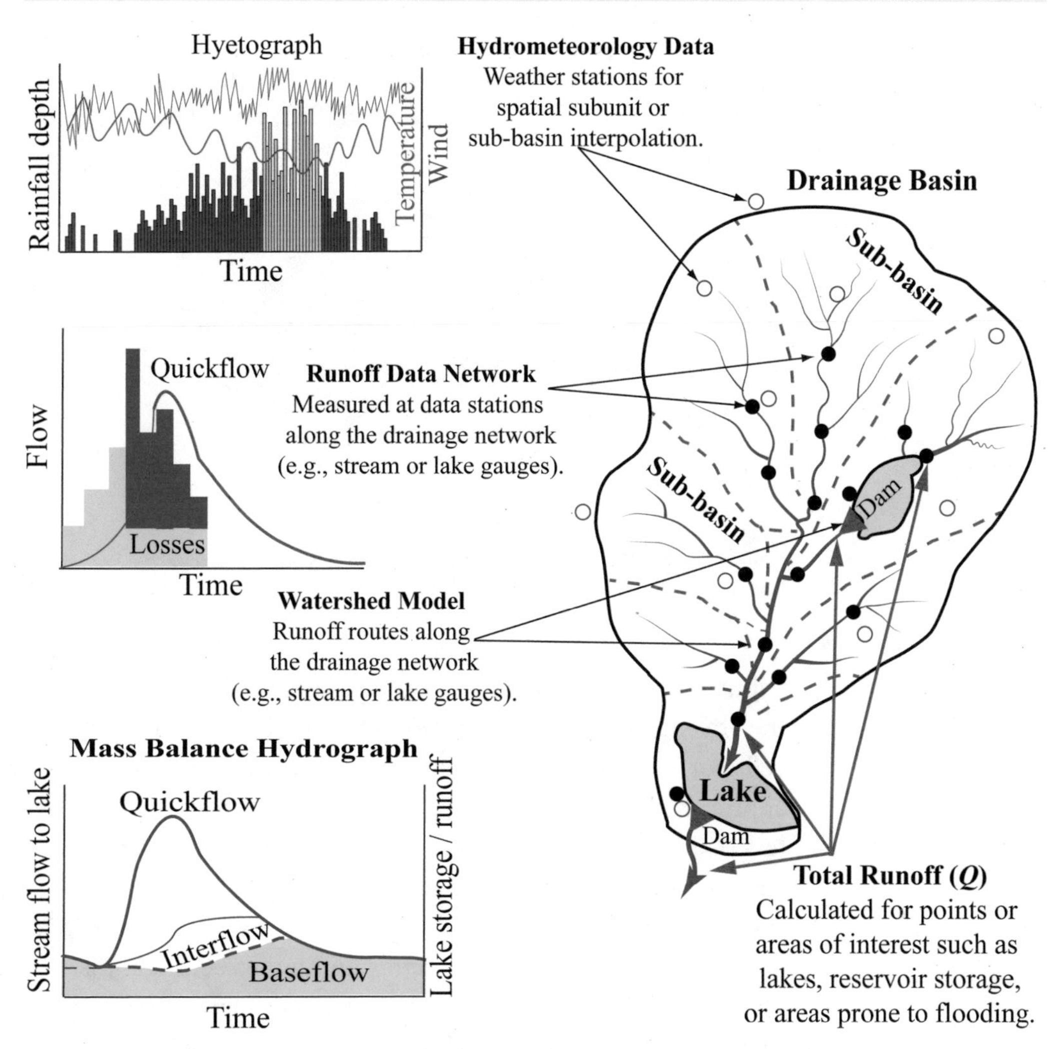

Figure 7.18. Runoff-routing process in a semidistributed (node-link) hydrologic model applied to lakes and streams. Modified from Ball et al. (2016).

each subarea receives an excess rainfall input that is converted to a runoff hydrograph at the node representing the subarea. The hydrographs are then routed successively through the links representing the drainage network to form the hydrograph at the catchment outlet. For more information on these types of models, see (Ball et al. 2016).

7.5.2d. Distributed Hydraulic Models

Distributed models (Figure 7.19) use a more spatially explicit approach, usually based on a grid that may be of a consistent size and shape across a study area or may be varied adaptively. Dis-

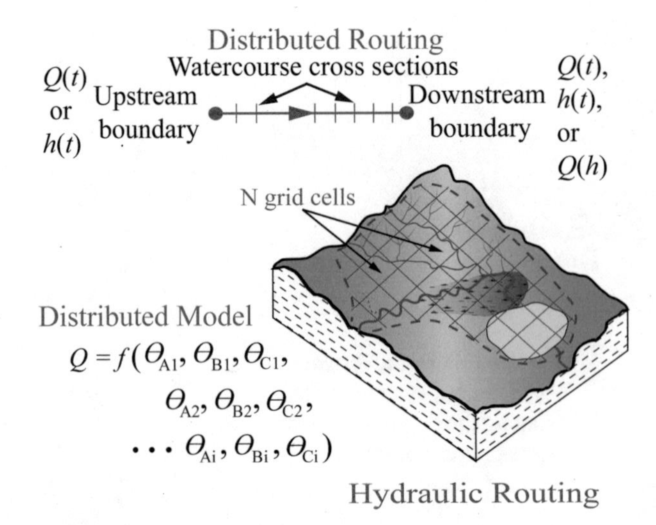

Figure 7.19. Spatial representation of hydraulic distribution storage/flow routing model. Modified from Ball et al. (2016).

tributed models require inputs and parameters for each grid cell; the advantage is that results can then be produced for each grid cell. For example, two-dimensional unsteady hydraulic routing approaches are commonly applied to grids to create spatially detailed information on flow depths, velocities, and flood hazard in rural and urban areas (Ball et al. 2016).

As repeatedly emphasized, the actual runoff formation processes in a watershed are complex and an exceedingly distributed process as the flow rate, velocity, and depth vary in space throughout the watershed. Hydraulic models are based on partial differential equations, such as the *Saint-Venant equations* for one-dimensional flow that allow the flow rate and water level h to be computed as functions of space and time, rather than time alone as in the lumped models (Chow, Maidment, and Mays 1988). The calculation of water level is important for determining the height of such structures as bridges and levees or for the delineation of the flood plain.

For flood design modeling, the current approach is to combine semidistributed hydrologic models with the hydraulic distributed models. Inputs generated from the semidistributed model for the upper watershed are incorporated into the hydraulic model to generate outputs for spatial flood mapping. The hydrologic model uses a semidistributed approach to deal with losses and runoff generation. Hydrologic routing is used for flow down slopes and from the upper reaches of the stream channel system. Hydraulic routing characterizes flow both within channels and overbank areas where detailed information on depths and extents are required (Ball et al. 2016).

An alternative to combining a semidistributed hydrologic model with a distributed hydraulic model is *direct rainfall* or *rainfall on grid* models. These types of models use a distributed approach to both hydrologic and hydraulic methods by gridding an entire catchment and simulating the runoff-routing process for each grid cell. Rain falling on a grid cell is converted to runoff, after allowing for losses, and this is added to any existing flow and hydraulically routed downstream using an unsteady two-dimensional approach (Ball et al. 2016). For more detailed information on direct rainfall models, see Ball et al. (2016, book 4, chapter 3).

7.6. Summary

The process required to describe mass balance of surface water from various areas of the drainage basin to a lake system suggests that highly complex methods and models would be required to represent the large number of pathways, flow speeds, and storage characteristics. However, surprisingly, simple mathematical approaches can be used to represent the movement of water along the different catchment pathways. Catchment response is usually highly damped so that short-term fluctuations in rainfall have little influence on the streamflow or lake-level hydrograph, and individual pathways do not need to be explicitly modeled but can be adequately estimated with the conceptual methods described throughout this chapter depending on the level of surface water representation needed. For more complex modeling requirements, see Ball et al. (2016); Chow, Maidment, and Mays (1988); and Maidment (1993) as well as the authors cited in chapter 11.

CHAPTER 8

Streamflow to Lakes

8.1. Introduction

The hydrology of flow in streams and river channels is a discipline unto itself in fluid mechanics. As such, this chapter is a brief overview of basic stream characteristics and streamflow as it pertains to discharge into or out of a lake system within a drainage basin. The basic principles of flow that are developed and discussed in chapter 9, such as the equation of continuity, Newton's second law of motion, Bernoulli's equation, fluid potential, and friction losses, all apply to streamflow.

8.1.1. Lotic System Framework

Lotic systems are defined as unidirectional flowing water systems (streams or rivers) imposed by gravity (Kalff 2002). They are ecosystems unto themselves that reveal catchment geomorphology, climate, and land use and are integrated into the overall catchment ecology and hydrology essential to the study of lake systems. Ward (1989) described lotic systems as a *four-dimensional structure* consisting of three spatial dimensions—*longitudinal* (length), *lateral* (width), and *vertical* (depth)—and one temporal (Figure 8.1).

Longitudinal aspects capture the upstream to downstream direction, typically consisting of three zones within the stream's morphology as it changes through the watershed:

1. *Headwaters*: where the slope is generally the steepest and volume the lowest; erosion is greater than deposition.
2. *Transfer zone*: where the slope decreases and more flow from the catchment enters the stream; erosion and deposition become more significant.

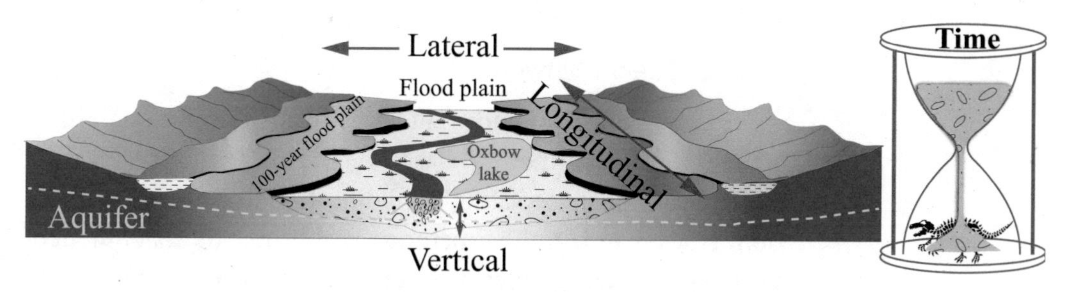

Figure 8.1. Four-dimensional lotic (stream) system as defined by Ward (1989).

3. *Downstream zone*: where slope is minimal and flow is at its highest; sediment deposition exceeds erosion.

Lateral aspects include across the channel, floodplains, and hillslopes. Although significant variation arises among stream types, a typical pattern includes the channel, the deepest part of which is called the *thalweg*; low floodplains that are flooded frequently; higher floodplains that are rarely inundated (e.g., 100-year or 500-year); terraces, which are former floodplains that a downcutting stream no longer floods; and hillslopes or other upland areas extending upgradient to the watershed boundary (Ward 1989).

The vertical dimension, or depth, includes the interactions between surface waters and groundwater. It is imperative to recognize that water bodies are not purely surface features; rivers and streams constantly interact with groundwater aquifers and exchange water, chemicals, and even organisms. As a stream travels through the watershed, it can often vary between *influent* reaches, where surface water leaks downward into the aquifer, and *effluent* reaches, where the stream receives additional water from the aquifer.

Finally, the temporal dimension results from a temporary response to evolutionary change. This aspect of time is important because rivers and streams are continuously changing. Structure as described in the other three dimensions should never be considered permanent, and watershed managers should always think of structure not just as what is there now, but in terms of the structural changes in progress and their rates of occurrence (Ward 1989).

8.1.2. Stream Networks

Horton's 1947 quantitative study of stream networks established a stream ordering method based on vector geometry. Strahler (1952, 1957) proposed a slight modification to Horton's method. Both Horton and Strahler methods established the assignment of the lowest order, number one, starting at the river's headwaters, which is the highest elevation point and correlates to topology, gravity, and trace flow downstream. Horton's method, modified by Strahler, is now referred to as *Strahler stream ordering system*.

The headwaters are the first order and downstream segments are defined at confluences (two streams converging into each other). At a confluence, if the two streams are not of the same order, then the highest numbered order is maintained on the downstream segment. At a confluence of two streams with the same order, the downstream segment gets the next highest numbered order. Figure 8.2 clarifies how the downstream numbering of stream order is done.

Divergences like braided streams maintain the same order all the way through the braid, just like it was a single stream; however, divergences that are not braided streams keep the upstream order number and follow the normal hierarchy further downstream. For example, stream 1 drains to stream 2, which drains into stream 3, and in turn drains into another common outlet (i.e., stream 4).

Empirically, Horton (1945) found the bifurcation ratios (see table in Figure 8.2) of stream channels were relatively constant from one stream order to the next (Chow, Maidment, and Mays 1988). However, efforts to establish a relationship between streamflow and indices of network structure such as runoff have not been very successful (Maidment 1993).

The Horton-Strahler methodology is important for defining the extent of a stream network, particularly in identifying fingertip tributaries and the location of the stream source as well as determining if they are ephemeral or flowing, and for the discovery of potential areas of saturation, or *variable source areas* (see section 6.2).

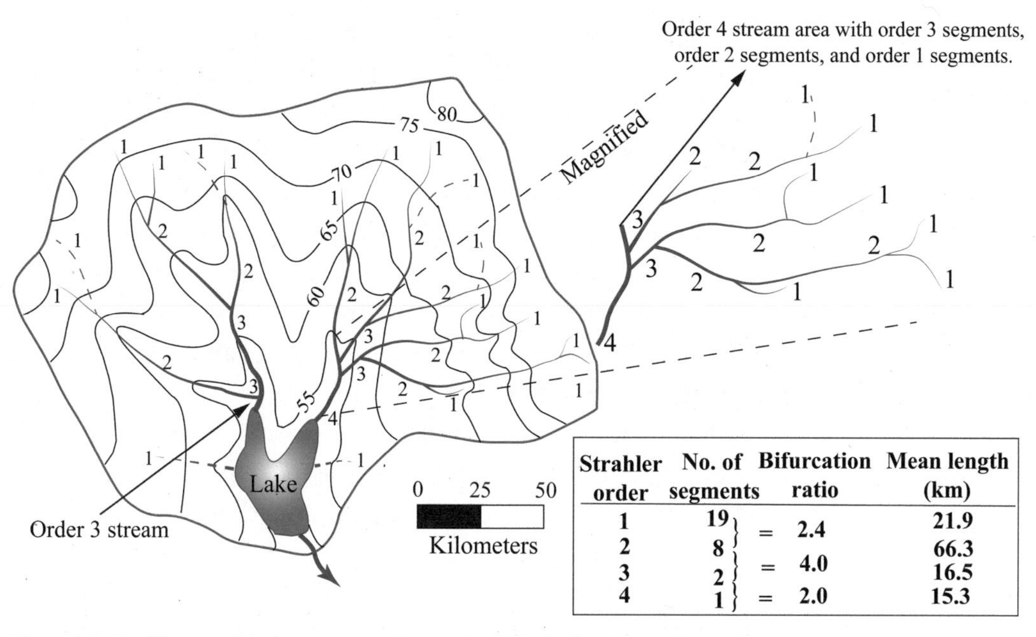

Strahler order	No. of segments	Bifurcation ratio	Mean length (km)
1	19	= 2.4	21.9
2	8	= 4.0	66.3
3	2	= 2.0	16.5
4	1		15.3

Figure 8.2. Strahler stream ordering system for stream networks flowing to a lake outlet within a hypothetical watershed.

8.1.3. Flow to Lakes

In a drainage basin, streamflow is the main form of surface water flow, and all other surface flow processes contribute to it. Therefore, determining flow rates in stream channels is a central task of surface water hydrology (Chow, Maidment, and Mays 1988). Combination of baseflow, interflow, and saturated overland flows generate the streamflow (Maidment 1993). A *streamflow rate* integrates all the hydrologic processes and storages upstream of a specific point on a drainage system at an exact time. These flow rates can be used as an integrated index for basin responses to the rainfall. Therefore, as discussed in chapter 7, it is important to analyze streamflow hydrographs, which explain the functional relationship between flow rates and time at a given location on the stream.

Open lakes typically have stable levels that do not fluctuate because input is always matched by outflow to rivers downstream. If more water enters an open lake than was previously leaving it, then more water will leave the lake. The drainage from an open lake, like that from ordinary rivers, is referred to as *exorheic* (from the Greek *exos*, outside, and *rhein*, to flow).

Exorheic (open) lakes drain into a river or other body of water that ultimately drains into the ocean, and they are characterized by stream and or groundwater (seepage) into or out of a lake (Kalff 2002). An exorheic lake is a lake where water typically flows out constantly and under almost all climatic circumstances. Because most of the world's water is found in areas of highly effective rainfall, most lakes are open lakes whose water eventually reaches the sea. For instance, the Great Lakes flow into the St. Lawrence River and eventually into the Atlantic Ocean (McMahon et al. 1992).

Streamflow is the flow rate or discharge of water $[L^3t^{-1}]$ along a defined natural channel. On a regional or river basin scale, it is the component of the hydrologic cycle that conveys water, originally falling as precipitation onto a catchment area from the land surface to the oceans.

When examining lake mass balance, the primary interest in lotic systems is flow rate or discharge (Q) in terms of volume into or out of the lake system over time $[L^3t^{-1}]$. The closer the streamflow measurements are to the lake system, the better the estimate of lake mass balance from this source. However, when stream gauges are located upstream, then the methods discussed here for determining streamflow inputs to a lake, such as the routing models, will be needed.

To accurately measure discharge, the velocity $[Lt^{-1}]$ is also of prime importance. Due to the complexity of the stream channel, velocity will vary both in time and space and provide the basis of standard flow classifications discussed next.

8.2. Velocity Distribution and Uniform Flow within Stream Channels

Section 9.6 discusses how a velocity profile varies due to the frictional losses at flow boundaries. Correspondingly, over the cross-sectional area of an open channel, the pattern of velocity distribution depends upon the character of the stream channel, including bottom and bank sediments, vegetation, and channel profile. The maximum velocities tend to occur just below the stream surface and away from frictional influences of the banks and bed of the stream channel as well as any drag due to vegetation.

Figure 8.3a shows an aerial view of a section of hypothetical meandering stream channel (with stream measurement points A,

(a) Aerial view of stream channel discharging to lake

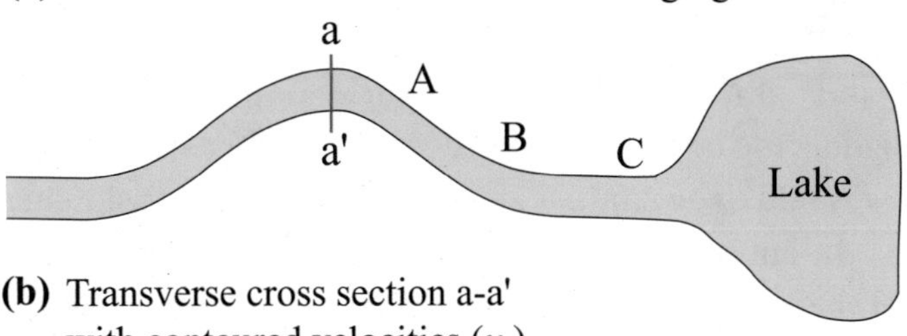

(b) Transverse cross section a-a' with contoured velocities (u_i) and vertical velocity profile

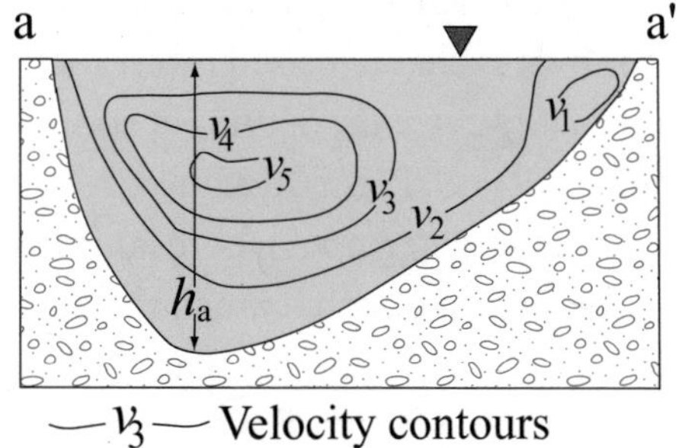

(c) Uniform and nonuniform flow

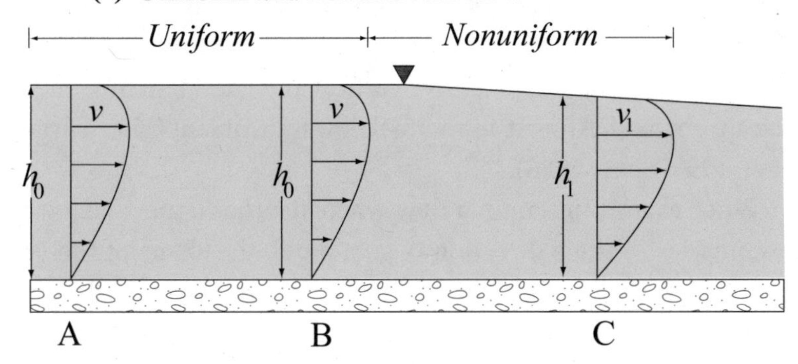

Figure 8.3. Types of flow patterns in a stream channel.

B, and C) feeding a lake. In Figure 8.3b a transverse cross section (a–a′) at the bend in the stream shows contours of equal velocity where the maximum velocities occur in the deeper part of the stream typical of conditions at the outside bend of a stream or river. Adjacent to the transverse section is a plot of the velocities in the vertical section corresponding with depth h_a.

Uniform flow is when the velocity distribution pattern does not change in the direction of flow, as shown in Figure 8.3c from A to B where the stream depth of flow (h_0), or *normal depth*, is constant. Nonuniform flow occurs between B and C where the normal depth (h_1) decreases, resulting in a change of velocity distribution.

8.2.1. Laminar and Turbulent Flow

Laminar flow is when fluid particles or parcels move such that neighboring layers of fluid slide by each other smoothly, thus the flow is said to be laminar, or *streamline* (see discussion in section 9.4). In laminar flow, each fluid parcel follows a smooth path, and these paths do not cross over or intersect one another. As the velocity of flow and depth increases, the moving fluid gains kinetic energy, and eventually inertial forces govern the flow patterns. *Turbulent flow* by contrast is characterized by erratic, small whirlpool-like circles called *eddy currents*, or simply eddies.

The *Reynolds number,* where $\mathbf{R} = \rho URH/\mu$ (modified from equation 9.15 in section 9.4), is used to determine where laminar flow transitions into turbulent flow. In an open channel, the diameter (D) for a tube is replaced by the hydraulic radius (RH) of the stream channel, and U is the average velocity. For channel flow, the critical **R** value for laminar flow is less than 500, transitioning to turbulent flow between 500 to 1,000. As the average velocity and depth increases, **R** increases, and the flow becomes increasingly turbulent, resulting in extensive vertical and lateral mixing in the stream channel. Almost all channel flow is turbulent (Hornberger et al. 1998; Shaw 1988).

When examining laminar flow, where flow has upper and lower boundaries in the z direction (e.g., a pipe), the shape of the velocity is parabolic. For a channel under laminar flow conditions, a similar profile would be expected with the highest velocities at the surface (e.g., cutting the pipe in half). In turbulent flow, the lower velocity changes much more rapidly due to the increased mixing (eddies) of fluid near the channel bed. Thus, the velocity distribution profile is logarithmic in shape as opposed to parabolic.

8.3. Calculating Channel Flow

Like lakes, the potentiometric surface or upper boundary of stream channel flow is open to the atmosphere and is designated the *free surface.* For a fluid of constant density, discharge Q $[L^3t^{-1}] = vA$ at any point in the stream channel. Because the vertical velocity at any point in the stream channel will vary (as illustrated in Figure 8.3) we are more interested in the average velocity U of flow where $Q = UA$, or $U = Q/A$, when describing overall discharge or velocity at a specific location of a stream, where A is the cross-sectional area [L^2] of flow. Due to the conservation of mass, the equation of continuity states that for a steady, uniform flow of constant-density fluid, the discharge Q remains constant such that $U_1A_1 = U_2A_2 =$ constant.

When describing flow in a small section or *reach* of a stream channel, changes in elevation and frictional losses can be initially overlooked. Thus, discharge (Q) at some point in a stream, is the product of the cross-sectional area (channel width w multiplied by water depth h) of the flowing water and its average velocity (U) and can be expressed as

$$Q = AU = Uwh = Uwh \tag{8.1}$$

where

Q = discharge [L^3t^{-1}],
A = area [L^2],
w = width [L],
h = depth [L], and
U = average velocity [Lt^{-1}].

The average velocity, as well as depth and width, increase as discharge increases.

Stream channel flow is primarily due to the gravitational pull on the weight of the water in a downhill direction. Hence, the stream gradient or slope is a significant hydraulic factor. Countering the downhill force is the resistance or drag of the banks and streambed that act to impede the flow. The velocity is dependent on the water surface gradient, which is the depth and slope of the stream channel, and inversely dependent on the boundary resistance.

The Bernoulli equation for flow in a stream channel is

$$\frac{U^2}{2g} + \frac{p}{\rho g} + z = h_T \tag{8.2}$$

where the total head h_T is a constant, with components velocity, elevation, and pressure head, respectively, and losses not considered due to friction. However, as flow moves downgradient, head losses are observed as a result of dissipating energy where mechanical energy is converted to thermal energy due to friction at the flow boundaries. Similarly, in a sloping stream channel

(Figure 8.4a), h_T decreases in the downstream direction and is dependent on the slope of the bed and *friction factor f* such that

$$\frac{dh_T}{dx} = -S_f \tag{8.3}$$

where the *friction slope* S_f is the slope of the total energy grade line.

Assuming uniform flow and a small angle of slope head loss (h_L),

$$S_f = \frac{h_L}{L} = S. \tag{8.4}$$

Head loss is dependent on the velocity of the fluid, the diameter and length of the cylinder, and the viscosity of the fluid (see section 9.8). The roughness of the cylinder or boundary wall must also be considered when determining head loss. This is expressed as the dimensionless friction factor *f*, which is a function of the streambed roughness (e.g., sand versus gravel) and the Reynolds number **R**. The friction factor is determined experimentally from a variety of streambed conditions.

The head loss (h_L) equation known as the *Darcy-Weisbach equation* is dependent on the length (L) and diameter (D) of a tube, viscosity of its fluid, velocity of the flow, acceleration due to gravity (g), and the friction factor of the tube wall:

$$h_L = f\frac{U^2}{2g}\frac{L}{D}.$$

This equation can be applied to channel flow by substituting the tube diameter with the dimensions of the open stream channel, or *hydraulic radius* (R_H), where for a circular cross section D, R_H is equal to the diameter of the tube divided by 4 (Hornberger et al. 1998) such that the Darcy-Weisbach equation then becomes

$$\frac{h_L}{L} = f\frac{U^2}{2g}\frac{1}{4R_H} = -S_f \tag{8.5}$$

The hydraulic radius (R_H) is defined as the ratio of the channel cross-sectional area (wh) to the wetted perimeter ($2h + w$) (Figure 8.4b) and can be described as

$$R_H = \frac{wh}{2h+w}. \tag{8.6}$$

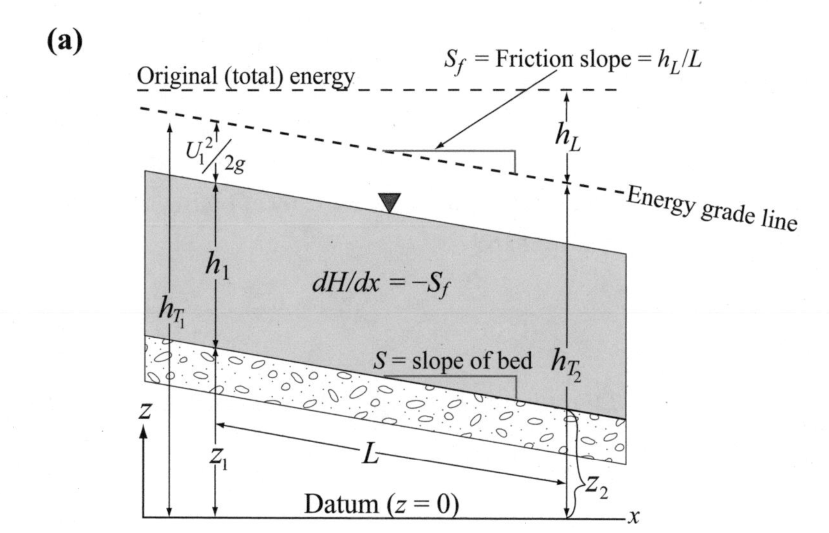

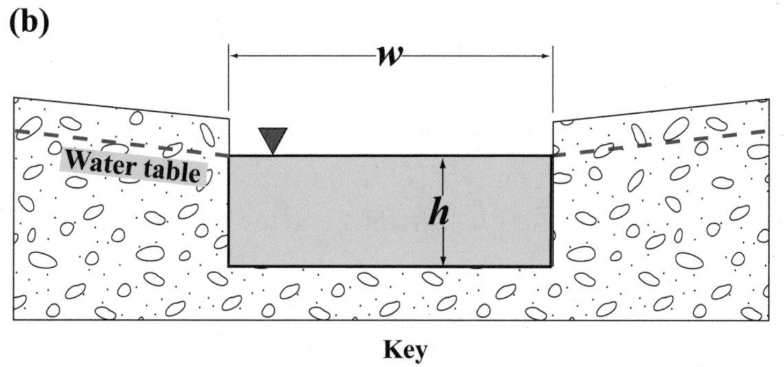

Key

▼ = Indicates *free surface*; the upper boundary of an open channel flow between the water and the atmosphere.

Datum = Convenient reference elevation; often mean sea level.

Figure 8.4. Stream channel flow showing (a) head loss (h_L) because of streambed slope and friction and (b) cross section of a rectangular channel with area $A = wh$. Modified with permission from Hornberger et al. (2014).

Assuming uniform flow where $h_L/L = S$, equation (8.4) can be rearranged (Hornberger et al. 1998) to solve for the average velocity (U) as a function of slope, hydraulic radius, and friction factor where

$$U = \sqrt{8g/f}\,\sqrt{SR_H} = C\sqrt{SR_H}. \tag{8.7}$$

This equation was first proposed by Antoine Chézy (1775) when he demonstrated thru empirical experimentation that the velocity of a river varies with its surface slope. This calculation is known as the *Chézy formula* where the constant C is referred to as the *Chézy number*. This expression simply states that velocity increases as the square root of depth and the square root of slope with the smoothness of the channel boundaries. This formulation presented hydrologists with a new basis for understanding flow of water in streams and, with minor changes, is still in use today.

The Chézy equation can be simplified by examining the effects of channel roughness only and replacing C by applying the *Manning equation* (Manning 1891)

$$C = \frac{R_H^{1/6}}{n} \tag{8.8}$$

where n is the *Manning roughness coefficient* and stream average velocity (U) is

$$U = \frac{R_H^{2/3} S_f^{1/2}}{n}. \tag{8.9}$$

When solving for this equation, Sf and n are dimensionless, and the units for velocity [Lt^{-1}] and the hydraulic radius [$L^{2/3}$] are not dimensionally homogenous (see section 1.4). Therefore, a constant k can be introduced with units of [$L^{1/3}t^{-1}$] where $k = 1\ m^{1/3}s^{-1}$ for SI units and $1.486\ ft^{1/3}s^{-1}$ for English units, and the equation becomes

$$U = k\frac{R_H^{2/3} S_f^{1/2}}{n}. \tag{8.10}$$

The Manning resistance coefficient n has been empirically determined for a variety of boundaries and types of channels (see Table 8.1). Many hydraulic factors that contribute to the loss of flow energy in a stream channel are incorporated into the resistance coefficient n. The major factor is channel surface roughness, which is determined by size, shape, and distribution of the sediment grains that make up the bed and banks of the channel (wetted perimeter). Five other key factors are channel surface irregularity, channel shape variation, obstructions, type and density of vegetation, and degree of meandering. Other general factors include depth of flow, amount of suspended material, channel obstructions, bed load, and changes in channel configuration due to deposition and scouring (Cowan 1956).

Assessment of individual hydraulic factors for roughness can be difficult to quantify. When coupled with the interaction of two or more contributing factors contributing to energy loss, a quantitative estimate of n can be largely subjective in nature as based on the

Table 8.1. Base values of the Manning's roughness coefficient *n*

Channel type and bed material	Median size or range of bed material (mm)	Base *n* value		
		Benson and Dalrymple (1967)[a]	Chow (1959)[b]	Bray (1979)
Sand channels				
(upper regime flow only)	0.2	0.012	—	—
	0.3	0.017	—	—
	0.4	0.020	—	—
	0.5	0.022	—	—
	0.6	0.025	—	—
	0.8	0.026	—	—
Stable channels				
Concrete	—	0.012–0.018	0.011	—
Rock cut	—	—	0.025	—
Firm earth	—	0.025–0.032	0.020	—
Coarse sand	1–2	0.025–0.035	—	—
Fine gravel	—	—	0.024	—
Gravel	2–64	0.028–0.035	—	—
Coarse gravel	—	—	0.028	—
Very coarse gravel	32–64	—	—	0.032
Small cobble	64–128	—	—	0.036
Cobble	64–256	0.03–0.050	—	—
Boulder	>256	0.04–0.070	—	—
Natural stream channels				
Clean, straight	—	—	0.030	—
Clean, winding	—	—	0.040	—
Winding with weeds and pools	—	—	0.050	—
With heavy brush and timber	—	—	0.10	—

Source: Modified from Coon (1998, table 1).
[a] Straight uniform channel.
[b] Smoothest channel attainable in indicated material.

experience of the stream researcher over a wide range of hydraulic factors (Coon 1998). As a result, a variety of methods have been developed for estimating *n* coefficients including estimates that are largely subjective in nature based on the experience of the stream researcher over a wide range of hydraulic factors (Coon 1998).

The US Geological Society (USGS) has published numerous publications in which the Manning coefficient has been computed based on field observations of depth of velocity, vegetation, slope, grain size, and a variety of other hydraulic factors and include statistical analysis and modeling for many river and stream systems. For further information, see USGS publications by Aldridge and Garret (1973); Aracement and Schneider (1987, 1989); Barnes (1967); Bjerklie, Dingman, and Bolster (2005); Coon (1998); Limerinos (1970); Jarret (1985); and Riggs (1976).

8.4. Streamflow Hydrographs and Field Measurements for Determining Streamflow

Streamflow hydrographs measure streamflow rate at specific stream location as a function of time and are used to observe the relationship between rainfall and runoff within a specific drainage basin. Streamflow hydrographs can be used to determine runoff effects to streams that flow into and out of open, or exorheic, lake systems. Most of the concepts and examples discussed next have been developed for streamflow hydrographs; however, these general principles are applicable for lake systems by substituting streamflow gauge data for variation in lake stage and subsequent change in area or volume as a result of storm events.

Stream gauge hydrograph data are measured at a stream or river gauge station that measures stream height. The simplest method uses a float system located in a protected structure (stilling well) hydrologically connected to the stream. As the river surface elevation changes so does the float, and these movements are recorded (Figure 8.5). Pressure transducers can also be used to measure the change in pressure as the river stage rises and falls.

From the gauge data, the stream velocity (Q), or rate of discharge [Lt^{-1}], is calculated based on such physical river characteristics as flow depths, channel shape, and cross sections and other features such as velocity measurement profile measured at certain

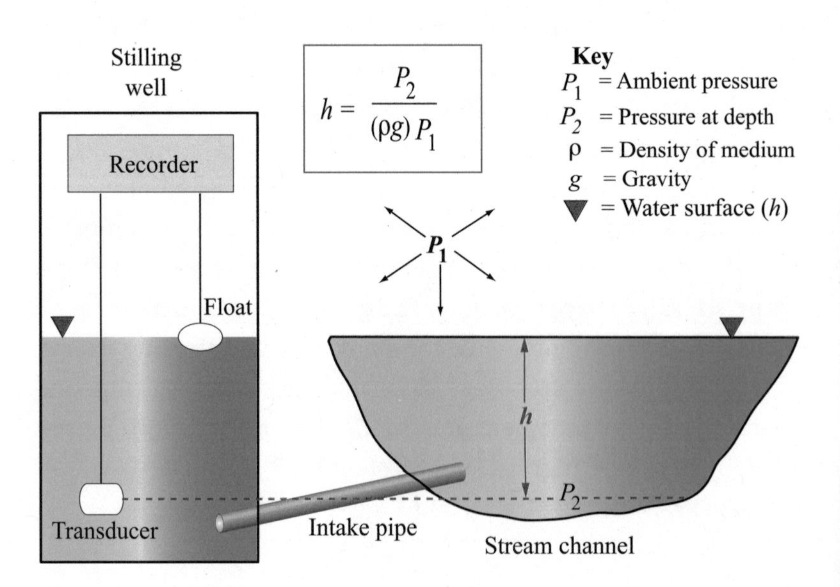

Figure 8.5. Stream gauge measurement using a float or pressure transducer.

points in the stream at the location of the gauge station. A velocity profile is constructed by measuring water depth and velocity at regular intervals across the channel. Velocity at each interval can be determined using a *current meter* or an *Acoustic Doppler Current Profiler.*

8.4.1. Current Meter

One of the most common methods of measuring stream discharge [Lt^{-1}] is the mechanical current meter (Figure 8.6). For this technique, the stream cross section is divided into numerous vertical subsections. Subsection width and depth are determined by a variety of methods depending on the condition of the stream. For example, width is commonly ascertained with a steel tape, cable, or similar measuring device. Depth can generally be determined by hanging a weight from a calibrated cable-and-spool system from a boat or bridge. The discharge in each subsection is calculated by multiplying the subsection area by the measured

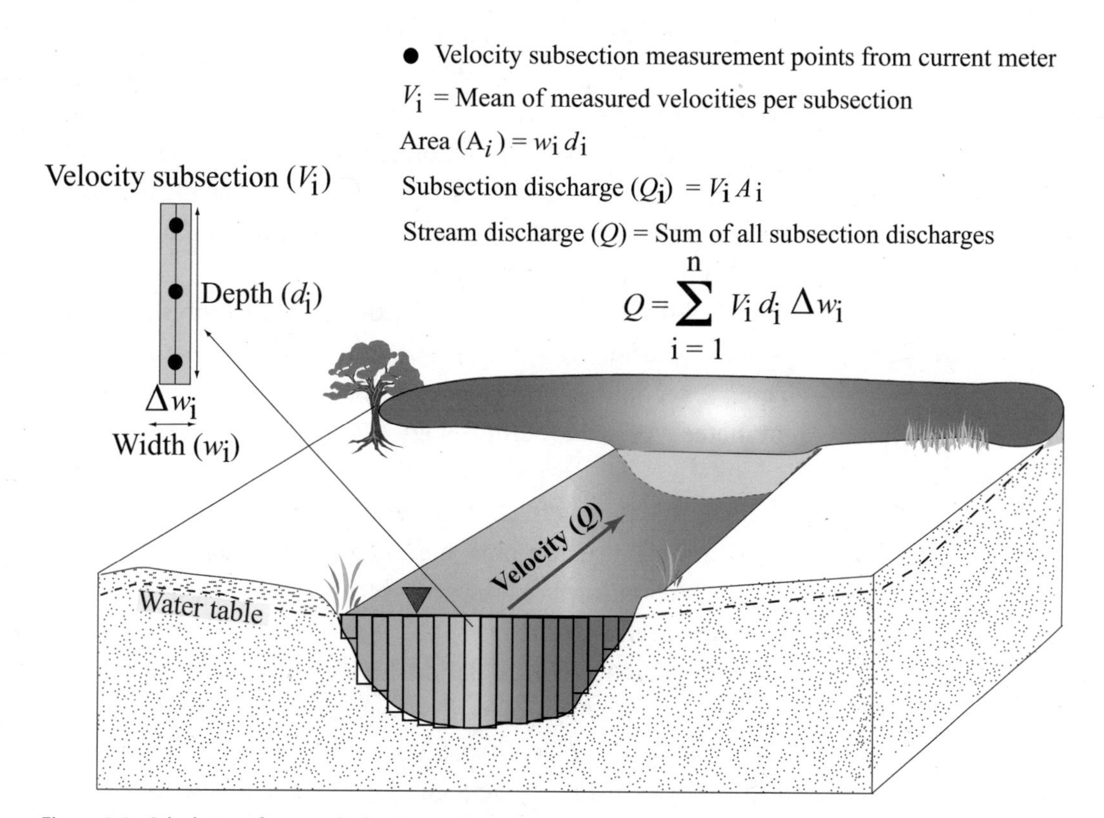

Figure 8.6. Calculation of stream discharge or velocity from current or velocity meters.

velocity. The total discharge is then computed by summing the discharge of each subsection.

8.4.2. Acoustic Doppler Current Profiler

Another method for determining water velocity measurements is the Acoustic Doppler Current Profiler (ADCP). The *Doppler effect* (or *Doppler shift*), named after the Austrian physicist Christian Doppler, is the change in frequency of a wavelength (or another periodic event) for an observer moving relative to its source. A common example of Doppler shift is the change of pitch heard when a vehicle sounding a horn or siren approaches and recedes from an observer. Compared to the emitted frequency, the received frequency is higher during the approach, identical at the instant of passing by, and lower during the recession (Possel 2017).

The ADCP uses the Doppler effect to determine water velocity by sending a sound pulse into the water and measuring the change in frequency of that sound pulse reflected back to the ADCP by sediment or other particulates being transported in the water. The change in frequency, or Doppler shift, that is measured by the ADCP is translated into water velocity. The sound is transmitted into the water from a transducer to the bottom of the river and receives return signals throughout the entire depth. The ADCP also uses acoustics to measure water depth by measuring the travel time of a pulse of sound to reach the river bottom and back to the ADCP (US Geological Survey, n.d.).

The following sources from Merkley (2004, lectures 3 and 4), US Geological Survey (n.d.), and US Bureau of Reclamation (2001, chapter 2, sections 10 and 11) are excellent for understanding current meter types and methodology:

"Current Metering in Open Channels": https://digitalcommons.usu.edu/ocw_bie/2/ (pdf download)
"How Streamflow is Measured": https://water.usgs.gov/edu/measureflow.html
"Energy Balance Flow Relationships" (2.10): https://www.usbr.gov/tsc/techreferences/mands/wmm/index.htm
"Hydraulic Mean Depth and Hydraulic Radius" (2.11): https://www.usbr.gov/tsc/techreferences/mands/wmm/index.htm

8.4.3. Determining Discharge across Stream Cross Section

The discharge (Q) at a cross section of area A is calculated by

$$Q = \iint VdA \tag{8.11}$$

The integral is estimated by summing the incremental discharges calculated from each subsection of measurement where $i = 1, 2, 3, \ldots, n$ (see Figure 8.6). The measurements are an average value over the interval of width Δw_i of the stream cross section such that

$$Q = \sum_{i=1}^{n} V_i d_i \Delta w_i. \tag{8.12}$$

From these calculated flow data, a *rating curve* (Riggs 1985) is constructed (Figure 8.7). Rating curves are prepared over a period of time and continually checked so as keep an accurate relationship between the stream channel characteristics and gauge, which change over time. The channel shape, size, slope, and roughness at the stream gauge determine the stage-discharge relation and are different at every stream gauge.

Stage-discharge hydrographs can be used to determine increase in volume of flow to streams during a rainfall event. For example, increases in stream gauge height and corresponding discharge for Peachtree Creek during a two-day storm event in Atlanta, Georgia, is shown in Figure 8.8. These types of data are important for flood control and estimating discharge to lakes, reservoirs, and wetlands.

Changes in stream channels occur due to deposition of sedi-

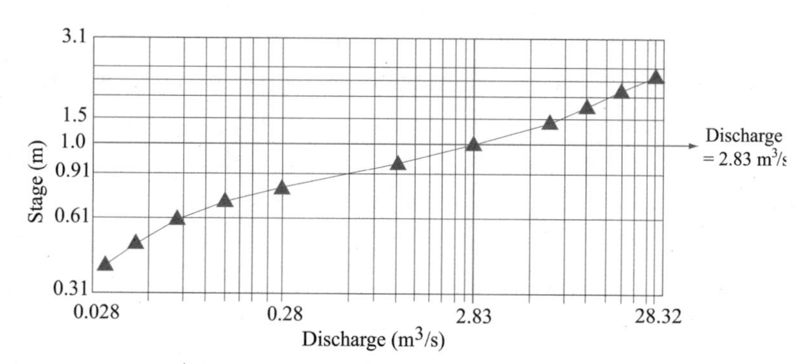

Figure 8.7. Hypothetical rating curve showing the relationship between stream stage and discharge. Modified from US Geological Survey (n.d.).

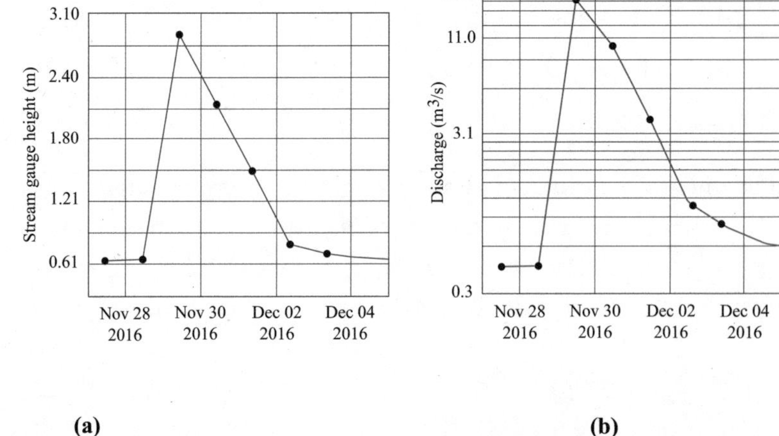

Figure 8.8. Peachtree Creek hydrograph showing change in river stage and discharge from a 70-mm two-day storm event in Atlanta, Georgia. Data from US Geological Survey (2016).

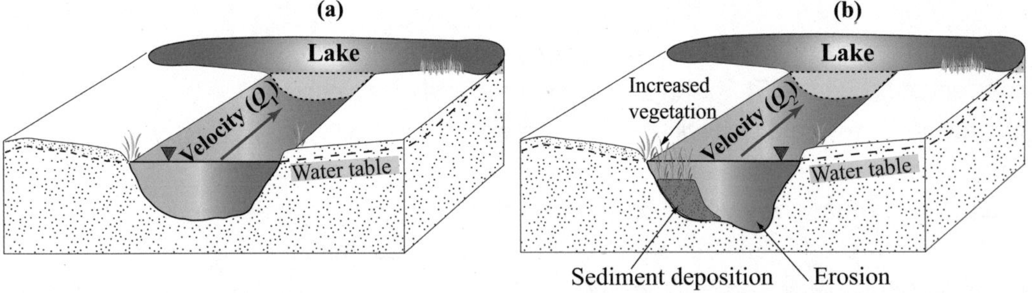

Figure 8.9. Changes in stream cross-sectional area and discharge (*Q*) before (a) and after (b) erosion (scouring), sediment deposition, and vegetation growth.

ments or debris, erosion (e.g., scouring), seasonal vegetation growth, ice, channel shape, and other similar factors (Figure 8.9). New discharge measurements plotted on an existing stage-discharge relation graph would show this, and the rating could be adjusted to allow the correct discharge to be estimated for the measured stage.

The USGS operates over 7,000 stream gauges nationwide. These stream gauges provide streamflow information for a wide variety of uses, including flood prediction, water management and allocation, engineering design, research, operation of locks and dams, and recreational safety and recreational use.

CHAPTER 9

Groundwater Flow

9.1. Introduction

To understand lakes, one must comprehend the flow of water into and out of these systems via the underlying subsurface in relation to physical hydrodynamic forces. The equations of fluid mechanics provide the basics for the quantitative description of groundwater flow. This chapter covers only the most rudimentary hydrodynamics of fluid flow through a porous media. For a more comprehensive investigation, readers are directed to the references at the end of this book.

Groundwater by definition occupies the zone of subsurface and, as depicted in the lake hydrologic cycle diagram (Figure 9.1), moves downward through the soil by percolation and then toward a lake, wetland, or stream channel as *seepage*. The subsurface boundary between the saturated and unsaturated zone, where the gauge pressure (measured relative to atmospheric pressure) is zero, is defined as the water table. Water will flow into an excavation, lake, or well up to this level and as such is an expression of a *free surface* (e.g., lake level).

Groundwater flow to and from lake systems will mainly involve saturated laminar flow conditions where flow occurs along a gradient from high to low energy (see Section 9.3.1) and will form the basis of the discussion throughout this chapter.

In general, lake and wetland systems are surface expressions of the water table and are in direct hydraulic connection with the water-table aquifer. Lakes receive inflow from sources where the hydraulic head is greater than the lake level and discharges flow to groundwater (or streams) where the hydraulic gradient is lower than the lake level. Lakes can be affected by underlying confined aquifers as well, dependent upon the potentiometric surface and especially if confinement is leaky or breached. Lake Jackson, in

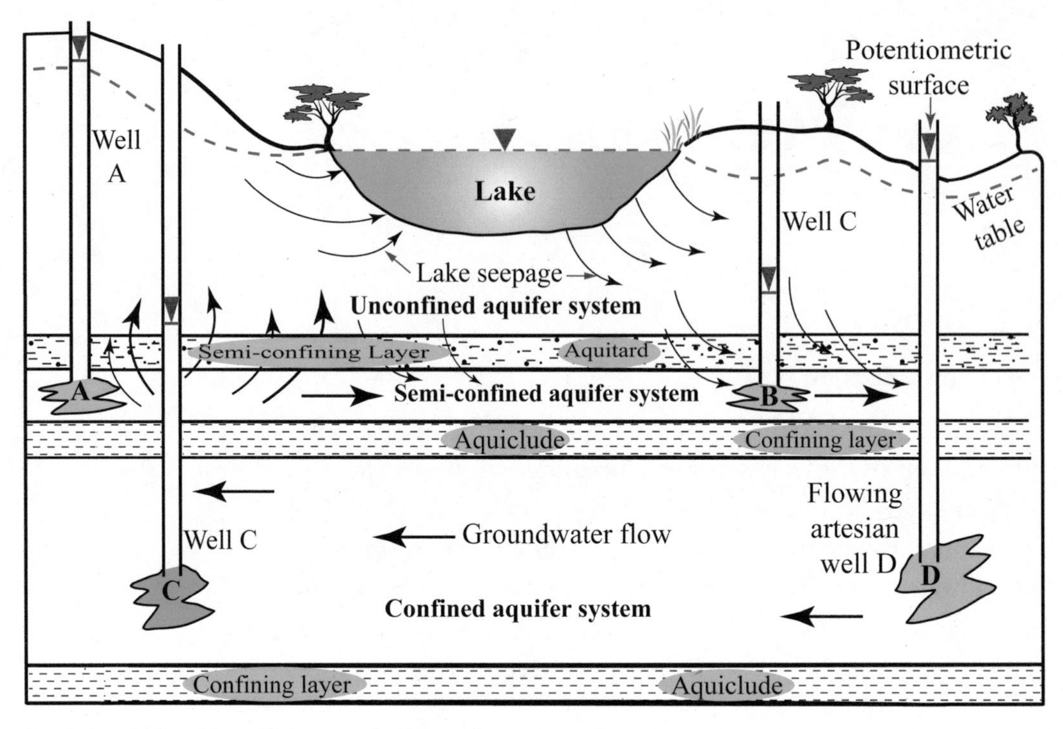

Figure 9.1. Hydrogeology of lake system and groundwater interactions.

Leon County, Florida (Figure 9.2), is an example of such hydrostratigraphic conditions where the confinement is intermittently breached when sinkholes open up and cause the lake to periodically drain through the *ponor* (karst term for "swallow hole"). NGVD refers National Geodetic Vertical Datum of 1988, which is a standard reference to altitude [L] above mean sea level.

9.2. Groundwater Systems

9.2.1. Aquifers, Aquitards, and Aquicludes

Before beginning a discussion on groundwater flow as it applies to lakes, the definition of naturally occurring groundwater systems must be established. Freeze and Cherry (1979) define an *aquifer* as a saturated permeable geologic unit that can transmit significant quantities of water under ordinary hydraulic gradients. An *aquiclude* is defined as a saturated geologic unit that is incapable of transmitting significant quantities of water under ordinary hydraulic gradients. An *aquitard* is a saturated geologic unit that is somewhere between an aquifer and aquiclude where the groundwater is sufficient to study the groundwater flow patterns

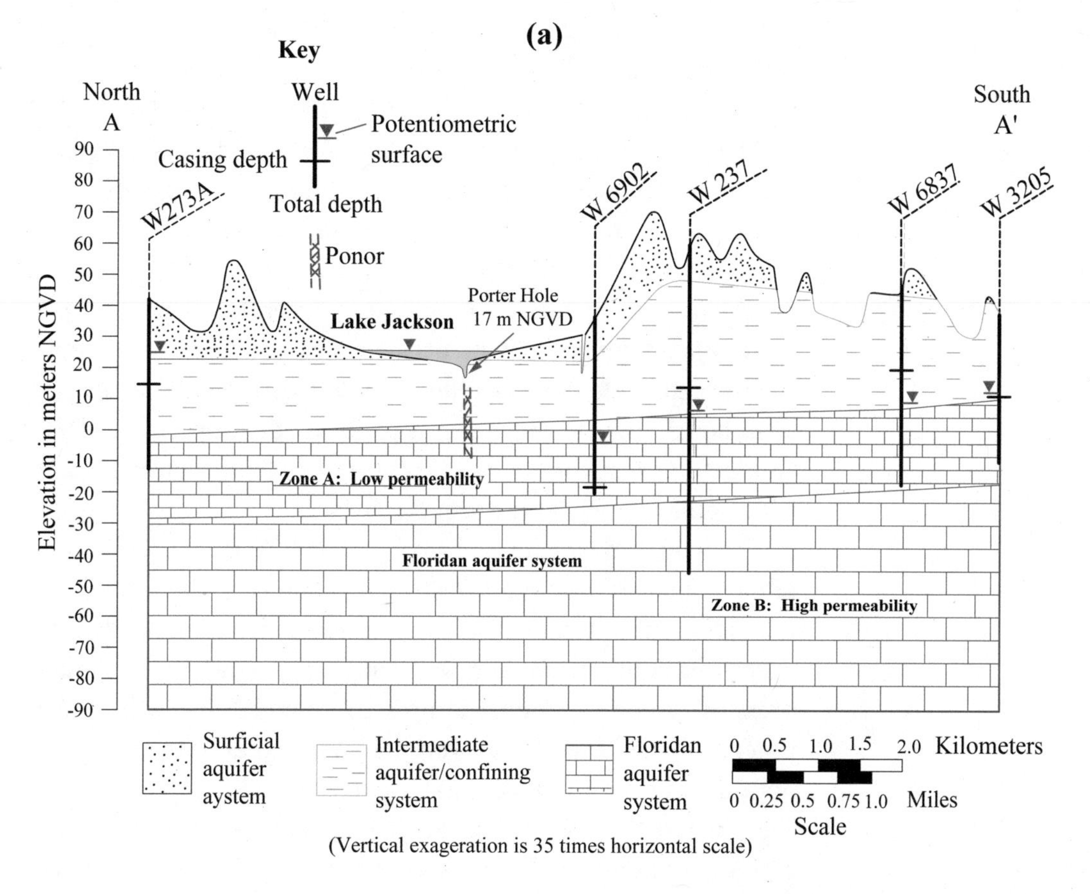

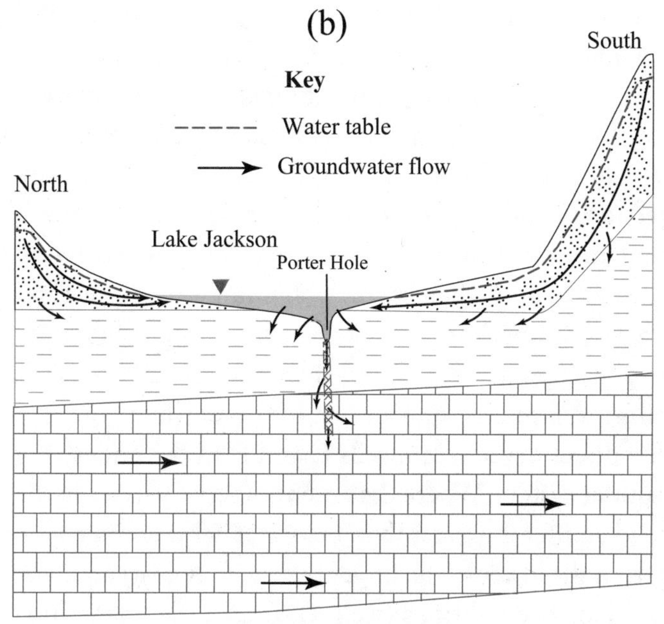

Figure 9.2. Regional cross section (a) of the hydrostratigraphy of Lake Jackson, Leon County, Florida, with a localized view (b) showing the flow of Porter Hole ponor into the underlying Floridian aquifer system. Modified from Wagner (1984) and Evans (1996).

but not able to produce water for economic use. The lithology or geologic makeup of the aquifer system is what defines its permeability characteristics. Coarse sands and gravels would generally make up a high yielding aquifer while shales and clays would be examples of the lithology found in aquitards or aquicludes (see section 9.7 for discussion on permeability of geologic lithology).

Aquifers can be unconfined, confined, or semiconfined. A *confined aquifer* is an aquifer confined between two aquicludes; a *semiconfined aquifer* may be between two aquitards or an aquitard or aquiclude. An *unconfined aquifer*, or *water-table aquifer*, is an aquifer where the water table forms the upper boundary. Confined or semiconfined aquifers occur at depth, whereas water table aquifers are near the ground surface.

The height to which water rises in a well is its *hydraulic head*, often defined as the *piezometric* or *potentiometric surface* and denoted throughout text figures as an upside-down triangle (▼). The water in a confined aquifer (and semiconfined aquifer) is under pressure, and a well penetrating a confined aquifer will have a potentiometric surface at a vertical level higher than the upper boundary of the confined aquifer (wells C and D in Figure 9.1). Wells where the potentiometric surface is higher than the upper aquifer boundary are known as *artesian wells*. When the potentiometric surface of an artesian well is above ground surface it is defined as a *flowing artesian well* (well D in Figure 9.1).

By plotting and contouring the potentiometric surfaces of wells on a map within a particular aquifer system, one can determine the direction of groundwater flow from the gradient of high hydraulic head to low hydraulic head. As seen in Figures 9.1 and 9.2, flow direction can vary in different overlying aquifers and between aquifers depending upon their spatial potentiometric surface and confining relationships. For example, by connecting the two potentiometric surfaces in wells A and B with a straight line, flow occurs from A to B in this semiconfined system, while flow is in the opposite direction between wells C and D in the underlying confined aquifer system. Where the potentiometric surface in the semiconfined aquifer is greater than the water-table aquifer, flow can occur from the lower to the upper (e.g., well A area) aquifers. Where the opposite conditions (well B) occurs, the unconfined aquifer acts as a source of water to the semiconfined aquifer.

In certain geologic environments where layers of impermeable

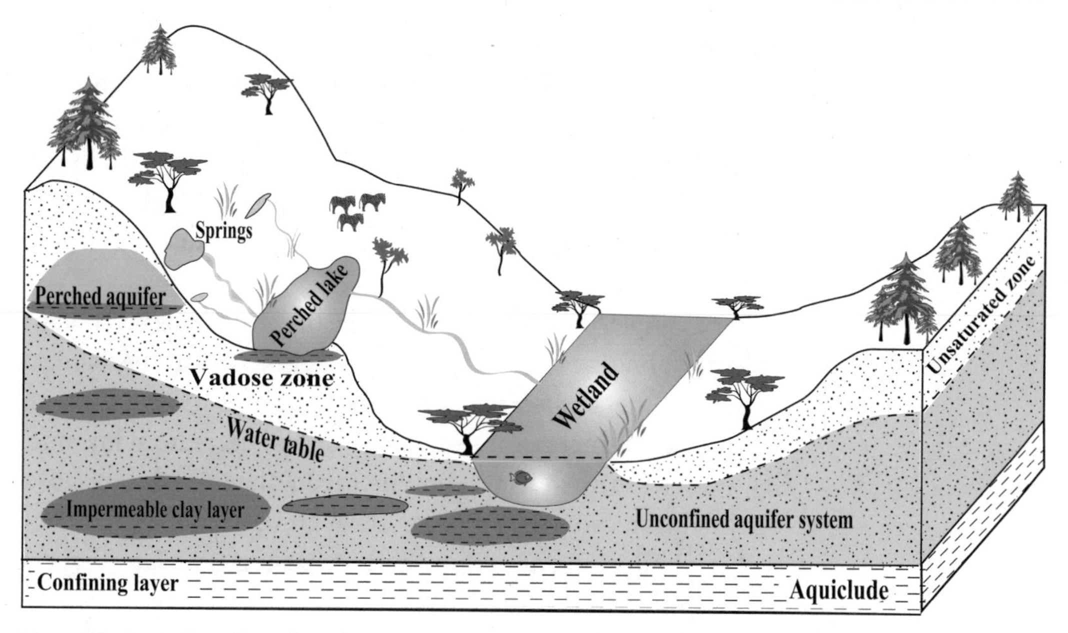

Figure 9.3. Perched aquifer and perched lake within a water-table aquifer system.

sediments such as clays (as in an *aquitard* or *aquiclude*) were deposited within more permeable sediments such as sands or gravel, a *zone of saturation* within an overall *vadose zone* (unsaturated zone) can form. These are known as *perched aquifers*, and in environments where they are close to the ground surface, perched aquifers can produce springs, lakes, or wetlands above the water-table aquifer (Figure 9.3). These perched systems in general are not very large and are subject to widely fluctuating potentiometric surface or lake stage levels dependent upon rainfall.

9.2.2. Elasticity, Transmissivity, and Storativity of Aquifers

Flow into and out of lake systems are influenced by the physical characteristics of the underlying aquifer systems and are a result of the aquifers ability to transmit and store water. In the saturated zone of an aquifer, the potentiometric surface or hydraulic head creates pressure (see Section 9.3.1 on fluid pressure) that can affect the arrangement of the lithologic material making up the aquifer, especially as it applies to the water in the voids. This gives a malleable nature to the aquifer, known as *elasticity*. For example, if the pressure increases, the lithologic aquifer skeletal structure will expand; if it drops, the structure will contract. Similarly, an increase in pressure will cause the water to contract and expand if the pressure decreases. If the head in an aquifer declines, the aquifer skeleton compresses, which in turn decreases the effective porosity and expels water. Additional water is released as the pore water expands due to lower pressure. Such external forces as barometric pressure, tidal influences, and the pressure of the weight of large objects (e.g., trains, buildings) can also cause aquifers to expand or compress.

For example, in wells near the ocean, the head levels will fluctuate corresponding to the tidal cycle. This is known as mechanical loading of the aquifer, and water levels in wells near the ocean will noticeably increase during high tide and decrease at low tide. As wells get further away from the ocean, the inland transfer of the tidal-induced pressure wave through the aquifer will occur with a diminishing amplitude and increasing time lag (Figures 9.4 and 9.5).

An inverse relationship exists between barometric pressure and water levels in wells. An increase in barometric pressure transmitted to the confined aquifer is similar to tidal loading; however, the well has a direct connection to the atmosphere, and because the atmospheric load is partially supported by the aquifer "skeleton," the net effect is a decrease in water level during increased barometric pressure and an increase in water level during decreased barometric pressure (Figure 9.5).

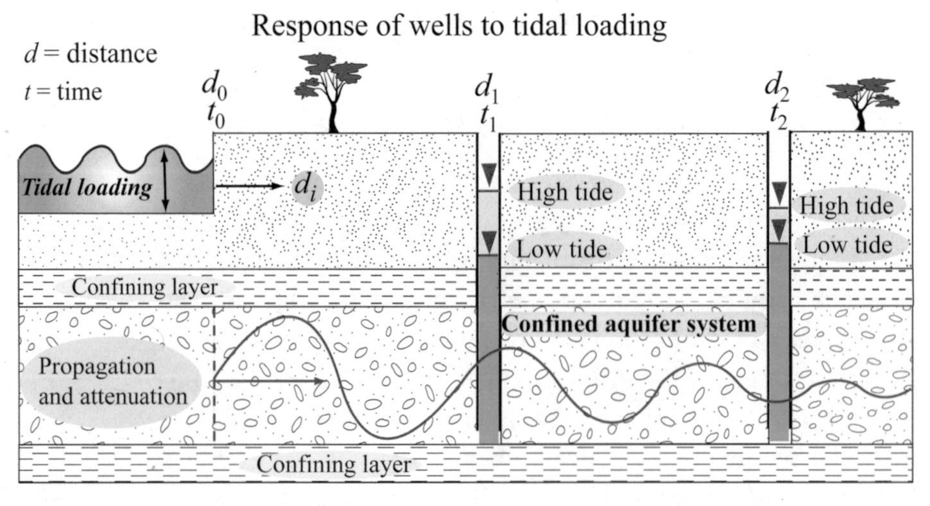

Figure 9.4. Response of wells penetrating a confined aquifer to oceanic tidal loading. Modified from Enright (1990) and Duncan, Evans, and Taylor (1994).

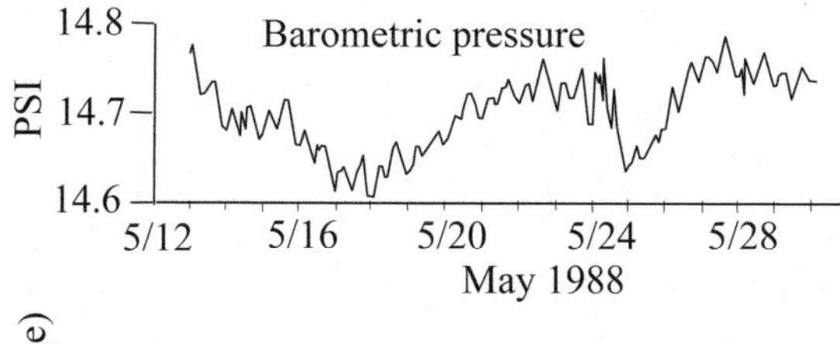

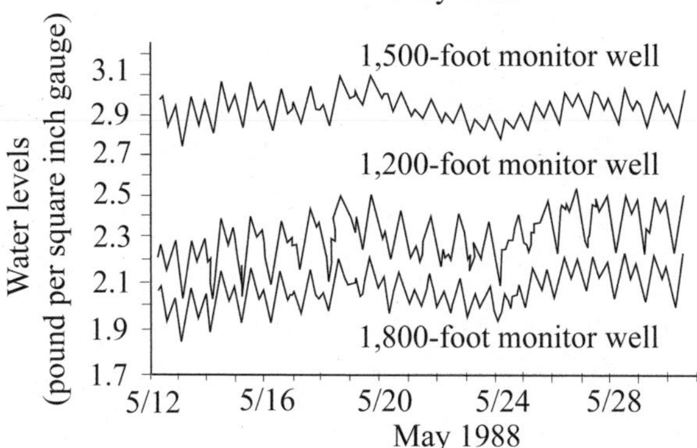

Figure 9.5. The effect of oceanic tidal loading and barometric pressure loading on water levels in the D. B. Lee injection and monitor wells system in Melbourne, Florida. Modified with permission from Duncan Evans, and Taylor (1994).

Elasticity influences the ability for aquifers to transmit and store water. The ability of an aquifer to transmit water is known as its *transmissivity* (T) (the units of which are [L^2t^{-1}]). Transmissivity is the product of a fully saturated aquifer, its thickness b [Lt] and the water-conducting character of the aquifer lithology, which is called the *hydraulic conductivity* (K) [Lt^{-1}], or velocity, such that

$$T = bK. \tag{9.1}$$

Transmissivity flow is assumed to be in a horizontal direction dependent upon a hydraulic gradient that is greater than zero (e.g., the flow direction arrows in Figure 9.1). For an unconfined aquifer, b is defined by the top of the water table above the top of underlying aquitard that bounds the aquifer.

The amount of water per unit volume in a saturated aquifer that can either be stored or released from storage due to the elasticity of the aquifer's lithologic structure, or skeleton, per unit change in head (h) is called the *elastic storage coefficient* or *specific storage* (S_s) where

$$S_s = \rho_w g(\alpha + n\beta). \tag{9.2}$$

S_s has the dimensions of [LL^{-1}] where

ρ_w = density of water,
g = the acceleration of gravity,
α = the aquifer compressibility,
n = porosity, and
β = the compressibility of water.

In a confined aquifer, *storativity* (S_a), a dimensionless coefficient, is the volume of water that a permeable aquifer will accumulate or discharge from a storage per unit surface area per unit change in head and is the product of specific storage (S_s) and the aquifer thickness (b) where

$$S_a = S_s b. \tag{9.3}$$

Freeze and Cherry (1979) define the storativity of a saturated confined aquifer of thickness b as the volume of water that an aquifer releases from storage per unit surface area of aquifer per unit decline in the component of the hydraulic head (h) normal to that surface (well A in Figure 9.6).

The volume of water stored in an unconfined aquifer is dependent upon the water table elevation, and as it changes so does aquifer thickness and the amount of water stored (well B in Figure 9.6). Thickness b is the saturated thickness of the aquifer or perpendicular height of the water table above the underlying confining unit that bounds the aquifer. Because of the variability of the water table, transmissivity and storage is not as well defined as in a confined aquifer. *Specific yield* (S_y) is the term used for storage in an unconfined, or water-table, aquifer.

As the water levels change in an unconfined aquifer, water drains or accumulates within the pore spaces and is related to the specific yield, such that

$$S_a = S_y + S_s b \tag{9.4}$$

where S_a is the storativity, S_y is the specific yield, b is the thickness of the saturated zone, and S_s is the specific storage. The volume

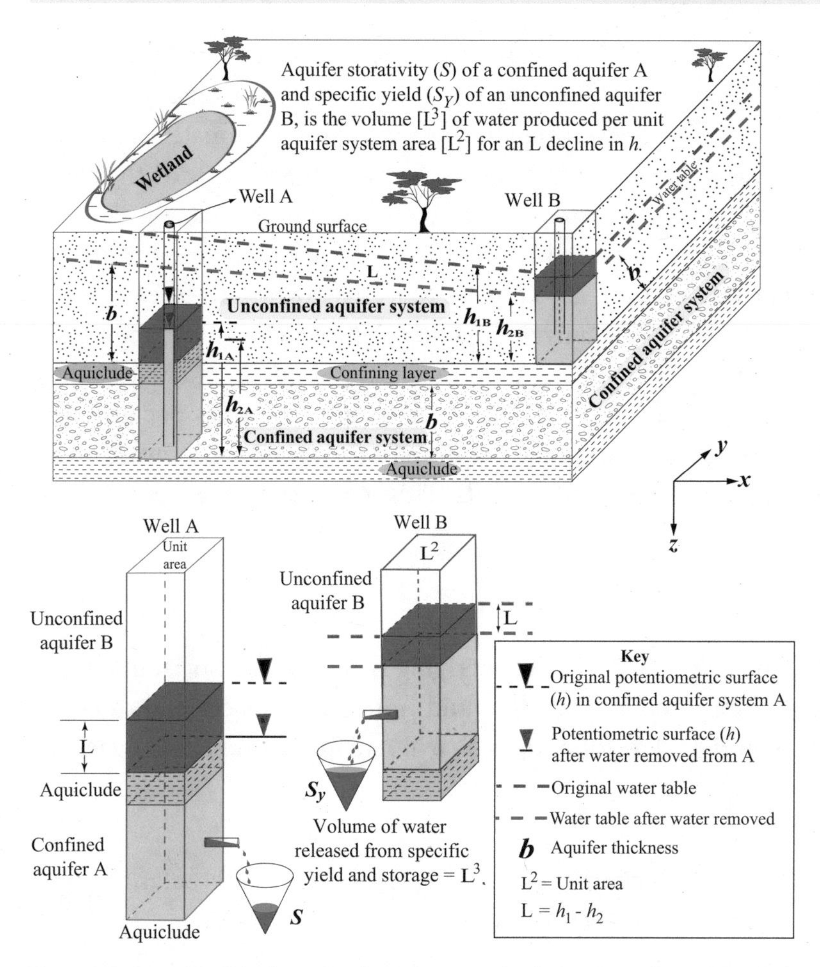

Figure 9.6. *Storativity* (S_a) of a confined aquifer system (well A) and *specific yield* (S_y) of an unconfined system (well B). The confined aquifer material is not drained and remains saturated; hence, b is constant. For the unconfined system, the water table can be variable and b can vary spatially.

(V_w or [L^3]) of accumulated or discharged water within an unconfined aquifer is

$$V_W = S_a A \Delta h \tag{9.5}$$

where A is the horizontal area and Δh is the decline in head. Therefore, similar to storativity, specific yield can be defined as the volume of water that an unconfined aquifer (well B) releases from storage per unit surface area of aquifer per unit decline in the water table (Δh).

The mechanisms of water expansion and aquifer compression that drive storage and specific yield are discussed further in sections 9.3.1 and 9.5. These physical processes can be summarized by treating a confined aquifer as a three-dimensional plane. The addition of the weight of the overlying lithologic material (including matrix water) would be a downward force acting on the plane. The downward force divided by the area of the plane is the total stress (σ_T). Because the plane is not in motion, Newton's second law (see section 9.5) conveys that this downward force must be balanced by an equivalent opposing force. Or, simply, the weight of the overlying material must be supported or held up by something. For a confined aquifer, the opposing forces per unit area (stresses) would be the pressure of the water and upward stress applied by the aquifer solids. This is known as the *effective stress* (σ_e) such that

$$\sigma_T = p + \sigma_e. \tag{9.6}$$

Thus, the weight of the overlying material on a horizontal plane is supported by both the fluid in the pore spaces and the solid materials. The total stress σ_T will remain constant over time, or

$$d\sigma_T = dp + d\sigma_e = 0. \tag{9.7}$$

Hence, any change in pressure must be counterbalanced by a change in effective stress, or

$$dp = -d\sigma_e. \tag{9.8}$$

When water is being pumped from a well within a confined aquifer, the hydraulic head (h) and, therefore, the fluid pressure (p) are decreasing at that point. Because water is slightly compressible, the fluid will expand somewhat, and water is released from storage in the aquifer. This additional volume of water produced (water mass does not change) may flow to the well and be withdrawn by pumping. This is one process by which water is being released from storage.

As fluid pressure and hydraulic conductivity are being reduced at the well, equation (9.6) states that there will be an increase in effective stress as part of the weight is being transferred from the fluid to the solid. Just as decreasing fluid pressure resulted in a slight expansion of the fluid (water), an increase in effective stress will result in compression of the confined aquifer material. This is analogous to squeezing a sponge for producing water to the well and is the second process where water is removed from storage.

In general, because aquifer solids and water are not very compressible, storativity values tend to be lower than values of specific yield. The dewatering of saturated spaces or pores of the aquifer matrix in an unconfined aquifer also occurs as a third mechanism, resulting in much higher values of storage. Freeze and Cherry (1979) gave storativity ranges from 0.005 to 0.00005 versus 0.01 to 0.30 for specific yield. If the aquifer systems are bounded by semiconfining units, the aquifer is referred to as a *leaky aquifer* and results in additional storage from the bounding aquitards.

9.2.3. Well Hydraulics, Specific Yield, and Aquifer Storativity

Aquifer systems and patterns of groundwater flow in a natural state are assumed to be in a relative steady state, whereas recharge and discharge balance out. This is especially true in the case of confined aquifers, which have a much slower response to recharge from precipitation and surface water effects. In contrast, water from rainfall and surface water systems (overland flow, wetlands, and lakes) infiltrates the soil and migrates downward, resulting in an almost immediate recharge response and increase of the water table elevation. Examples of relatively immediate output from an unconfined aquifer and decrease in the water table elevation would be discharge to wetlands and losses from evapotranspiration. Hence, the specific yield is a hydrological parameter that determines and represents the dynamic response of the water table aquifer to changes in inputs and outputs.

Observation wells and subsequent well hydrographs that show the variation in the water table levels through time are important tools for observing the behavior of unconfined aquifers over time and space. Observation wells in confined aquifers that monitor the potentiometric surface serve a similar purpose (for example, Figure 9.5 as an example of a hydrograph that shows the effects of tidal loading and barometric pressure on confined aquifers).

When examining lake systems, the unconfined, or water-table, aquifer is usually the primary source or sink to the lake system. Aquifer water levels and corresponding lake stage are a result of the amount of water in storage. Human interaction with the system, such as groundwater mining by well pumping, can dramatically alter the natural flow gradient of an aquifer and lake system. Pumping produces a decrease in hydraulic head and subsequent increase in hydraulic gradient at some point toward the well. *Drawdown* is the term used to describe the change in water level at the pumping well or neighboring observation wells. The greatest amount of drawdown is at the pumping well, and it decreases with distance from the pumped well. The pattern drawdown produces is typically cone-shaped and designated the *cone of depression* (Figure 9.7). The extent and shape of the cone of depression within an unconfined aquifer is dependent upon its transmissivity, specific yield, pumping rate, and pumping time. The volume of the cone of depression is the volume pumped by the well divided by the specific yield:

$$V_{cone} = \frac{V_{pumped}}{S_y}. \tag{9.9}$$

Pumping a well near a lake can reverse the natural flow gradient between the water-table aquifer and lake system and negatively impact lake stage.

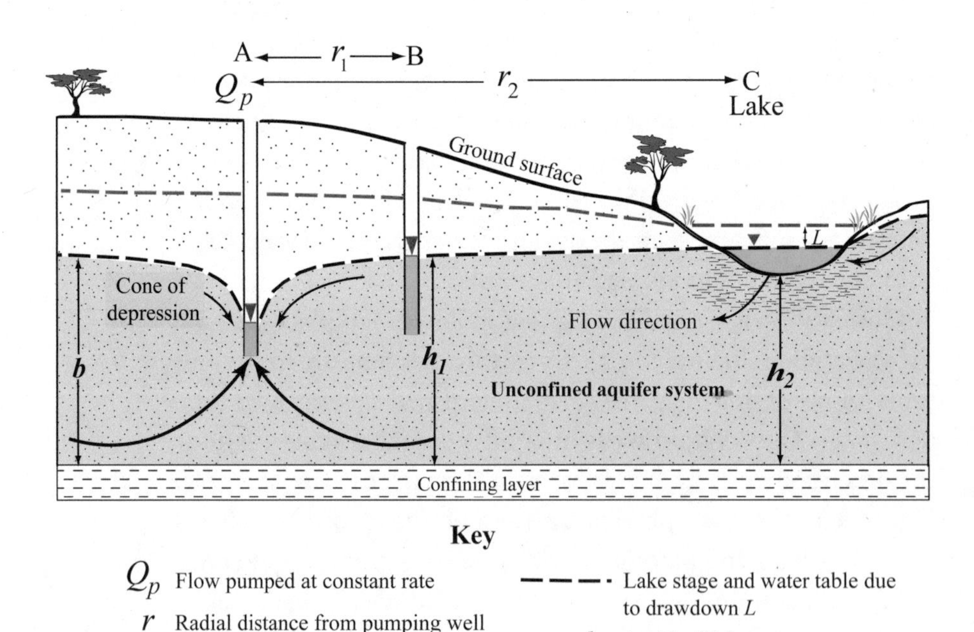

Figure 9.7. Well-pumping cone of depression in an unconfined aquifer system causing lake level drawdown.

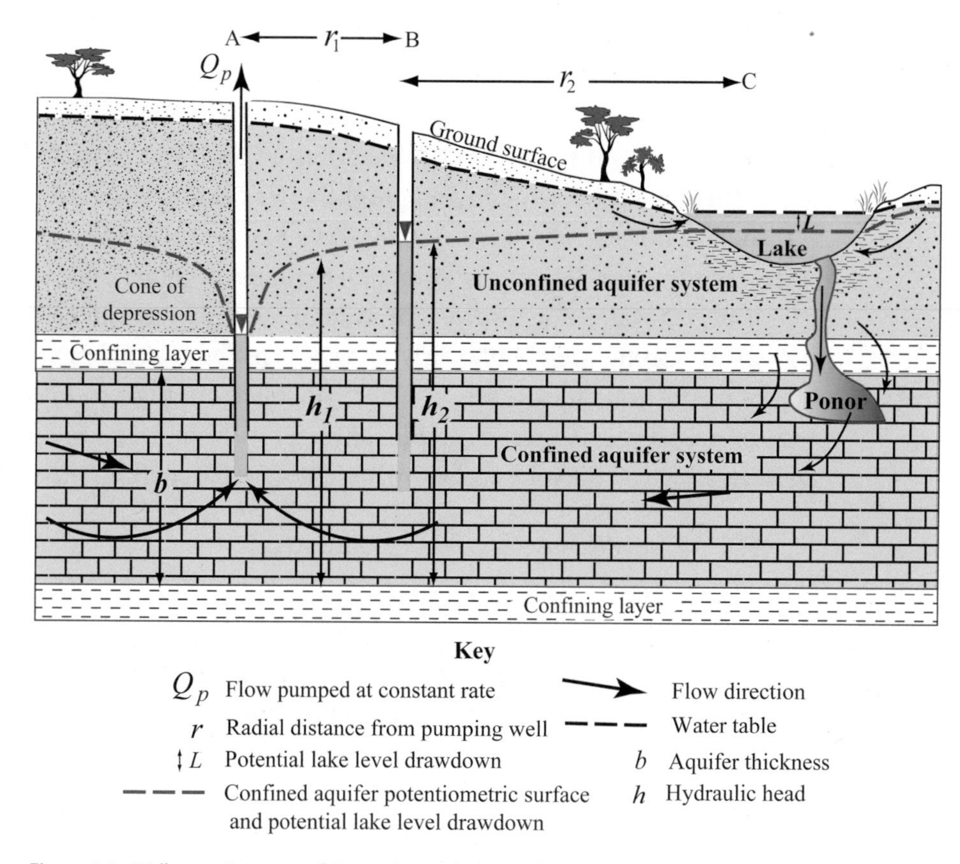

Figure 9.8. Well-pumping cone of depression within a confined aquifer system causing a drawdown of lake level due to an active lake bottom ponor connected to the underlying confined aquifer system.

The cone of depression generated by pumping a confined well is similar in concept to an unconfined aquifer with the volume of the cone in a confined system dependent upon and replaced by aquifer storativity instead of specific yield in equation (9.9). As stated earlier, storativity values are much less than specific yield, thus, given the same rate and time of pumping, the drawdown of the potentiometric surface will be much greater in a confined aquifer. In confined aquifers that are heavily pumped over extensive time periods, the cone of depression can be quite extensive, covering hundreds of kilometers with tens of meters of drawdown.

In general, pumping from confined aquifers will not impact lake systems, however, if the lake is associated with underling karst or sinkhole features (common to Florida lakes), or if the overlying confinement is leaky, then long-term pumping could impact lakes and wetlands (Figure 9.8).

9.3. Groundwater Hydraulics

When considering the flow of groundwater (or any fluid), it is necessary to discuss the forces acting on the fluid particle. The three main forces acting on groundwater causing it to move through a porous medium are *gravity*, *external pressure*, and *molecular attraction*. Gravity as a function of the earth's gravitational pull moves water downward. The combination of atmospheric pressure and the weight of overlying water creates external pressure in the zone of saturation where flow is from high to low pressure. The third force, molecular attraction, is where fluid molecules adhere to the solid surface materials of the lithologic matrix.

9.3.1. Fluid Pressure and Static Conditions

The variation of pressure with depth in a fluid at rest is essential in understanding the behavior in a lake system. The equation describing the variation in pressure at any thin parcel or point in a body of liquid (Figure 9.9a) is called the *hydrostatic equation*, where

$$dp / dz = \rho g \tag{9.10}$$

or

$$dp = -\rho g d \tag{9.11}$$

indicates that the rate of decrease in pressure p with the distance as we proceed upward relative to gravity is ρg, the unit weight (density $\rho \times$ gravity g) of the overlying fluid. In the case of water near the earth's surface, we assume that density to be constant and then integrate equation (9.11) where pressure is now a function of depth. Placing the parcel of water at the surface where $z = 0$, we obtain

$$\int_{p}^{ps} dp = -\rho g \int_{-d}^{0} dz \tag{9.12}$$

where ps is pressure at the surface, or atmospheric pressure. Integrating

$$p - ps = \rho g d \tag{9.13}$$

this equation becomes a form of the hydrostatic equation that can determine the absolute pressure at any point within a static fluid

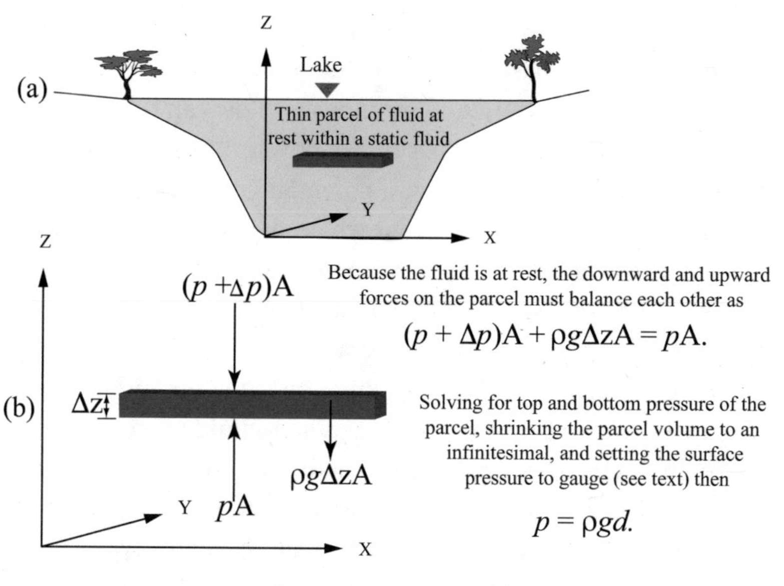

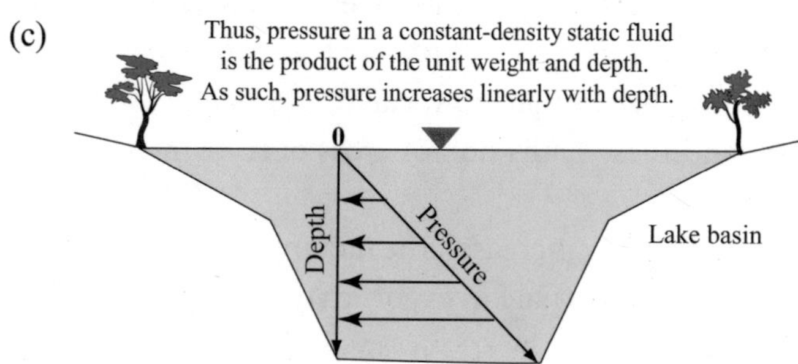

Figure 9.9. Variation of pressure with depth in a static water body.

(Figure 9.9b). By convention, pressure at the surface is *zero gauge pressure* ($ps = 0$), so what is actually calculated is pressure relative to the atmospheric pressure. To convert gauge pressure to absolute pressure, atmospheric pressure would be added to the gauge pressure. Throughout the remainder of this text, convention implies that p is the gauge pressure. Therefore, equation (9.13) becomes

$$p = \rho g d. \tag{9.14}$$

Hence, the pressure in a constant-density static fluid is the product of the unit weight and the depth, and pressure increases linearly with depth (Figure 9.9c).

9.4. Fluids in Motion: Laminar and Turbulent Flow

The study of fluids in motion or fluid flow is called *hydrodynamics*. This is a complex and vibrant field of study where many aspects are poorly understood, especially as it applies to groundwater flow and lake systems. However, with certain simplifying assumptions and knowledge of basic physics of fluid flow, a good understanding of the hydrodynamics of groundwater flow can be attained.

To understand the analysis of the physical processes of fluid flow, it is essential to recognize a potential gradient analogous to heat flow or an electrical current where flow is from a higher to lower energy state. For these two comparable processes, the temperature and voltage are potential quantities and the rates of flow are proportional to the potential gradient.

There are two major types of fluid flow: *laminar* and *turbulent* (Figure 9.10; see also section 8.2.1). When a fluid parcel begins to move due to force or gradient acting on the parcel, it must overcome the internal friction or viscosity of the fluid. Thus, slowly moving fluids are dominated by viscous forces. This type of flow is smooth such that the neighboring layers of fluid slide by each other and the flow is said to be *laminar* (Latin for "layers"), or a streamline flow. In laminar flow, each fluid parcel follows an orderly path, and these paths do not cross over or intersect one another.

As the velocity of flow increases, the moving fluid gains kinetic energy and eventually inertial forces govern the flow patterns. Turbulent flow is characterized by erratic, small whirlpool-like circles

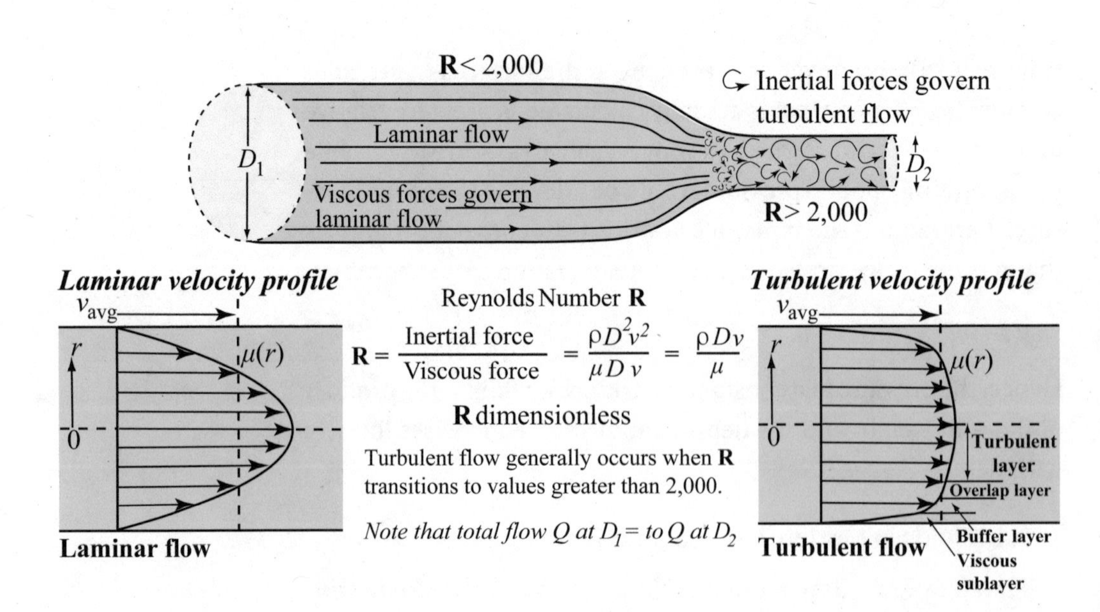

Figure 9.10. Laminar flow versus turbulent flow in a circular conduit of varying diameter.

called eddy currents, or simply eddies. Because of the rotational motion in turbulent flow, eddies absorb a great deal of energy. Although a certain amount of internal friction (viscosity) occurs in laminar flow, it is much greater in turbulent flow.

Osborne Reynolds, a professor of engineering, studied through a series of experiments the conditions in which the flow of fluid in pipes transitioned from laminar flow to turbulent flow. From these experiments came the dimensionless *Reynolds number* (**R**) that examines inertial forces (e.g., proportional to mass times acceleration expressed as $(\rho)(L^2)(v^2)$) versus viscous forces (shear stress times area) to determine whether flow will be laminar or turbulent, such that

$$\mathbf{R} - \frac{\textit{Inertial Forces}}{\textit{Viscous Forces}} = \frac{\rho v^2 D^2}{\mu v D} = \frac{\rho v D}{\mu} \tag{9.15}$$

where μ is the viscosity of the fluid, ρ is the fluid density, v is the fluid velocity, and D is the diameter of the conveyance the fluid is flowing through. Through a series of experiments of flow in circular conduits, Reynolds determined that laminar flow will generally transition to turbulent flow at **R** values greater than 2,000 (Figure 9.10).

Under most natural groundwater conditions where there is no fracturing, flow velocity is sufficiently low to where viscous forces govern. With the exception of ponor flow, movement into and out of lake systems via groundwater seepage will be treated as laminar.

To complicate matters, flow pathways through a porous medium, such as sand, is complex, convoluted, and tortuous as individual particles of water wind their way through the pore openings and cannot be completely quantified. However, because of the small size of the pores, it is assumed the flow is laminar and that frictionless flow is nonexistent, therefore, the hydraulic head is the driving force of flow.

One way to describe or represent a simplified version of a system such as flow through porous media is to construct a *conceptual model* of the specific area of investigation. The conceptual model consists of a set of assumptions that reduce the real problem and the real domain to a simplified version acceptable to the objectives of describing the system. For example, the position, size, and shape of individual sand grains and path of flow between them leads us

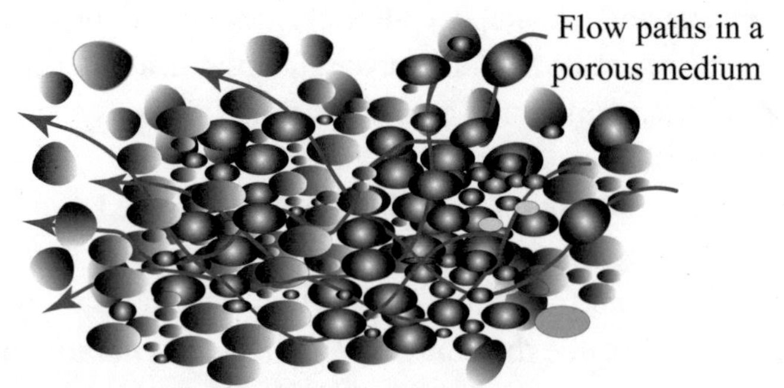

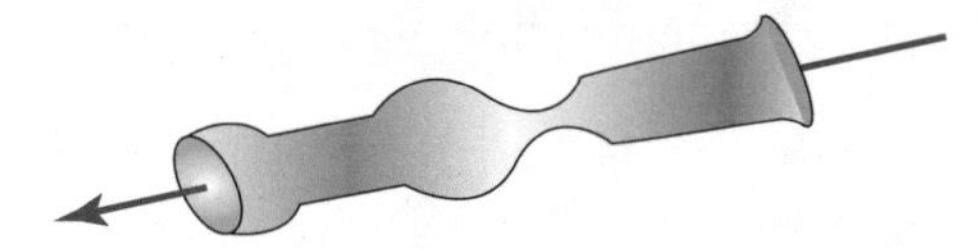

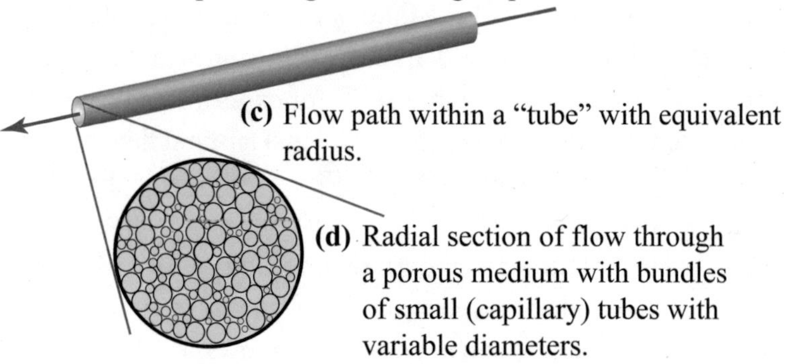

Figure 9.11. Conceptual model analogous to laminar flow through a series of capillary tubes. Modified with permission from Hornberger et al. (2014).

to a conceptual model of flow through a series of capillary tubes of varying diameters (Figure 9.11). Next the conceptual model is expressed in the form of a *mathematical model*, where the precise language of mathematics provides a powerful mechanism for describing a system in a compact, simple, and efficient way, which is examined in the next section.

9.5. Molecular Attraction, Fluid Viscosity, Friction, Head Loss, and Laminar Flow

Molecular attraction of flow through a porous medium results in the restriction of fluid movement as a result of shear stresses acting tangentially to the solid surface and normal stresses acting perpendicularly to the surface. Collectively we can characterize these restrictive forces as "friction" where this shearing resistance is the viscosity of the fluid (Rumer 1969). Mechanical energy is transformed to thermal energy because of the frictional resistance between the fluid viscosity and the particle boundary. Different fluids have different molecular attractions and viscosity. Syrup is more viscous than water, grease is more viscous than engine oil, and liquids, in general, are more viscous than gases.

Viscosity can be visualized by an idealized situation known as *Couette flow* (Couette 1890), which is laminar flow between two parallel plates (Figure 9.12). An external force F is applied to one plate, allowing it to move relative to the other at a constant velocity u where the speed of the plate is small enough that the fluid particles move parallel to the plate. The fluid in direct contact to the plates is held to the plate surface due to the adhesive force between the molecules. Thus, the upper surface of the fluid moves with the same speed as the moving plate, whereas the fluid in contact with the stationary plate remains stationary. The stationary layer retards the layer just above it, this layer retards the flow of the next layer, and so on. The particle speed will value linearly from zero at the stationary bottom plate to a maximum u at the top.

Each fluid layer will move faster than the one just below it, and the friction between them will result in a force resistant to their relative motion. The magnitude F is proportional to the speed u and the area A of each plate, and it is inversely proportional to the depth of the fluid between the two plates z where

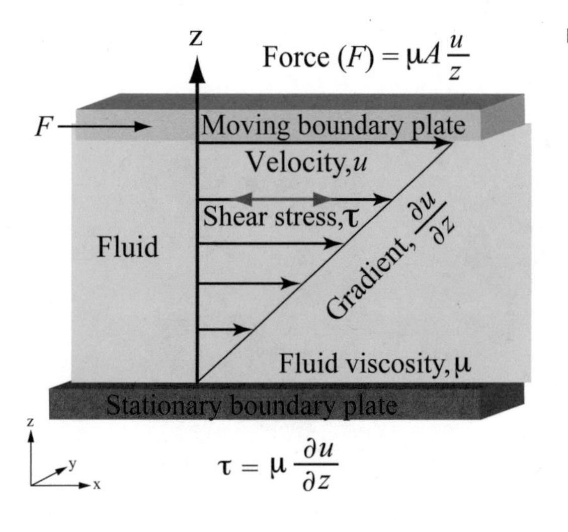

Figure 9.12. Couette flow.

$$F = \mu A \frac{u}{z}. \tag{9.16}$$

The constant of proportionality, μ, is the viscosity of the fluid. The ratio u/z is rate of shear deformation or shear velocity and is the derivative of fluid speed perpendicular to the plates, expressed by Sir Isaac Newton as

$$\tau = \mu \frac{\partial u}{\partial z}. \tag{9.17}$$

This concept can be applied to laminar flow through a pipe (analogous to flow between particles in a porous media in the conceptual model from section 9.4), where the wall of the pipe (or particle surface) acts as a stationary boundary (Figure 9.13). As a result, there will be velocity gradient away from the wall within a frictional flow with zero velocity at the boundary and maximum velocity at the center. This *parabolic velocity gradient* within a frictional flow results in a dissipation of energy, convert-

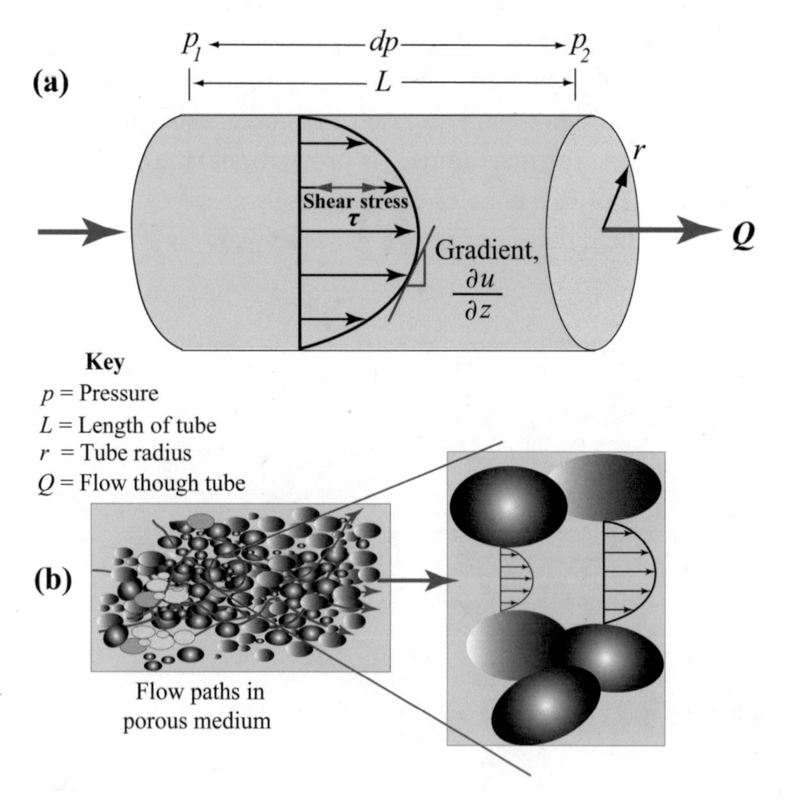

Figure 9.13. Parabolic velocity gradient of a viscous fluid within a pipe and between particles in a porous medium.

ing mechanical energy (head) to thermal energy, and is measured as head loss (h_L).

Because of viscosity, a pressure difference between two ends of the tube (or pore) is necessary to sustain a steady flow of any real fluid. The rate of fluid flow in a round tube depends on the viscosity of the fluid, the pressure gradient, and the dimensions of the tube. French physicist and physiologist J. L. Poiseuille (after whom the term "poise" is named after) first determined how the variables affect flow rate of an incompressible fluid undergoing laminar flow through a cylindrical tube. The result is known as *Poiseuille's equation*, which is

$$Q = \frac{\pi r^4 (p_1 - p_2)}{8\mu L} \tag{9.18}$$

where r is the inside radius of the tube, L is the tube length, and $p_1 - p_2$ is the pressure difference between ends of the tube (Figure 9.13a). This equation applies to laminar conditions only and is not applicable to turbulent flow.

Poiseuille's equation states that the flow rate is directly proportional to the pressure gradient: $p_1 - p_2$ (or dp) / L. Q also depends on the fourth power of the tube radius r. This means that if the pressure gradient is held constant, and the tube's radius is halved, then the flow rate is decreased by a factor of 16. Or, conversely, the pressure to maintain a constant rate of flow can be greatly affected by only a small change in radius, which is another demonstration of the complexity of flow through a porous medium (e.g., Figure 9.13b).

Recall that the conservation of mass (or equation of continuity) is where $Q = vA$ in frictionless flow at any point in the tube. However, when examining parabolic distribution of velocities, we are now more interested in the average velocity U of flow where $Q = UA$ or $U = Q/A$. One can then derive Poiseuille's equation (9.18) in conjunction with experimental relationships, determined by examining the Reynolds number **R** of viscous forces in a smooth conduit, such that

$$U = -\frac{dp}{dx}\frac{D^2}{32\mu} \tag{9.19}$$

where D is the diameter of the tube and the negative sign depicts flow from high to low pressure (Debler 1990). Flow (Q) through the tube is a product of the average velocity (U), and a cross-sectional area of the tube is expressed as

$$Q = AU = \frac{\pi D^2}{4}\left[-\frac{dp}{dx}\frac{D^2}{32\mu}\right] = -\frac{\pi R^4}{128\mu}\frac{dp}{dx} \tag{9.20}$$

This form of Poiseuille's law for flow of a viscous fluid through a capillary tube states that flow is directly proportional to the pressure gradient (dp) and the fourth power of the radius of the tube, and inversely proportional to the fluid viscosity. Given a constant viscosity and pressure gradient, flow through a tube (or pore opening) with a 10 mm radius will be 10^4 times greater than flow through a radius of 1 mm.

9.6. Darcy's Law

Henri Darcy (1856) was a French hydraulic engineer who, when designing a water filtration system in city of Dijon, first empirically described fluid flow through a porous sand medium by conducting a series of experiments on the basis of gradients that he expected to influence flow. To do this, Darcy filled cylindrical iron pipes with sand and systematically measured various factors that he hypothesized would define flow through the apparatus.

The apparatus Darcy designed (similar to Figure 9.14) consisted of a hollow cylinder filled with sand of a cross-sectional area A [L^2] and length l; it was stoppered at both ends with inflow and outflow tubes and had two manometers positioned along the cylinder.

The way it works is water is introduced to the system until the sand is saturated and the inflow rate Q [L^3] is equal to outflow rate. Setting an arbitrary datum of elevation $z = 0$, the elevations of the manometer intakes within the cylinder are z_1 and z_2. The sum of the elevation head (Dz) and pressure head ($p/\rho g$) is represented as h [L] and varies from h_1 to h_2 along the cylinder. A manometer measures the hydraulic head along the cylinder. By varying l ($l_2 - l_1 = \mathrm{D}l$) and the differences in hydraulic head ($h_2 - h_1 = \mathrm{D}h$), total discharge Q varies in direct proportion to A and Dh and inversely to Dl. Thus, flow rate Q is a function of the flow area, elevation, fluid pressure, and a proportionality constant K (with dimensions of LT^{-1}), which can be expressed as

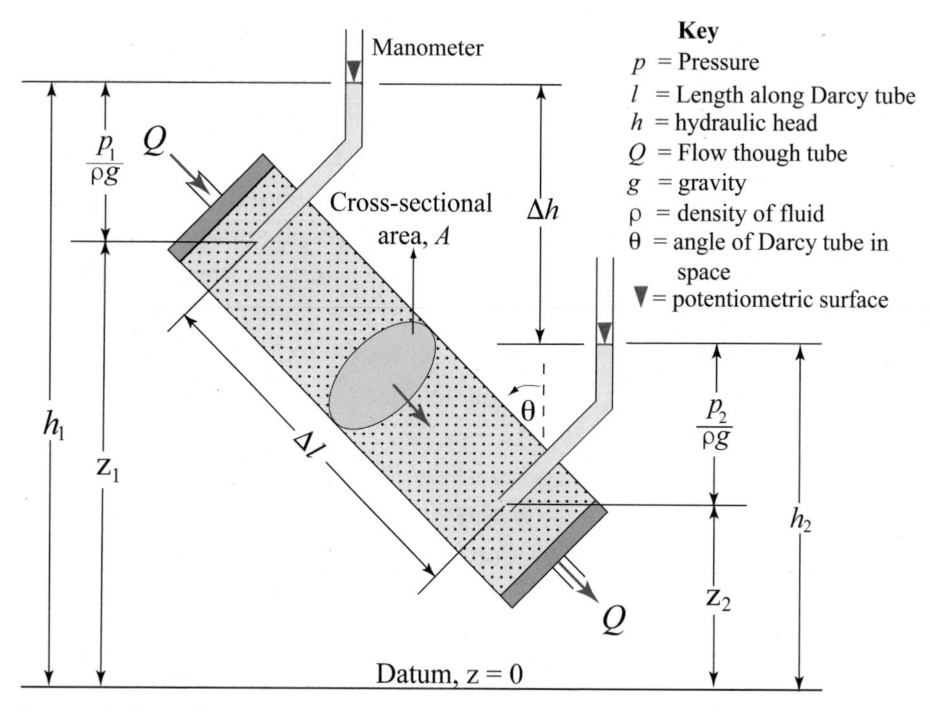

Figure 9.14. Sand-filled cylindrical apparatus illustrating Darcy's law.

$$Q = -KA\frac{\Delta h}{\Delta l} \tag{9.21}$$

$$\frac{Q}{A} = -K\frac{\Delta h}{\Delta l} \tag{9.22}$$

$$q = -K\frac{dh}{dl} \tag{9.23}$$

where $q = Q/A$ is the *specific discharge*, h is the *hydraulic head*, and dh/dl in differential form is the *hydraulic gradient*. Note that the negative sign associated with the hydraulic gradient indicates that positive specific discharge corresponds with a negative hydraulic gradient, where flow is in the direction of decreasing head proportional to the hydraulic gradient. K, the proportionality constant with dimensions of $[Lt^{-1}]$, is known as the *hydraulic conductivity* and is the function of the medium (e.g., sand) in the cylinder and the fluid moving through the medium. For example, gravel would have a higher K than sand, which would have a greater K value than clay (see Table 9.1 in section 9.9), and water would have a higher K than molasses.

If the hydraulic gradient is held constant, then q is proportional to K and has the dimensions $[Lt^{-1}]$ of a velocity or flux. In reality, because of the tortuosity of water particles traveling through a porous media (Figures 9.11 and 9.13), q is not a true velocity (i.e., flow of a particle in a linear path from points A to B) and is correctly referred to as *specific discharge* and not velocity. It should be noted that in some hydrogeology texts (e.g., Freeze and Cherry 1979; Fetter 1988) specific discharge is denoted as v. So as not to confuse it with velocity v used throughout the text, the q notation is preferred, as is also characterized in the hydrology literature (e.g., Maidment 1993).

If the hydraulic gradient (dh/dl) and K are held constant, and q is independent of the angle θ, then Darcy's law is valid for groundwater flow in any direction in space. This is true where θ is greater than 90° and flow through the cylinder is against gravity.

To better describe flow through a porous medium in terms of Darcy's equation (9.23), the pressure gradient (dp/dx) in equation (9.19) can described in terms of the hydraulic head as derived in equations (9.17) through (9.20), such that the average velocity of flow

$$U = -\frac{dh}{dl}\frac{\rho g}{\mu}\frac{D^2}{32} \tag{9.24}$$

and

$$q = -K\frac{dh}{dl} \tag{9.23}$$

where K depends on the properties of the fluid and porous lithologic material. This *Darcy-Poiseuille analogy* suggests that $D^2/32$ describes the variation in K in terms of pore diameter and that the $\rho g/\mu$ variation in K is a condition of fluid density and viscosity. As discussed, the diameters of the pores in lithologic material are not well defined or measurable; however, the grain size, type of lithologic material, and depositional environments can give some indication of pore size and conductivity (e.g., glacial till gravel is more porous than beach fine-sand deposits).

Darcy's law demonstrates that the volumetric flow rate is a function of the flow area, elevation, fluid pressure, and a proportionality constant K. It may be stated in several different forms depending on the flow conditions. Since its discovery, it has been found valid for any Newtonian fluid. Likewise, while it was established under saturated flow conditions, it may be adjusted to account for unsaturated flow and multiphase flow (e.g., fractures).

9.6.1. Hydraulic Conductivity: Darcy's Proportionality Constant *K*

Hubbert (1956) conducted numerous experiments varying fluid density, viscosity, and the geometric properties of sand and determined that K, commonly known as hydraulic conductivity, is a function of the properties of both the porous medium and the fluid. If one varies the viscosity μ and density ρ of the fluid (e.g., water versus molasses) flowing through the porous media, then it should be no surprise that Darcy's proportionality constant will also vary accordingly. By varying both the porous medium and the fluid run through a Darcy apparatus under a constant hydraulic gradient dh/dl, Hubbert reported the following proportionality relationships:

$$K \propto \frac{1}{\mu} \tag{9.25}$$

$$K \propto pg \tag{9.26}$$

$$K \propto d^2 \tag{9.27}$$

where d is the diameter of the grain size. When examining flow on microscopic scales, Hubbert (1940) realized another dimensionless proportionality constant C needed to be included as properties of the porous medium that affect flow other than grain size, such as distribution of grain size, porosity, shape (sphericity), roughness, and the nature of their packing or special orientation. Therefore, K can be rewritten as

$$K = \frac{Cd^2\rho g}{\mu}\frac{dh}{dl} \tag{9.28}$$

where Cd^2 is the function of the medium and pg/μ is the property of the fluid.

Cd^2 is known as *specific* or *intrinsic permeability*, or simply *permeability*, and has been assigned the notation of k with the dimensions of $[L^2]$ where

$$K = \frac{k\rho g}{\mu} \tag{9.29}$$

9.7. Hydraulic Head and Hubbert's Classic Treatise on Fluid Potential

Darcy's law is indispensable in the understanding of groundwater flow as it applies to hydraulic head gradients where flow is from higher to lower head values. However, this law is based only on empirical evidence and, as stated by Hubbert (1940, 795), "to adopt it empirically without further investigation would be like reading the length of mercury column of thermometer without knowing that temperature was the physical quantity being indicated."

To reiterate, the physical processes that involve flow necessitate the acknowledgment of a potential energy gradient where flow is from higher to lower energy values. Hubbert (1940, 794) defines *potential* as "a physical quantity, capable of measurement at every point in a flow system, whose properties are such that flow always occurs from regions in which the quantity has higher values to those which it has lower, regardless of the direction in space."

Hubbert recognized that two potential quantities for flow are *elevation* and *fluid pressure*. If the Darcy cylinder were positioned vertically ($\theta = 0°$), flow would occur from high elevation to low elevation in response to gravity. Conversely, if the cylinder were placed horizontally ($\theta = 90°$), flow would be induced by creating a pressure gradient from one end of the cylinder to the other. These processes are further defined by Hubbert in terms of Newton's laws of motion and the mechanical energy (e.g., work) required to move a fluid parcel from any two points in a flow system as well as the energy loss of the unit mass during this process.

Recall from basic physics that work is done by a force when the force moves an object through a distance d. Energy (kinetic and potential) can be defined as the ability to do work. *Kinetic energy* (KE) is energy of motion: a body of mass m and speed v has translational kinetic energy of ½ mv^2. *Potential energy* (PE) is associated with the position or configuration of bodies, such as gravitational energy, where $PE_g = mgz$ and z is the object above an arbitrary reference point (e.g., z_0). The law of conservation of energy states that energy can be transferred from one state to another but the total energy remains constant. It is valid when friction is present, as the heat generated is another form of energy. In the absence of friction and other nonconservative forces, the total mechanical energy is conserved such that KE + PE is constant. When nonconservative forces are involved, $W_{NC} = \Delta KE + \Delta PE$, where W_{NC} is the work done by the nonconservative force (e.g., friction).

Simply stated, fluid flow through a porous media is a mechanical process where the forces driving fluid movement must overcome the frictional forces, or *drag*, between the moving fluid parcel and the grains of the porous medium. As the parcel travels through the media, the mechanical energy is transformed into thermal energy due to the frictional resistance between the parcel viscosity and the media (see section 9.6). Hence, the direction of flow is from regions of higher mechanical energy to regions of lower mechanical energy, such that the mechanical energy per unit mass at any point in the system can be defined as the work required to move a unit mass of fluid from a randomly chosen standard state to the point in question. Or as specified by Hubbert (1940, 797), "The fluid potential for flow through porous media is therefore the mechanical energy per unit mass of fluid."

To describe mechanical energy in terms of elevation and pressure, an arbitrary standard state A is established where elevation $z = 0$, pressure $p = p_0$, and p_0 is atmospheric (Figure 9.15). At this point, a unit mass m of fluid with density ρ_0 will occupy a volume V_0 where $V_0 = 1/\rho_0$. The work required to lift the unit mass of fluid from the standard state A to some point B in the flow system at elevation z and fluid pressure p. At B the unit mass with density ρ occupies a volume $V = 1/\rho$. Given the fluid parcel is water and assuming incompressibility, the mass, density, and volume at point A and point B are equal. The fluid will have a velocity $v = 0$ at standard state and velocity v_B at point B.

The energy or work required to move a unit mass (m) from point A to B consists of three components. The first is the work (w) necessary to move the mass from elevation $z = 0$ to elevation z, which represents loss of potential energy:

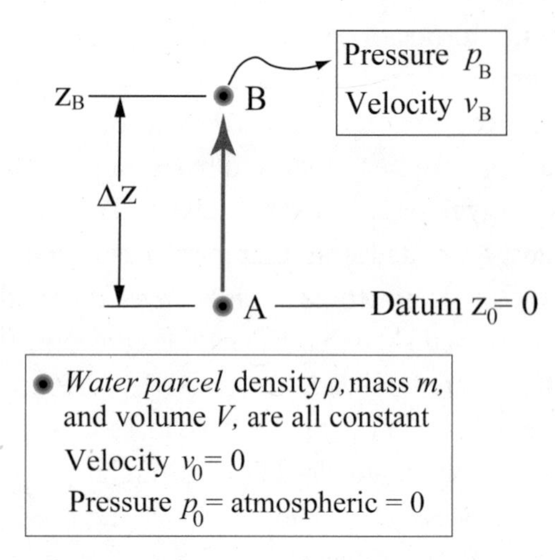

Figure 9.15. Movement of a water parcel from points A to B.

$$w_1 = mgz. \tag{9.30}$$

Second is the work required to accelerate the fluid from $v = 0$ to velocity v, which is the loss in kinetic energy:

$$w_2 = \frac{mv^2}{2}. \tag{9.31}$$

Third is the work required to raise the fluid pressure from $p = p_0$ to p where, as just stated, $V = 1/\rho$, which represents the loss of elastic energy, or $p - V$:

$$w_3 = m\int_{p_0}^{p} \frac{V}{m} dp = m\int_{p_0}^{p} \frac{dp}{\rho}. \tag{9.32}$$

The mechanical energy per unit mass, or *fluid potential* (Φ), is the sum of w_1, w_2, and w_3. Setting a unit mass of fluid to $m = 1$ and combining equations (9.30), (9.31), and (9.32), then

$$\Phi = gz + \frac{v^2}{2} + \int_{p_0}^{p} \frac{dp}{\rho} \tag{9.33}$$

which is the classic *Bernoulli equation* defining energy loss during fluid flow and derived in most college physics and fluid mechanics texts. Water in terms of its characteristics under conditions near the earth's surface can be considered an incompressible fluid (fluids with a constant density at varying pressures), thus ρ is not a function of pressure and

$$\Phi = gz + \frac{v^2}{2} + \frac{p}{\rho}. \tag{9.34}$$

where g is the acceleration due to gravity, z is the elevation of the base of the piezometer (see Figure 9.15), p is the pressure applied by the water column, v is the velocity head, and ρ is fluid density.

As flow velocities in porous media are extremely low, the second term can be negated. In addition, it is common in groundwater hydrology to set the atmospheric pressure (p_0) equal to zero and work in gauge pressures, thus equation (9.34) can be simplified to

$$\Phi = gz + \frac{p}{\rho} \tag{9.35}$$

From Darcy's experiments (see Figure 9.14), the hydraulic head (h) at any point in the cylinder can be defined as the sum of the elevation head (Dz) and pressure head ($p/\rho g$), hence:

$$h = \frac{p}{\rho g} + z \tag{9.36}$$

such that

$$p = \rho g(h - z) \tag{9.37}$$

and

$$\Phi = gz + \frac{\rho g(h - z)}{\rho}. \tag{9.38}$$

Then, after cancelling terms,

$$\Phi = gh. \tag{9.39}$$

From equation (9.39), the fluid potential Φ at any point P in a porous medium is simply the hydraulic head multiplied by the acceleration due to gravity. Gravity is nearly constant at the earth's surface, therefore, the hydraulic head h and Hubbert's fluid potential Φ are virtually perfectly correlated measurements where Φ is energy per unit mass and h is energy per unit weight.

By convention, ψ symbolizes the pressure head $p/\rho g$, and its relationship with elevation head z is fundamental in understanding groundwater flow and meets Hubbert's definition of a potential as a physical quantity, capable of measurement, and where flow occurs from regions of higher h value to regions of lower h value. This concept is demonstrated in Figure 9.16, showing the flow direction between two piezometers. A piezometer is analogous to Darcy's manometer and is the basic tool for measuring hydraulic head in the field. It consists of a pipe placed in the ground that is open to water flow at the bottom and to the atmosphere at the top. Water rises in a piezometer in direct proportion to the total fluid energy at point P at the bottom of the piezometer, which is open to the aquifer lithology. Thus, the point of measurement is at the base of the piezometer, not at the fluid surface level (▼). This point is used to determine the h at the corresponding elevation (z) of the groundwater reservoir. The common datum, by convention, is taken at sea level, which is zero elevation. The boundary between the water table aquifer and unit of confinement demarking a confined aquifer can also be used.

Hydraulic head distribution within a regional groundwater system is a three-dimensional hydrogeologic system. If large numbers of piezometers are placed strategically throughout the system, it is possible to contour the positions of equal hydraulic heads as center points forming an *equipotential surface*. Traces of the equipotential surfaces on a two-dimensional cross section are called

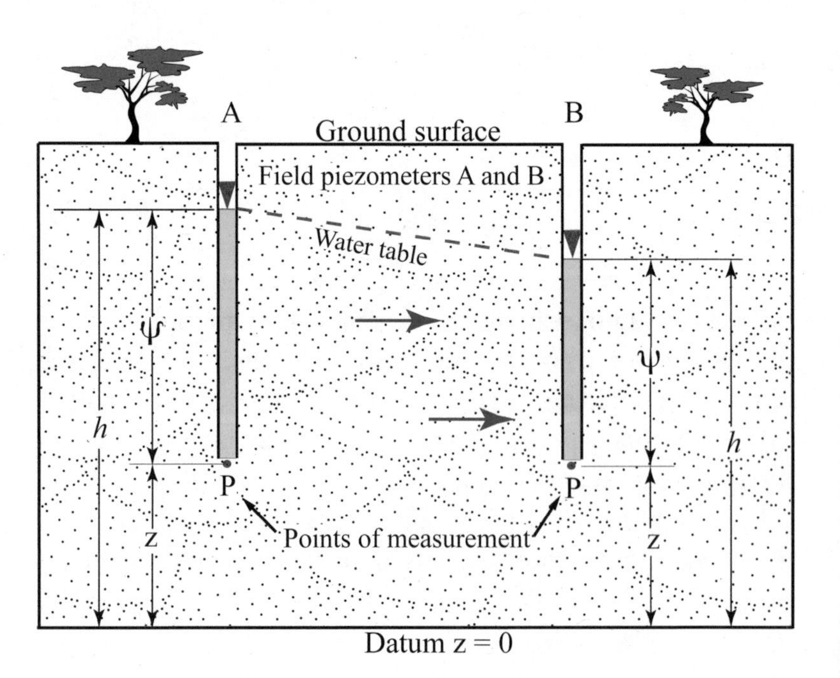

Figure 9.16. Hydraulic head (h) is equal to pressure head (ψ) plus elevation head (z) of two field piezometers, resulting in flow from the higher head (left) to lower head (right).

equipotential lines. If a pattern of equipotential lines can be established, then flow lines can be constructed perpendicular to the equipotential lines, establishing a *maximum potential gradient*. A set of intersecting equipotential lines and flow lines is known as a *flow net* (see section 9.10).

Returning to Bernoulli's equation (9.34) under conditions of steady flow, the total energy of an uncompressible fluid is constant at each point along a flow path (e.g., the water table boundary between piezometers A and B in Figure 9.16) in a closed system where g is the acceleration due to gravity, z is the elevation of the base of the piezometer, ψ is the pressure applied by the water column, v is the velocity, and ρ_w is fluid density, such that

$$\Phi = gz + \frac{v^2}{2} + \frac{p}{\rho_w}. \tag{9.40}$$

The total head is

$$h_T = h_z + h_y + h_p. \tag{9.41}$$

As stated earlier, the velocity head in groundwater systems is usually negligible, hence the total head h_T is equal to the hydraulic head h where h_z and h_p are expressed in units of length [L].

9.8. Head Loss

Head loss (h_L) can be demonstrated using Darcy's apparatus positioned horizontally ($\theta = 90°$) (Figure 9.17) and containing a homogeneous and isotropic porous medium. Several manometers are placed along the length to measure the head (h), and flow is induced by creating a constant pressure gradient at one end of the cylinder. The height of the of water at each manometer is a measure of the pressure at that point (p_i) within the cylinder and can be expressed as

$$p_i = \rho g h_i. \tag{9.42}$$

In the derivation of Bernoulli's equation, friction flow was assumed to be insignificant, which is valid in experiments involving flow over short distances or glass tubes with low frictional properties. For the experiment demonstrated in Figure 9.17, the Bernoulli equation would predict that pressure and velocity along the cylinder would be the constant.

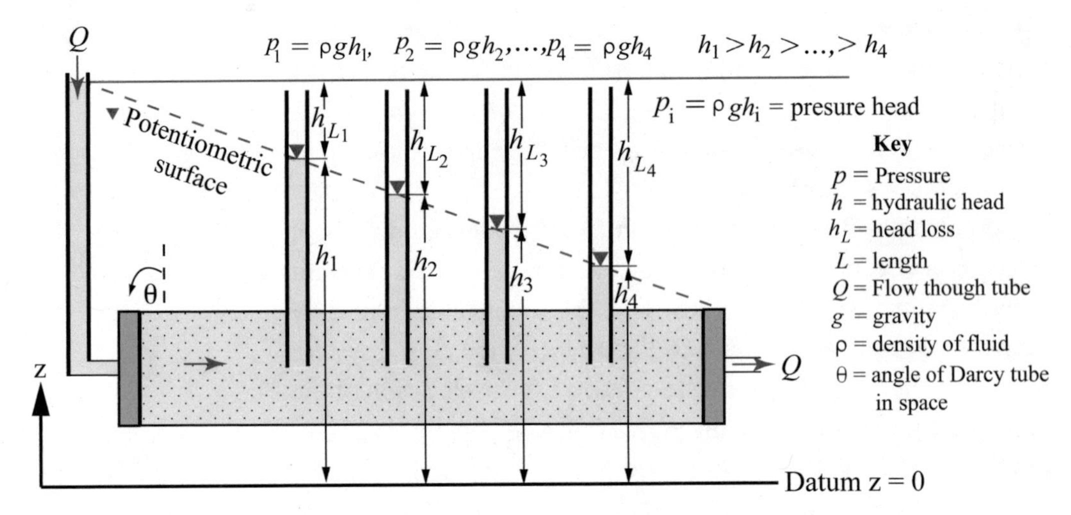

Figure 9.17. The height of water in each manometer decreases linearly along the length of the Darcy cylinder, demonstrating energy loss due to friction (head loss) as flow travels down gradient.

This would entail that $h_1, h_2, \ldots, h_i$ along the cylinder would be equal. However, empirical observations divulge that $h_1 > h_2 > \ldots > h_i$. Because the cylinder is horizontally placed and the diameter and flow are constant, the pressure along the cylinder must be continuously decreasing along the cylinder. Based on the conservation of mass (or volume), to account for energy loss as a result of the flow of fluids with viscous properties, the Bernoulli equation (as derived in equation 9.33) must be modified as

$$\Phi = gz + \frac{p}{\rho} + h_L = constant. \tag{9.43}$$

Head loss is dependent on the velocity of the fluid, the diameter and length of the cylinder, and the viscosity of the fluid, such that

$$h_L = f\frac{U^2}{2g}\frac{L}{D}. \tag{9.44}$$

Note that the roughness of the cylinder or boundary wall must also be considered. This is expressed as a dimensionless friction factor (f) and is determined experimentally for different materials usually associated with pipe flow under controlled laboratory conditions.

9.9. Hydraulic Properties of a Porous Medium

The lithology of rocks and sediments near the earth's surface usually contain empty spaces due to the physical and chemical weathering processes that continually decompose and disaggregate the rocks and sediments. Movements of rock mass near the surface can cause the rocks to crack or fracture. Dissolution of carbonate sediments can cause cavities and enhance already present fractures, resulting in karst features. Sediments are assemblages of individual grains that are deposited by water, wind, ice, or gravity. The openings between the sediment grains are called *pore spaces* and, along with the fractures, cracks, and voids of the various geologic materials deposited over time, are the means by which water is stored and transmitted within the subsurface.

The *porosity* of lithologic materials is the percentage of the rock or soil that is void of material. Porosity (dimensionless) is defined mathematically by

$$n = \frac{Vv}{Vt} \tag{9.45}$$

where

n = porosity (as a percentage),
V_v = volume of void space in a unit volume of earth material, and
V_t = total unit volume of earth material, including voids and solid material.

Fluids flowing through the pores, or void spaces, in rock and sediment may not move through the entire volume of void space due to noninterconnectedness between pores (as well as dead-end pores with no outlet). An example of this would be volcanic pumice, a highly vesicular, highly porous rock with little interconnectedness that is light enough to float on water.

Primary porosity refers to the void spaces between the grains in sediment or rock whereas *secondary porosity* is due to either fracturing or chemical dissolution of the mineral framework, such as dissolution of carbonates in karst environments. The *effective porosity* (n_e) is the porosity available for fluid flow. The *effective*

Table 9.1. Representative values of porosity, permeability, and hydraulic conductivity

Sediment or rock type	Porosity n	Permeability k (m^2)	Hydraulic conductivity K (m/day)
Clays	0.40–0.60	10^{-19}–10^{-15}	10^{-7}–10^{-3}
Silts	0.35–0.50	10^{-16}–10^{-12}	10^{-4}–10^{0}
Fine to coarse sand	0.15–0.45	10^{-14}–10^{-9}	10^{-2}–10^{+3}
Shales weathered	0.30–0.50	10^{-16}–10^{-12}	10^{-4}–10^{0}
Shales at depth	0.01–0.10	10^{-20}–10^{-16}	10^{-8}–10^{-4}
Sandstones	0.05–0.35	10^{-17}–10^{-12}	10^{-5}–10^{0}
Carbonate mud	0.40–0.70	10^{-13}–10^{-15}	10^{-3}–10^{-1}
Limestone	0.10–0.35	10^{-16}–10^{-12}	10^{-4}–10^{2}
Karst and reef	0.05–0.50	10^{-13}–10^{-5}	10^{-1}–10^{7}
Dolostone	0.001–0.15	10^{-16}–10^{-12}	10^{-4}–10^{0}
Chalk	0.15–0.45	10^{-15}–10^{-12}	10^{-3}–10^{0}
Salt	0.001–0.005	10^{-22}–10^{-20}	10^{-10}–10^{-8}
Unfractured igneous rocks	0.0001–0.01	10^{-21}–10^{-17}	10^{-9}–10^{-5}
Fractured igneous rocks	0.01–0.10	10^{-17}–10^{-13}	10^{-5}–10^{-1}
Basalts	0.01–0.25	10^{-18}–10^{-11}	10^{-6}–10^{-1}

Source: From Maidment (1993).

pore fraction (epf) is the ratio of the porosity available to flow to the total porosity and can be defined as

$$n_e = n(epf). \tag{9.46}$$

As discussed in section 9.4, hydraulic conductivity is a measure of fluid flow through the interconnected void spaces as a function of both the medium and the fluid properties. The permeability (k) is derived to separate out the effects of the medium only (equation 9.46) such that $n_e = k$.

Porosity n, permeability k, and hydraulic conductivity K for various lithologies and rock types have been measured in both the laboratory and the field. Representative values (meters and meters/day) of these three parameters are summarized in Table 9.1. These values have been obtained from numerous sources and are not meant to be substituted for site-specific applications.

9.9.1. Heterogeneity, Anisotropy, and Hydraulic Conductivity

The lithology of rocks and sediments contain a directional component due too various geological and sedimentation processes that formed them, including horizontal stratification, variable energies of deposition and bedding planes, structural movement of rock mass resulting in fracturing, dissolution of carbonates, and weathering.

For types of lithologies where hydraulic conductivities vary directionally in three-dimensional space (e.g., $K_x \neq K_y \neq K_z$) are as known *anisotropy.* For example, because of the plate shape and electrostatic charge of clay minerals, deposition usually occurs, resulting in K values much greater in the horizontal direction than vertical. *Isotropy* is when the hydraulic conductivities are the same in all directions ($K_x = K_y = K_z$). In horizontally bedded sedimentary deposits, it is common to have vertical conductivities that vary with horizontal conductivities such that $K_x = K_y \neq K_z$. This is known as *transverse isotropy*, and in general the permeability of these sediments are highest along the directional or horizontal plane of bedding stratification and least permeable perpendicular to the stratification plane.

A hydrogeologic unit where hydraulic conductivity is the equivalent from position to position, or independent of space, is said to be *homogeneous*. However, no unit is truly homogeneous because of the temporal and spatial variability of the geologic processes responsible for creating and modifying the placement of rocks and sediments.

Heterogeneity in geologic environments, in general, can manifest in a large range of scale, thus K is a function of the scale that is being examined. At the microscopic pore level, K involves variation in orientation and packing of sediment grains, and as the scale changes, K can be subjected to a multitude of geologic environments and processes. For example, secondary modifications (such as fracturing) and dissolution introduce their own distinct complexities for influencing hydraulic conductivity.

Estimations of K are based on the scale and system one is examining; as such, a variety of methods have been developed to estimate K. When examining an aquifer system, for example, pump tests can be conducted at different wells, giving a volume-integrated estimate of the system as a whole. Although K may vary spatially, such as not coinciding with particular bedding lithologies in an aquifer, the overall *effective hydraulic conductivity* for the system can be projected.

Lake bottom sediment deposition, structure, and other hydrogeology heterogeneities are typical for lake systems, thus spatial variation in lake hydraulic conductivity is not atypical (see section 10.6). Processes that result in lake bottom heterogeneity include edge effects, such as erosion and deposition of the lake shoreline sediments; karstic (e.g., ponors) and fluvial (streamflow) processes; biological interactions; and spatial and temporal effects due to lake hypsometry and seasonality impacts. These types of heterogeneities are discussed in more detail in section 10.6.

9.10. Continuum Concept and Representative Elementary Volume

All the processes involving flow through a porous medium described previously are based on flow pathways through a permeable lithology on a minute level. This is a complex, convoluted, and tortuous conveyance as individual particles of water wind their way through the various pore openings and configurations (Figure 9.18a). Specific discharge (q), although easily measured, represents a macroscopic average of a *representative continuum* concept of flow (Figure 9.18b) since flow is impossible to measure quantitatively on a microscopic scale.

Hubbert (1956) emphasized this continuum concept of a macroscopic control volume that is large with respect to the microscopic individual pores but small in terms of the space within which significant statistical variations of the macroscopic properties may be expected. A composite modified from Hubbert (1956) and Bear (1972) illustrates this point with a hypothetical plot of porosity (n) of a lithologic medium versus the increasing volume (V_1, V_2, . . . , V_i) measured at a point P within a porous medium (Figure 9.19). At the microscopic level, porosity values vary wildly. For example, the level may range from all solid ($n = 0$) to all pores ($n = 1$). As the volume increases, a statistical average smooths out the microscopic variations until there are no longer any variations with the size of the sample.

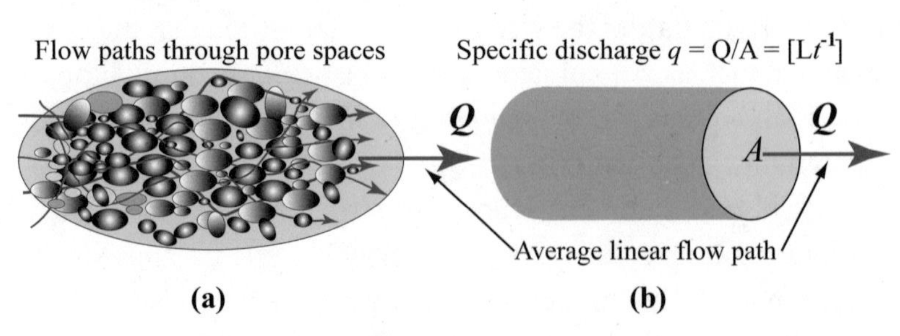

Figure 9.18. Microscopic (a) versus macroscopic (b) concepts of flow.

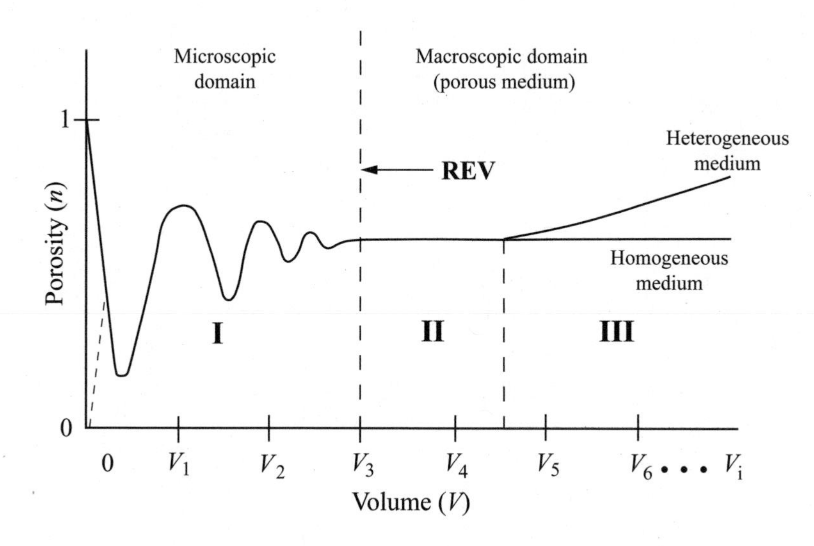

Figure 9.19. Microscopic and macroscopic domains of a porous medium and the point of representative elementary volume (REV) at V_3. Modified from Hubbert (1956) and Bear (1972).

Bear (1972) defined this limit as the *representative elementary volume* (REV), a volume of sufficient size such that there are no significant statistical variations in the value of a particular property with the size of the element.

9.10.1. Flow in a Fractured Medium

Fractures, joints, and faults serve as conduits for groundwater flow and can be the dominant secondary porosity feature in some geologic media. They are often the only method of water flow in such igneous and metamorphic rocks as granites, basalts, gabbro, quartzite, gneiss, and schist where the primary porosity is less than 1%. In areas of karst, they can have an enormous influence on lake and wetland systems.

When examining fracture flow in an aquifer system, the REV continuum concept can be used where the domain of the individual fractures is equivalent to the microscopic domain, and the macroscopic domain can be applied to the aquifer as a whole. If the fractured medium is adequately dense that the fractures behave similar to a granular porous medium, this approach is valid and Darcy's theory can be applied. If the fracture spacing is variable in a given direction, the media will exhibit trending heterogeneity. If fractures have a certain orientation more so in one direction, then anisotropy will be demonstrated (Freeze and Cherry 1979).

Fractured basalt flows can demonstrate how geologic medium parameters and REV vary over space (Figure 9.20). For example,

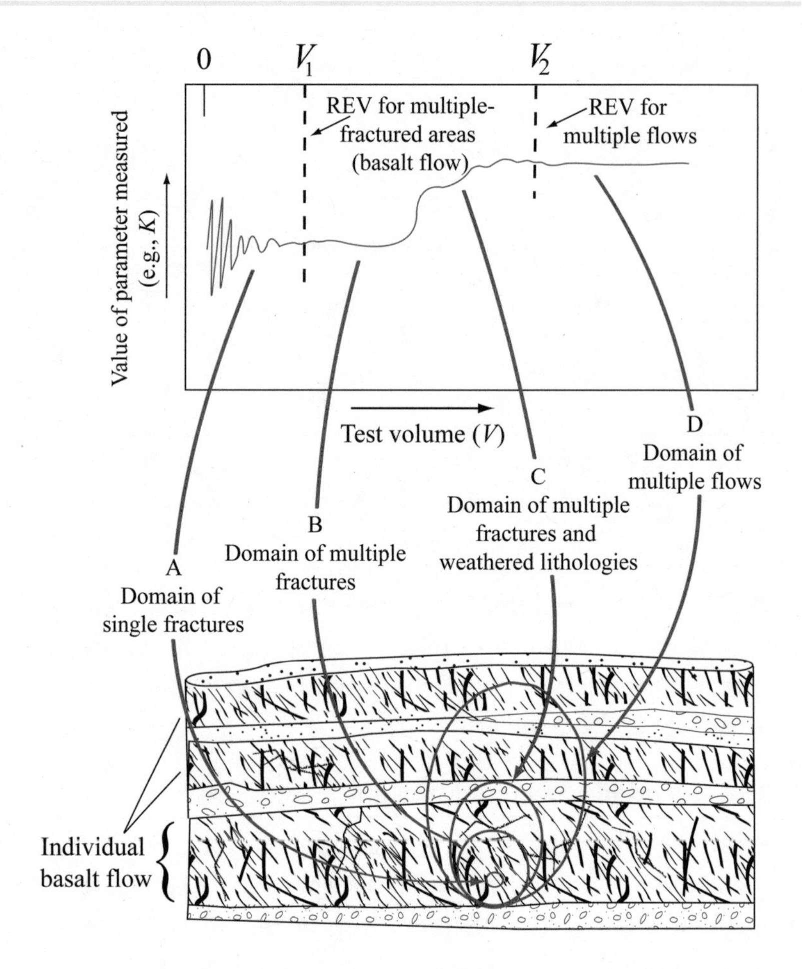

Figure 9.20. Representative elementary volume (REV) as applied to fractured basalt. Modified from the Department of Energy (1986).

a parameter such as hydraulic conductivity (*K*) can behave either homogeneously or heterogeneously depending on the scale investigated. A small sample volume, as demonstrated in domain A, can contain all openings or all solid rock and *K* will vary significantly. As the sample volumes increase, both fractures and solid rock will be included and dependent upon the scale examined, and the variation presumably decreases as the volume increases such that the REV is determined to be V_1 in domain B. But as the scale continues to expand, *K* will again vary as it encompasses both fractured basalt and weathered intermittent lithology deposited in between basalt flows, as seen in domain C. As the scale continues to increase, it again approaches a homogeneous permeability, and a second REV is established as the larger domain includes several basalt flows and discontinuities, as seen in domain D. Hence, a

REV can exist in both a porous and fractured medium. In a fractured medium, however, the continuum approach is scale dependent and may or may not be a feasible based upon the scope of the investigation.

If Darcy equations cannot be applied to a fractured rock domain, then flow must be described through single fractures, joints, or sets of joints or fractures (i.e., ponors). This requires detailed data on fracture or joint orientation, opening, size, and connectivity. In addition, problems of turbulent flow may exist, and hydraulic conductivity may vary within the fracture system.

9.11. Hydraulic Gradients, Boundary-Value Problem, and Direction of Flow

Throughout this discussion, direction of flow has been determined along a gradient from high to low energy values. In physics, a *field* is a physical quantity that has a value for each point in space and time (x,y,z,t) and represents a continuous distribution of vectors, scalars, or tensors described in terms of this special coordinate system and time. For example, on a weather map, the surface wind velocity (a vector quantity has magnitude and direction) can be assigned to each point on the map. The magnitude of the vector is the distance between the two points, and the direction refers to the direction of displacement. A simple example was described in moving a water parcel from point A to point B in Figure 9.15. A scalar field associates physical quantity to every point in space. Examples of scalar fields would include temperature distribution throughout space, pressure distribution in a fluid (such as pressure with depth in Figure 9.9c), and hydraulic head gradients in Figure 9.16.

Hydraulic head distribution within a heterogeneous regional groundwater system can vary spatially and is another example of a scalar field. The flow direction, or hydraulic head (h) gradient, can be determined by placing several piezometers throughout a region. Starting at any point represented by a piezometer, it is probable that the head value at each piezometer increases in some direction x, y, or z and decreases in others. Using a partial derivative, as head is a function of direction, h along the x-axis can be written as $\partial h/\partial x$. Similarly, the gradients of head along the y and

z axes can be expressed as $\partial h/\partial y$ and $\partial h/\partial z$, respectively. As such, Darcy's law can be defined in each direction as

$$q_x = -K\frac{\partial h}{\partial x} \tag{9.47a}$$

$$q_y = -K\frac{\partial h}{\partial y} \tag{9.47b}$$

$$q_z = -K\frac{\partial h}{\partial z} \tag{9.47c}$$

Freeze and Cherry (1979) discuss how analysis of fluid flow and direction through a porous media is based on a *boundary-value problem* involving partial differential equations to describe flow at any point in the system (see section 11.3.1a). They define a boundary-value problem as a mathematical model involving a four-step process:

1. Examine the physical problem.
2. Replace the physical problem with an equivalent mathematical problem.
3. Solve the mathematical problem with accepted techniques of mathematics.
4. Interpret the mathematical results based on the physics of flow.

Typically, when describing groundwater flow through a certain hydrogeological system, the following information is needed:

a. Size and shape or spatial boundaries of the region of flow
b. Equation of flow within the region
c. Boundary conditions at the regional boundaries
d. Initial conditions in the system of interest
e. Spatial distribution of the hydrogeological parameters that control the flow
f. A mathematical method of solution

If the magnitude and direction of flow velocity are relatively constant through time, the system is said to be in *steady state*. If the system of interest is in steady state, then initial conditions (item d) can be removed.

One of the simplest examples of a boundary-value problem is

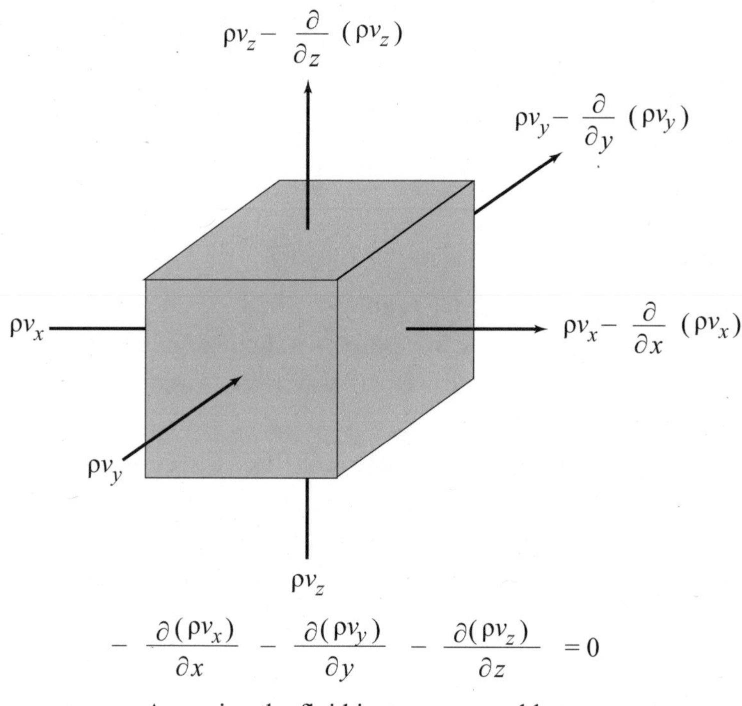

Figure 9.21. Elemental control volume illustrating steady-state flow through a saturated anisotropic porous medium. Rate of fluid mass in equals the rate of fluid mass out.

the *elemental control volume* concept discussed in section 2.3 on mass balance. The law of conservation of mass for steady-state flow through a saturated porous medium requires that the rate of mass flow into the volume is equal to the flow rate leaving the control volume and satisfies the boundary-value problem parameters of (a), (b), (c), (e) and (f) (Figure 9.21). Assuming that the fluid entering into the volume is incompressible, then the equation of continuity can be written as

$$-\frac{\partial v_x}{\partial x} - \frac{\partial v_y}{\partial y} - \frac{\partial v_z}{\partial z} = 0. \tag{9.48}$$

Replacing the parameters of Darcy's law for the velocity and satisfying condition (e),

$$\frac{\partial}{\partial x}\left(K_x \frac{\partial h}{\partial x}\right) + \frac{\partial}{\partial y}\left(K_y \frac{\partial h}{\partial y}\right) + \frac{\partial}{\partial z}\left(K_z \frac{\partial h}{\partial z}\right) = 0 \tag{9.49}$$

which is steady-state flow through a saturated anisotropic porous medium.

For a homogeneous isotropic porous medium where $K_x = K_y = K_z$, K is a constant and the steady flow can be reduced to

$$\frac{\partial^2 h}{\partial x} + \frac{\partial^2 h}{\partial y} + \frac{\partial^2 h}{\partial z} = 0. \tag{9.50}$$

This is known as the *Laplace equation*, which can be used to describe hydraulic head (h) at any point in a three-dimensional flow field. The solution of equation (9.50) describes flow of fluid along the x, y, and z axes where the $K_x \neq K_y \neq K_z$.

Since flow does not just occur in just three directions a scalar field is applied to determine the rate of change or gradient of head h in any direction with the most important being the maximum rate of change at any point in the system. The gradient (∇) h for the scalar field can be expressed as

$$grad\, h = \nabla h = i\frac{\partial h}{\partial x} + j\frac{\partial h}{\partial y} + k\frac{\partial h}{\partial z} \tag{9.51}$$

where i, j, and k are unit vectors in the x, y and z directions, respectively. The gradient of h (∇h) represents the maximum rate of change of head spatially as referenced by the Cartesian coordinate system in the x, y, and z directions. In the isotropic system where the $K_x = K_y = K_z$, Darcy's law can be defined as a gradient ∇ such that

$$q = -K\,\nabla h. \tag{9.52}$$

When K is equal in all three principle directions, as in equation (9.50), the direction of maximum head change is perpendicular to lines of equal head as it coincides with the direction of groundwater flow.

By using the Laplace equation in a two-dimensional context of an (x,z) coordinate system, one can construct a contoured flow net to describe flow in a certain direction with three or more piezometers in a horizontal plane and then apply the three-point problem for determining strike and dip in structural geology. The flow direction into a lake system can then be ascertained by two strategically placed piezometers monitoring the water table, and

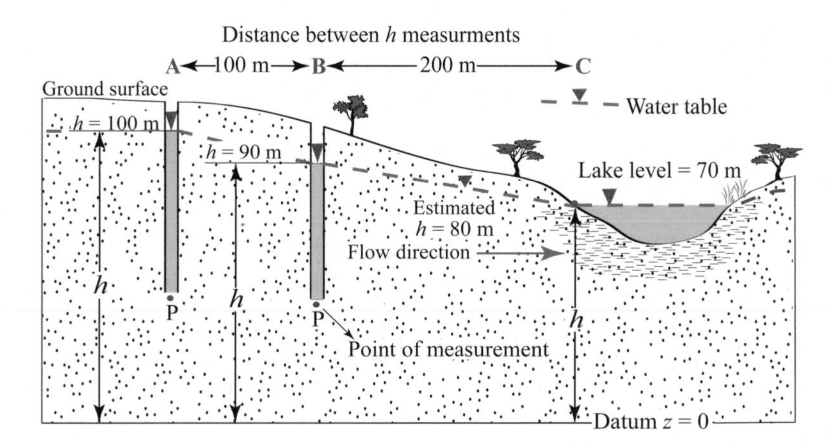

Figure 9.22. Flow direction from two piezometers and lake stage.

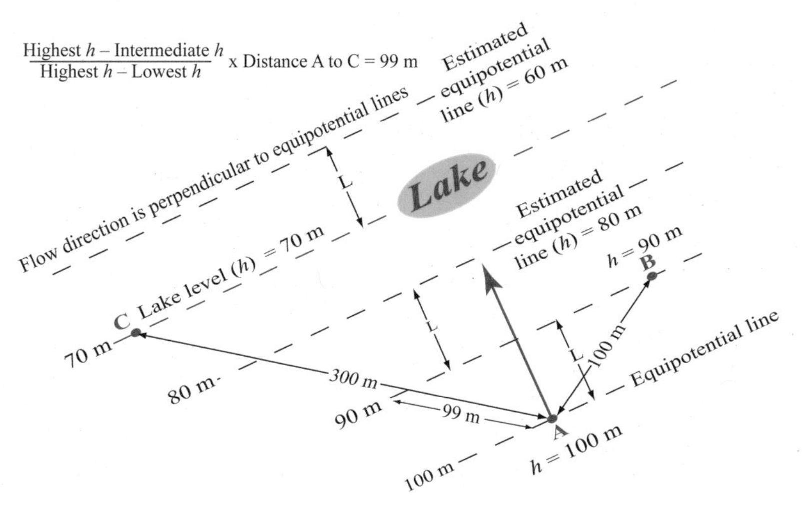

Figure 9.23. Determining flow direction into a lake system using two piezometers, lake level (Figure 9.22), and the three-point problem from structural geology in an areal view.

the lake level can be substituted for the third piezometer (Figures 9.22 and 9.23).

9.12. Field Mapping Equipotential Lines and Flow Nets

As stated earlier, a potentiometric map is a contour map of the potentiometric surface (elevation and pressure head summed) (see Figure 9.17). There are two types of potentiometric maps. The first uses the hydraulic gradients for a confined aquifer system that assumes a relatively constant aquifer thickness (b). The second is for determining flow direction of unconfined or water-table aquifers. The fact that the water table in the unconfined aquifer system represents the upper boundary of the region of flow complicates matters as the cross-sectional area or thickness of this aquifer can vary with the height of the water table.

9.12.1. Potentiometric Mapping for Confined Aquifer Systems

For a confined aquifer system of constant thickness b and in a steady-state condition, the specific discharge through the aquifer at boundary conditions where $x = \text{A}$ to $x = \text{C}$ (e.g., piezometers A and B and lake C in Figures 9.22 and 9.23) can be expressed with Darcy's law in the x direction:

$$q = -K\frac{dh}{dx}. \tag{9.53}$$

Rearranging and integrating over the distance between piezometers in the area of interest (e.g., A to C in Figure 9.24) of the confined aquifer, an expression of hydraulic head h can be attained by

$$dh = -\frac{q}{K}dx \tag{9.54}$$

$$\int_{h_C}^{h} dh = -\frac{q}{K}\int_{0}^{x} dx \tag{9.55}$$

$$h - h_C = -\frac{q}{K}x \tag{9.56}$$

$$h = h_C - \frac{q}{K}x. \tag{9.57}$$

Equation (9.57) states that the head decreases linearly from piezometer A to C. This is demonstrated in Figure 9.24 as the potentiometric surface (blue dashed line) sloping from A to C, establishing flow in this direction as well. When the equipotential lines have been constructed, the direction of flow, or *streamlines*, can be drawn in perpendicular to the equipotentials. Just as in a regular topographic map, the spacing of the contours indicates the slope of the flow, or *potentiometric plane*, within the boundary conditions.

Flow nets are created through a trial and error process, beginning with the boundary conditions as demonstrated in Figures 9.23 or 9.24 with three points of hydraulic head giving a direction of flow on the geographic plane. The boundaries are carefully

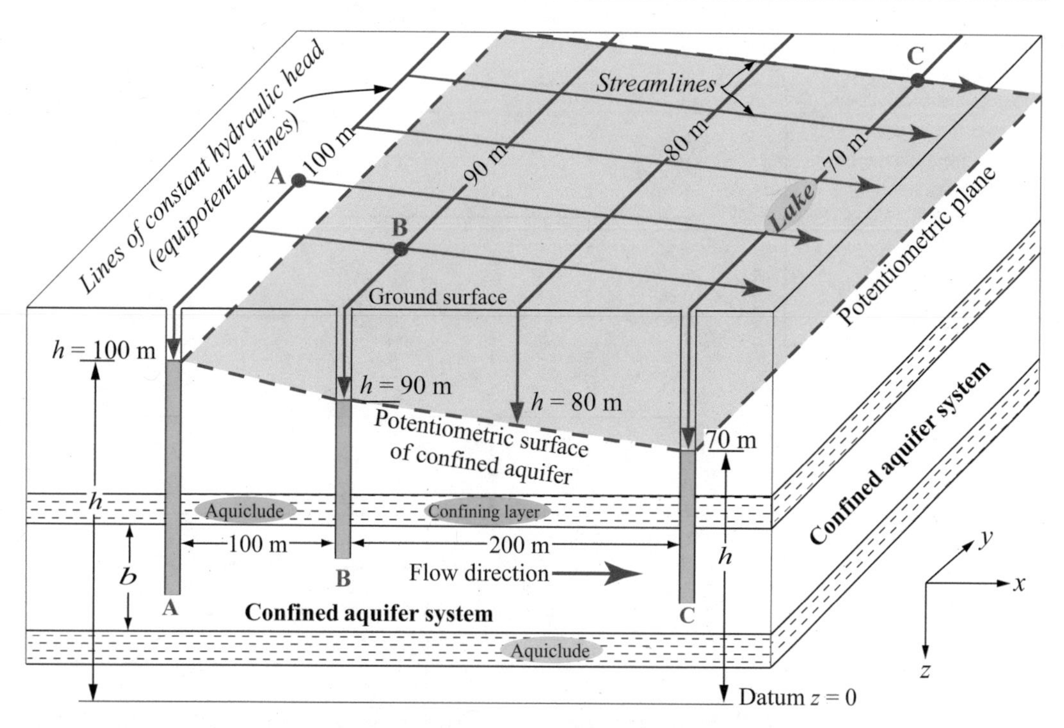

Figure 9.24. Three-dimensional drawing of simple flow net with lines of constant hydraulic head, or *equipotential lines*, in a confined aquifer with constant thickness *b* under steady-state conditions constructed from the three-point problem.

drawn to scale and initial streamlines sketched in on the map. The streamlines should be equally spaced across the width of the flow region and intersect the equipotential lines at right angles, and beginning and ending at equipotentials with at least two of the streamlines forming the boundaries of the flow region.

The final stage is to add the intermediate equipotential lines (Figure 9.24 at 80 meters and Figure 9.25 at h_2, h_4, h_5, and h_6 equipotential lines). They must intersect the streamlines at right angles, including the boundary streamlines, and the bound areas should be approximately square.

Flow nets can be used to estimate groundwater flow through confined aquifers under steady-state conditions, if the hydraulic conductivity K is known. The area between two adjacent streamlines is known as a *streamtube* or *flowtube*. If flow lines are equally spaced, then the discharge through each streamtube is the same. Using the confined aquifer system in Figure 9.24, a two-dimensional flow net in planar view can be constructed (Figure 9.25). Consider the flow through area IV in the planar view. If one

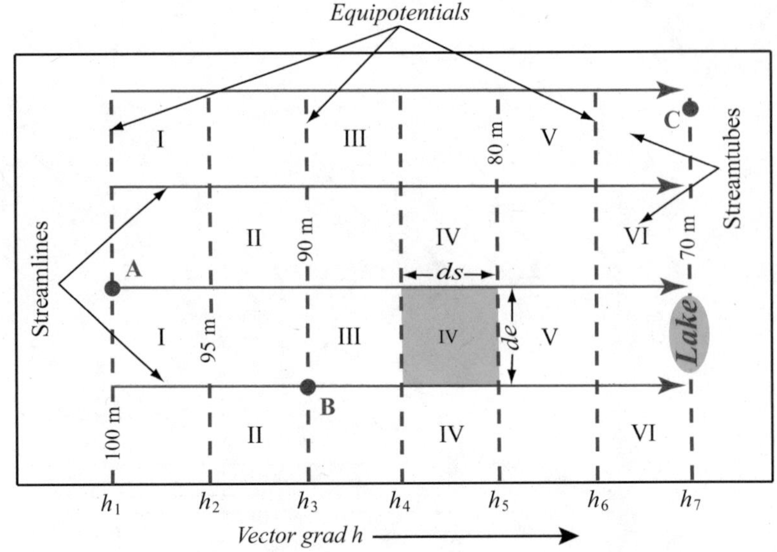

Figure 9.25. Plane view of the aquifer shown in Figure 9.24 showing how flow nets can be used to quantify groundwater flow in the area bounded by wells A, B, and C.

of the squares (e.g., IV) is isolated with sides *ds* (parallel to the streamlines) and *de* (parallel to the equipotentials), then Darcy's law can be used to calculate flow through the square:

$$Q_{IV} = qA = K(deb)\frac{dh}{ds} \tag{9.58}$$

and because $ds = de$, then

$$Q_{\mathrm{IV}} = Kbdh \tag{9.59}$$

where Q is total discharge through the streamtube $[L^3t^{-1}]$ and dh is the head difference in the square, or in the case of area IV, $h_4 - h_5$. Discharge for the set of flow in Q_{IV} can be determined by number of flow paths bounded by the adjacent pair of streamlines, which in the case of Q_{IV} is 4. Or for the system, summing the number of streamtubes defined under the boundary conditions in Figure 9.25. Q can also be calculated with a "partial" streamtube, such as on the outer boundaries of wells B and C in Figure 9.24 if the boundary conditions are known.

Equipotential lines are curvilinear in most aquifer systems except in the simplest of systems. Figure 9.26 is an example of a map of the potentiometric surface of the Floridan aquifer system in

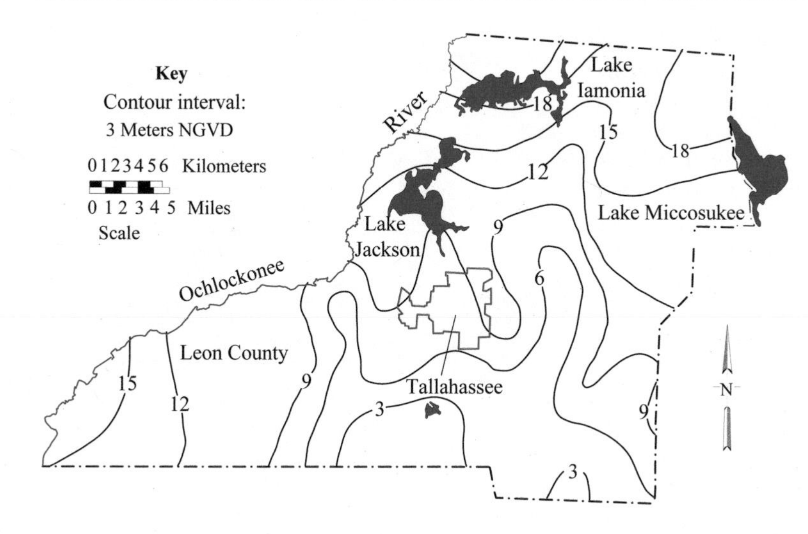

Figure 9.26. Potentiometric surface of the Floridan aquifer system in Leon County, Florida. Modified from Wagner (1989) and Evans (1996).

Leon County, Florida. This is an area with three large lake systems that drain periodically due to their karstic nature and breach of confinement to the underlying aquifers because of ponors in the lake bottom. (The hydrostratigraphy of Lake Jackson in Figure 9.2a depicts the general hydrogeology of the area.)

9.12.2. Flow through an Unconfined Aquifer

When trying to estimate flow into a lake system, the water table in the unconfined aquifer system represents the upper boundary of the region of flow, and the cross-sectional area or thickness of this aquifer can vary with the height of the water table. Figure 9.27a depicts steady-state flow through an unconfined aquifer system bound by an underlying confining lithology and a lake system to the right. The cross-sectional area of Figure 9.27a decreases from left to right, and based on continuity (as demonstrated by Bernoulli's equation and Darcy's law), we know that the hydraulic gradient and flow increases in the direction of flow.

Dupuit (1863) solved this problem by assuming that the hydraulic gradient is equal to the slope of the water table (dh/dx) in the x direction (or y direction, dh/dy) between h measurements, with no flow in the vertical, or z, direction. Similar to flow nets in a confined aquifer system, the streamlines are horizontal and the equipotential lines are vertical. Applying Darcy's law where q' is specific discharge through an unconfined system,

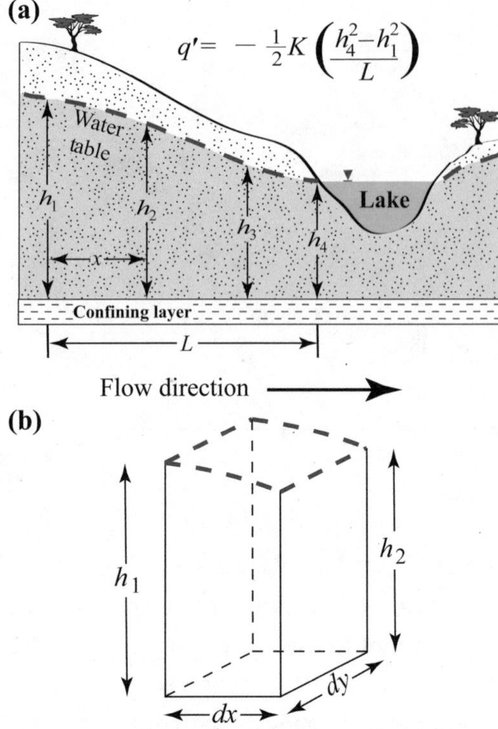

Figure 9.27. Steady-state flow through an unconfined aquifer system from h_1 to h_4 bordering a lake (a), and flow through an elemental control volume bound by a water table and underlying confining unit between h_1 and h_2 (b).

$$q' = -K\frac{dh}{dx}. \tag{9.60}$$

Setting up the boundary conditions based on Dupuit's assumptions then, the saturated thickness of the aquifer where $x = 0$ is h_1. At x = L then $h = h_4$. Integrating equation (9.60) with Dupuit's boundary conditions,

$$\int_0^L q'\,dx = -K\int_{h_1}^{h_4} h\,dh. \tag{9.61}$$

Integrating and substituting for x and h then

$$q'L = -\frac{1}{2}K\left(h_4^2 - h_1^2\right) \tag{9.62}$$

and rearranging gives the *Dupuit equation* of

$$q' = -\frac{1}{2}K\left(\frac{h_4^2 - h_1^2}{L}\right). \tag{9.63}$$

Total flow through an unconfined aquifer can be estimated in the x and y directions using the concepts of continuity of flow through a control volume (Figure 27b). Flow through the left face of the prism in the x direction where dy is the width of the prism face can be found by

$$q'_x dy = -K\left(h_1 \frac{\partial h}{\partial x}\right)_x dy. \tag{9.64}$$

Flow through the right prism face is

$$q'_{x+dx} dy = -K\left(h_2 \frac{\partial h}{\partial x}\right)_{x+dx} dy. \tag{9.65}$$

The change in flow rate between h_1 and h_2 in the x direction can be determined by

$$(q'_{x+dx} - q'_x)dy = -K\frac{\partial}{\partial x}\left(h\frac{\partial h}{\partial x}\right)dxdy. \tag{9.66}$$

Using the same process flow in the y direction, one can determine flow volumes and change in flow rate between h values, as such

$$(q'_{y+dy} - q'_y)dx = -K\frac{\partial}{\partial y}\left(h\frac{\partial h}{\partial y}\right)dydx. \tag{9.67}$$

9.12.3. Flow Maps Typical to Lake Systems Bounded by Unconfined Aquifer Systems

Examples of flow nets (in planar and cross-sectional views) common to lake systems are recharge lakes, where the lake is a source to the unconfined aquifer (Figure 9.28); discharge lakes, where the lake acts as a sink for the unconfined aquifer (Figure 9.29); and flow-through lakes, where the flow is in one direction (Figure 9.30). Streamtubes are mostly curvilinear and are “square” only in the simplest aquifer systems.

These are very basic examples of lake and aquifer systems. Hence, simplifying assumptions—such as the aquifers being homogenous, isotropic, and infinite in extent—must be made to solve for the flow equations given in the previous section. If enough simplifying assumptions are made, the descriptive equation can be solved analytically. However, this simplification can diminish the correctness of the system being investigated. Incor-

Figure 9.28. Flow nets of a recharge lake system (planar view and cross-sectional views).

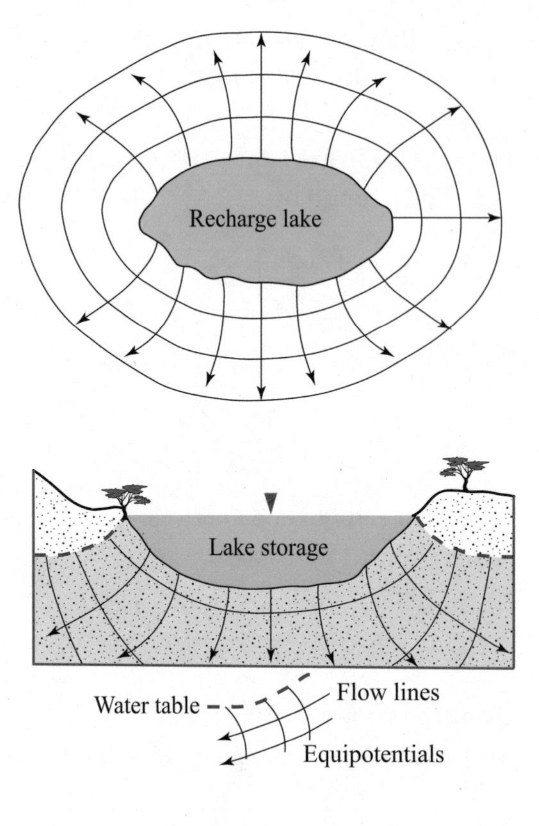

Figure 9.29. Flow nets of a discharge lake system (planar view and cross-sectional views).

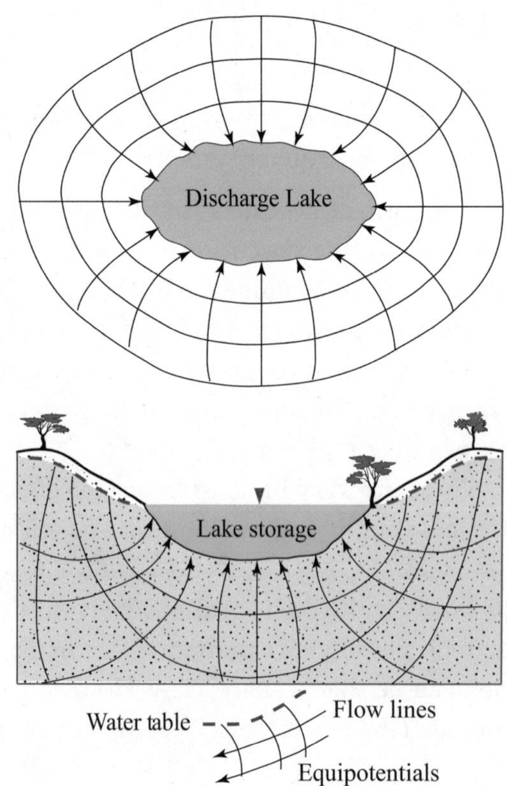

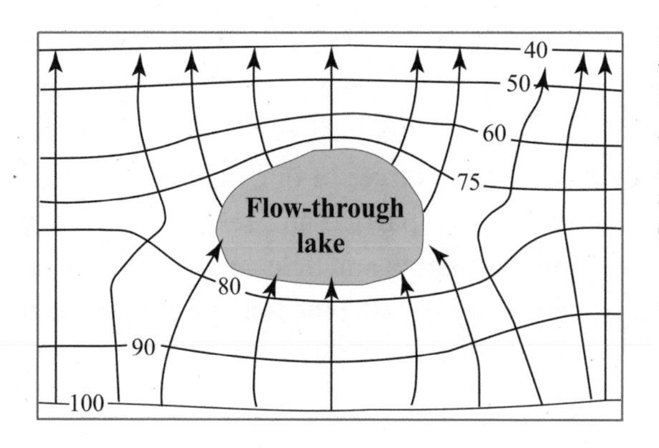

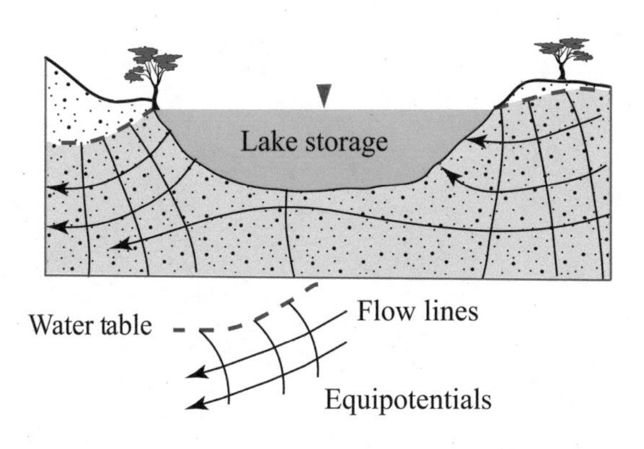

Figure 9.30. Example of flow nets of a flow-through lake system (planar view and cross-sectional views). Equipotential numbers are meters above National Geodetic Vertical Datum.

porating more complex equations into a model is often arduous and can prove to be a difficult solve.

9.13. Summary

When considering lake-groundwater interactions, the complex nature of subsurface conditions can rarely be described accurately by simple mathematical expressions. With the advent of the computer, numerical methods can be employed and are especially useful for analyzing complex lake-aquifer interactions with irregular boundaries; multifaceted hydrogeology; and impacts due to variable pumping rates, recharge, or the effects of karst or fractured geology.

Basic lake-groundwater interactions are discussed in more detail in chapter 10 (Lake Seepage). And chapter 11 systematically defines lake mass balance and hypsometry as well as numerical methods and other complex modeling concepts used to describe lake-aquifer systems.

This chapter has presented a large variety of topics on groundwater flow and aquifers as they apply to lake systems. Each section is a discipline unto itself and thus this chapter is meant only as an introductory overview on lake hydrodynamics and groundwater interaction. For more information, the references provided at the end of the text are all excellent sources of information depending upon the topic of interest. In particular, Freeze and Cherry (1979), Fetter (1988), and Domenico and Schwartz (1990) are all classic hydrogeology textbooks covering the topics herein as well as many other related subjects in more detail.

CHAPTER 10

Lake Seepage

10.1. Introduction

King (1899), in his theoretical field analysis regarding topographical and gravity influences on groundwater flow, noted that the contours of the groundwater level are replicated by the surface topography and that water moves from high areas to low areas. If nothing else is known about the hydrogeology of an area, from the topographical profile one can infer basic information on the uppermost portion of the water table aquifer and lake system, including probable areas of groundwater divides. The defining area where the surface water of a lake (or wetland) and the underlying unconfined aquifer intersect occurs along an obvious break in slope where the groundwater flow lines merge into the surface water of the lake system (Figure 10.1). Winter (2001) defines this as "an upland adjacent to a lowland separated by an intervening steeper slope."

Flow lines are initially parallel to each other, representing equal flow rates and moving from higher to lower potentiometric surfaces. As the flow lines approach the break in slope at the lake bed interface, they curve upward as flow moves into the lake or curve downward as flow moves out (Figure 10.2). Studies by McBride and Pfannkuch (1975) and Pfannkuch and Winter (1984) reported exponential decreases in hydraulic conductivity K with distance from the lake shoreline in lakes with relative homogeneous and isotropic conditions. The diverging flow lines indicate that rate of flow per unit area is decreasing (Winter, LaBaugh, and Rosenberry 1988). However, lakes and wetlands can form under a variety of conditions that include heterogeneous and anisotropic hydrogeology, and seepage patterns, if any, can vary spatially and temporally.

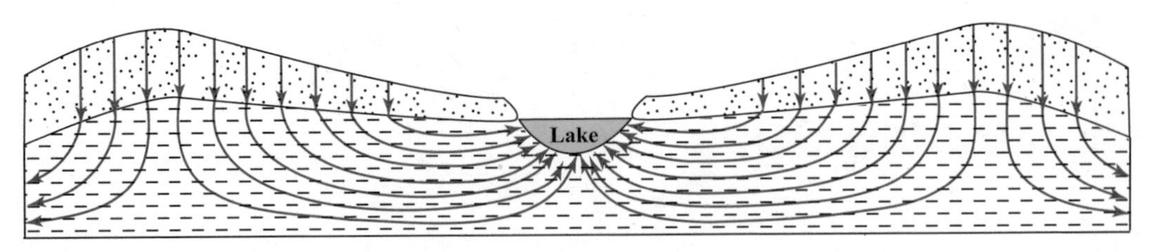

Figure 10.1. Groundwater flow in a watershed. Modified from King (1899).

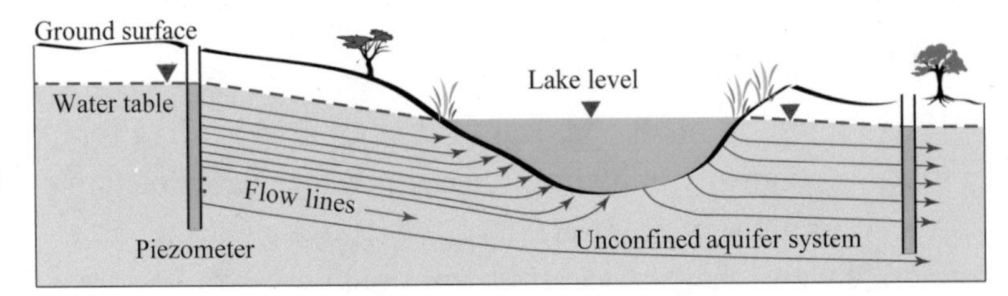

Figure 10.2. Seepage between an unconfined aquifer system and lake (or wetland) with an ideal isotropic and homogenous lithologic matrix.

10.2. General Lake-Groundwater Interactions

Regional hydrogeology strongly influences lake systems such that most large permanent lake systems are often discharge areas for regional groundwater systems (Stephenson 1971). However, lakes can form and are influenced by a wide variety of hydrogeological and environmental conditions. During a rainfall event, for example, a lake may be a temporary source of *recharge* to groundwater systems as the surface water runoff into the lake causes the lake stage to rise higher than the water table and the general flow direction will be away from the lake (Figure 10.3a). As the rain enters the soil and recharges the surficial aquifer system, the water table near the lake rises above the lake stage (Figure 10.3b), the flow direction reverses, and the surficial aquifer system again discharges to the lake. Therefore, a lake can be both a recharge or discharge system in response to seasonal or environmental changes.

With the advent of numerical modeling (see section 10.4), the complex nature of groundwater and lake interaction has been further examined. Winter (1976) used finite-difference modeling (section 11.3.1b) to investigate lake and groundwater interaction under a variety of hydrologic conditions. For example, Figure 10.4 demonstrates the effects of both a low (a) and high (b) hydraulic conductivity aquifer on lake-groundwater flow patterns.

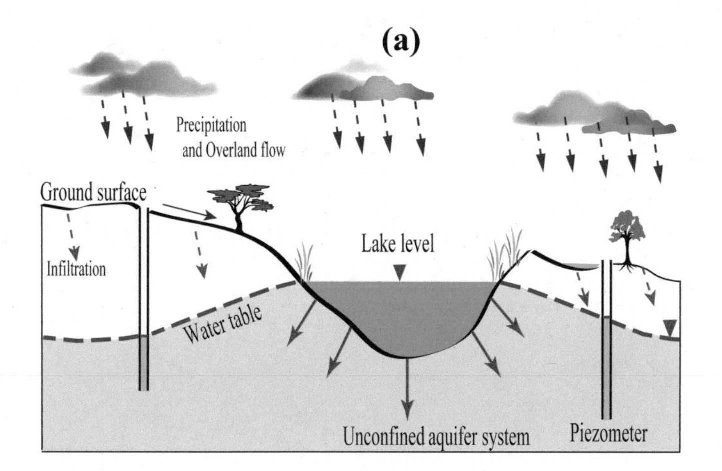

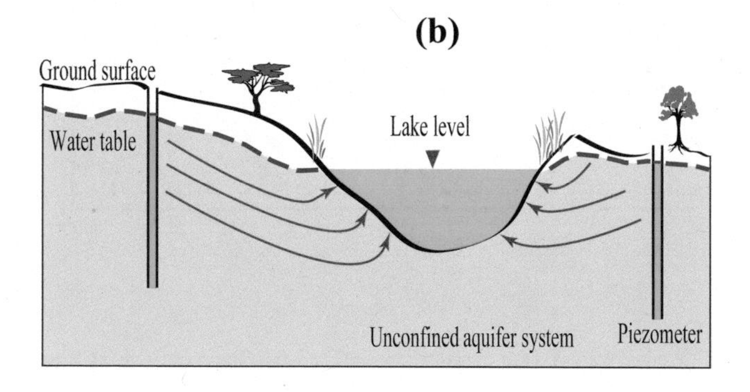

Figure 10.3. Flow direction due to a rainfall event goes from lake to aquifer (a) until the water table rises above the lake stage, causing flow to reverse direction (b).

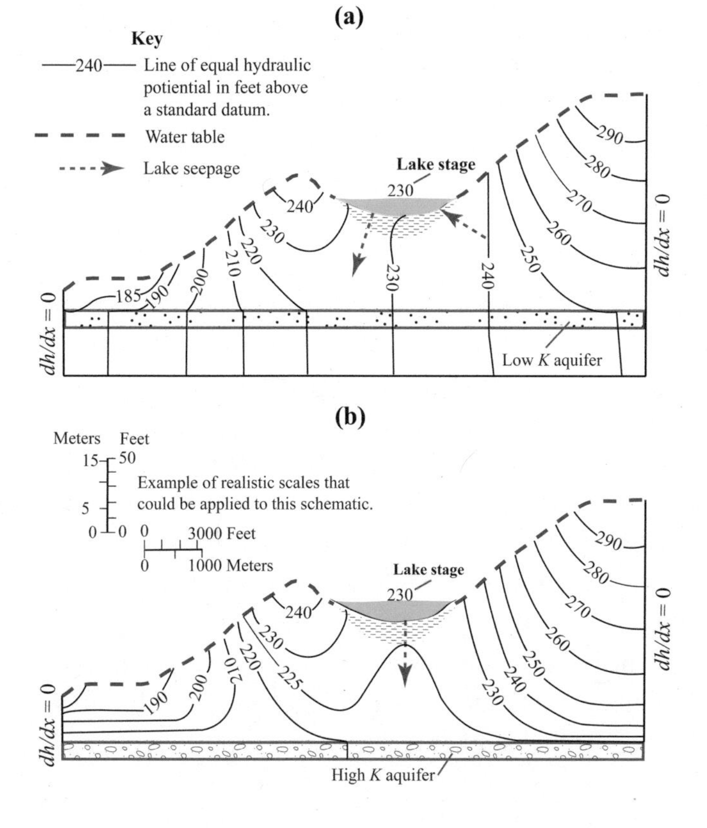

Figure 10.4. The effect of aquifers with low (a) versus high (b) hydraulic conductivity on lake and groundwater flow patterns. Modified from Winter (1976).

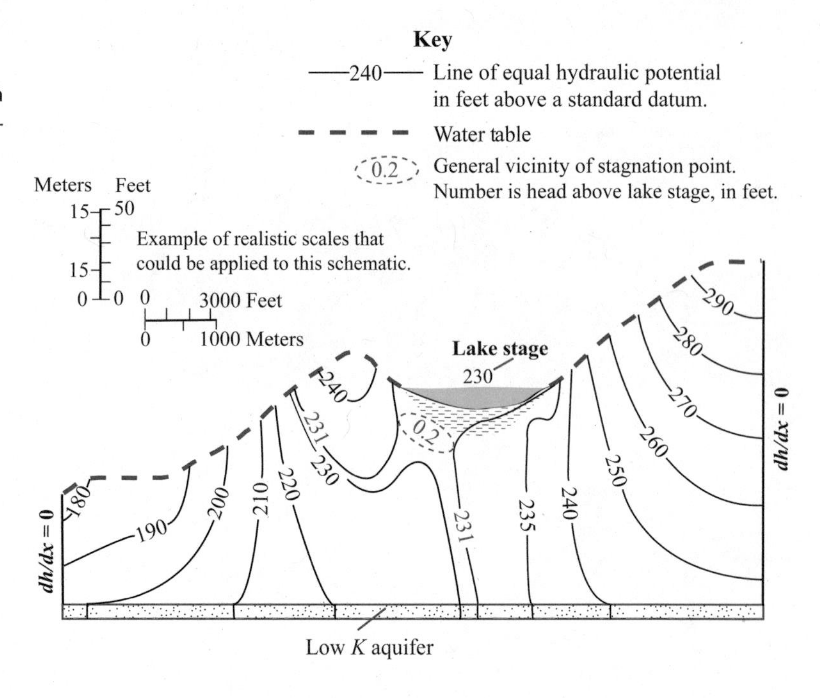

Figure 10.5. Example of a stagnation zone and hydraulic head distribution in a single-lake system. Modified from Winter (1976).

Toth (1963) observed the presence of *stagnation zones* at the juncture of flow systems when modeling regional groundwater systems. A stagnation zone is a point of minimum head along a groundwater subsurface divide separating one flow system from another (Figure 10.5). This zone is a function of head distribution and occurs where the vectors of flow are equal in reverse directions and therefore cancel each other, creating an area of no flow.

10.3. Determining Seepage

Seepage rates [Lt^{-1}] through lake bottom sediments and ponors are generally estimated by two methods. The first is to approximate seepage for the entire lake by subtracting the *potential evapotranspiration* (PET) from the change in lake levels over time (*dh/dt*). The second is to measure the hydraulic conductivity or seepage of lake bottom sediments. However, depending on the quality (e.g., pan location and methodology, etc.) and complexity of PET measurements, measuring seepage directly will result in better data. For example, two common estimates of seepage are using lake bottom seepage meters and/or collecting sediment cores and measuring the hydraulic conductivity in the laboratory using falling-head permeameters or other similar techniques.

10.3.1. Falling-Head Permeameter Hydraulic Conductivity

Hydraulic conductivities of the lake bottom sediments can be estimated by collecting lake bottom cores and analyzing the cores using a falling-head permeameter. Other techniques, such as a hydraulic potentiomanometer (Winter, LaBaugh, and Rosenberry 1988), can be used to estimate seepage. In the falling-head test, the head, as measured in a pipe of cross-sectional area a, can fall from H_0 to H_1 during time t (Figure 10.6).

The hydraulic conductivity (K) is calculated from

$$K = \frac{aL}{At} \ln\left(\frac{H_0}{H_1}\right) \mathrm{T_c} \tag{10.1}$$

where a is the area of the standpipe, A is the area of the sample, L is the length of the sample, and $\mathrm{T_c}$ is a viscosity correction factor derived from the water temperature [T] (Klute and Dirksen 1986).

Measuring seepage directly using seepage meters is preferred to using cores to estimate hydraulic conductivity as the process of core collection, subtracting PET from lake-level data, and similar methods will often result in additional error components. For example, disturbance of the sediment characteristics during coring

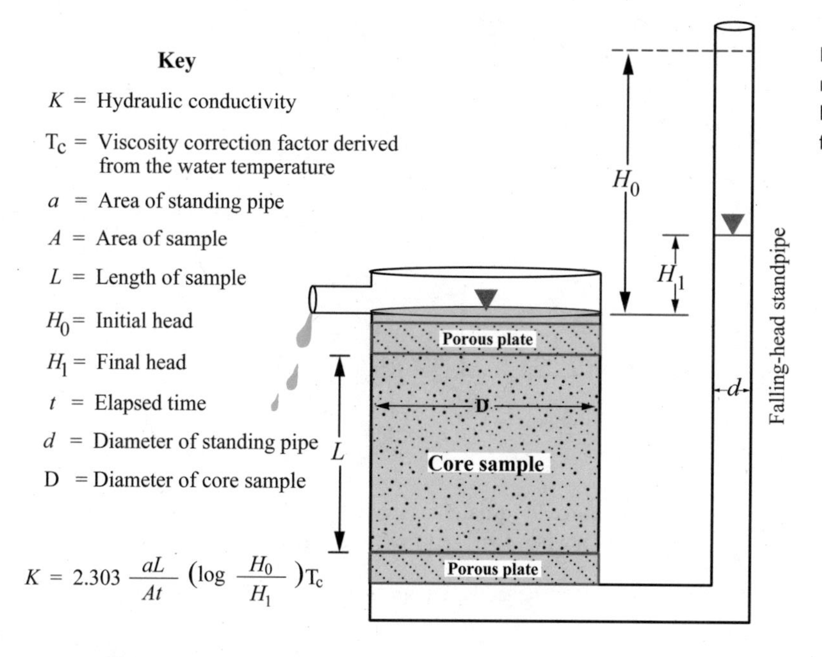

Figure 10.6. Measurement of saturated hydraulic conductivity (K) with a falling-head permeameter.

or increased flow along sides between the core and the falling-head apparatus can occur when indirectly measuring seepage in a laboratory.

10.3.2. Seepage Meters

The basic concept of a seepage meter is to enclose and isolate an area of the lake sediment to surface water interface with a cylinder that is open at its base and vented at the top to an enclosed flexible bladder with a known volume (Figure 10.7). A gain in water volume in the bladder indicates that flow is occurring from groundwater to surface water, while a loss indicates that flow is occurring from surface water to groundwater.

The change in the volume of water in the bladder over a measured time interval and area of the cylinder is used to determine the direction and rate of flow [Lt^{-1}] between surface water and

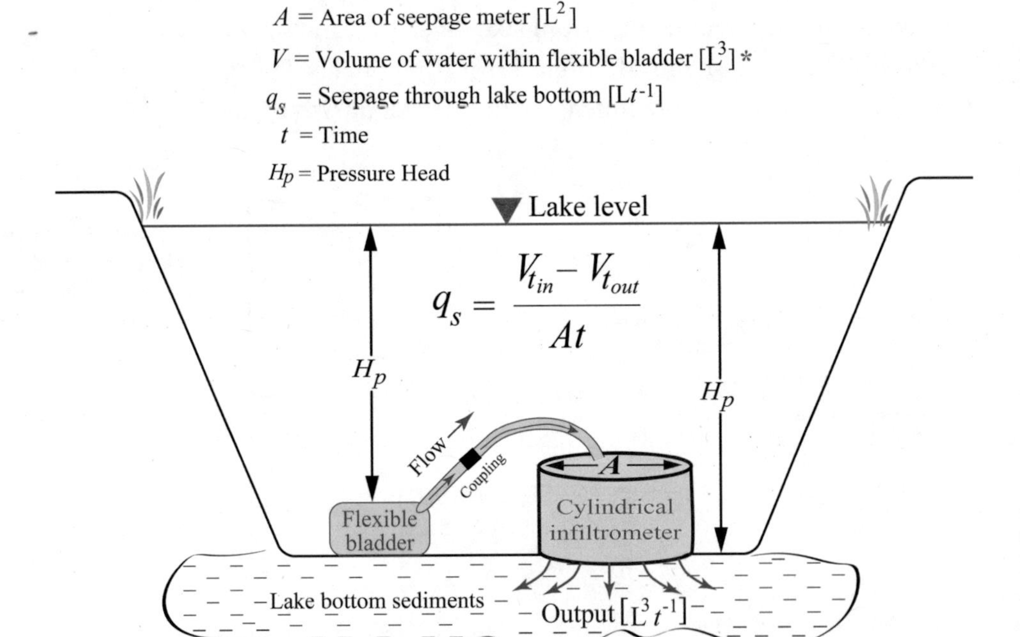

Figure 10.7. Basic design of seepage meter with bladder system. Modified from Evans (1996).

groundwater of the lake bottom lithology. Recall from section 9.6, equations (9.22) and (9.23), that this is a measure of specific discharge ($q = -K(dh/dl) = Q/A$) where K (from equation 9.23) would represent the hydraulic conductivity of the lake bed sediments and Q represents flow through the lake bottom (see section 10.4 for further discussion).

Seepage (q_s) can be empirically measured by

$$q_s = \frac{(V_{tin} - V_{tout})}{At} \tag{10.2}$$

where A is the area of the cylindrical infiltrometer and t is the elapsed time ($t_{in} - t_{out}$) between initial placement and removal of the seepage bladder (Figure 10.7). If the volume at t_{in} is greater than the volume at t_{out}, then seepage is coming into the lake from the underlying aquifer. See section 10.4 for further discussion.

Seepage meters have an advantage over other methods of measuring groundwater to surface water exchange since flow measurements can be made directly without measurements of the hydraulic conductivity of the sediment. Seepage meters are particularly useful when many measurements are needed to define the exchange in different areas of a lake.

10.4. Seepage and Average Linear Velocity

As discussed in section 9.6, seepage as specific discharge is not a true velocity but an apparent velocity representing that at which water would move through the underlying lake lithology as if this matrix were an open conduit (see Figure 9.18). However, water can only move through pore spaces of the matrix through which saturated flow occurs.

To find the velocity that water is essentially traveling, the seepage is divided by the effective porosity (see section 9.9) to account for the actual open space available for flow. For a cross section A, the true velocity can be ascertained representing the rate at which the water moves through the pore spaces by modifying equation (10.2) such that

$$\overline{q_s} = \frac{Q}{n_e A} = -\frac{Kdh}{n_e dl} \tag{10.3}$$

where $\overline{q_s}$ is the *average linear velocity* (also termed seepage velocity) and n_e is the effective porosity.

One of the uses of average linear velocity is to use it as the basis to predict the average linear rate of movement of a solute (dissolved substance) via groundwater that may impact a lake or wetland system. These solutes may be natural constituents, artificial tracers, or contaminates. The process by which solutes are transported is by the bulk motion of flowing groundwater known as *advection*. Because of advection, nonreactive solutes are carried at an average linear velocity of the groundwater. These solute plumes, however, tend to spread out from the projected path due the phenomenon of *hydrodynamic dispersion*, which in general is a result of diffusion (at low velocities; see Fick's equation, section 4.4.1) and mechanical mixing during fluid advection. Hence, equation (10.3) will require an empirical coefficient (n_{ed}) based on lithologic properties of the aquifer:

$$\overline{q_s} = \frac{Q}{n_{ed}A} = -\frac{Kdh}{n_e dl} \tag{10.4}$$

Reactive solutes are substances that undergo chemical, biological, or radioactive change and require further modeling processes to predict travel time and concentrations, all beyond the scope of this text. For further information on solute transport see, Freeze and Cherry (1979), Fetter (1988), and Domenico and Schwartz (1990).

10.5. Construction and Placement of Seepage Meters

Seepage meters can be made from a variety of inexpensive materials and specific designs. Cylindrical infiltrometers can be made by cutting sections from 55-gallon drums (Lee 1977) or welded cylinders whose size can be tailored to the lake study area. Other infiltrometers documented in the literature (Carr and Winter 1980) include inverted plastic trash cans (Rosenberry, LaBaugh, and Hunt 2008), lids from desiccation chambers (Duff et al. 1999), fiberglass domes cemented to a limestone bed (Shinn, Reich, and Hickey 2002), and even galvanized stock tanks (Landon, Rus, and Harvey 2001).

Like infiltrometers, a large variety of types and sizes of the seepage-measuring bladders have been used in seepage studies (Carr and Winter 1980). Bladders as large as fifteen-liter trash bags (Erickson 1981) and four-liter sandwich bags to as small as condoms have been used to measure seepage volume (Fellows and Brezonik 1980; Duff et al. 1999; Isiorho and Meyer 1999; Schincariol and McNeil 2002).

Infiltrometers with larger diameters have the advantage of more accurately measuring lake bottom sediments with low seepage rates and better integrate small-scale spatial variability in seepage flux (Rosenberry, LaBaugh, and Hunt 1980). However, larger infiltrometers can be unwieldy and especially difficult to manage in small boats, and they may be harder to insure a good seal in lake bottoms laden with vegetation, debris, or coarser sediments such as gravel. Hence, infiltrometers should be selected based on a variety of conditions, such as hydraulic conductivity, bottom sediments, available materials, and ease of use.

Evans (1996) used SCUBA (self-contained underwater breathing apparatus) for placing varying-sized cylindrical infiltrometers (one-meter diameter, including coffee cans) on the lake bottom by rotating the cylinders slowly into the substrate to minimize sediment disturbance. The latitude-longitude locations were determined and marked by floating buoys attached to cinderblocks. As recommended by Lee (1977), the elevated venthole and flexible tubing were left open for at least forty-eight hours to allow gas and invertebrates from the sediments to escape. Medical IV bags (500–1000 ml each) were used as flexible bladders and attached by using quick-connect spring-operated valve couplers that automatically shut off flow when the couplers were disconnected. This ensured that no water entered or exited the bladder or infiltrometers during installation or removal. The bladders were secured to the lake bottom to ensure maximum head effect and limit movement.

Rosenberry and LaBaugh (2008) discuss a "memory effect" caused by the manufacturing process that results in a slight pressure created in a plastic bag as it moves into a more relaxed position, which can result in measurement errors. To reduce any potential errors due to the memory effect, the IV-bag bladders were filled with a known volume of water prior to attaching one to each infiltrometer.

Removal of infiltrometers can be problematic if large or placed in adhesive-like clay or silt and left over long periods. In this case, second-stage SCUBA regulators were used to place air into the infiltrometers, allowing the positive buoyancy force to aid in removing the infiltrometer out of the sediments. A similar method of placing a tube under the cylinder and blowing air into the infiltrometer also works to float the infiltrometer.

Case Study 10.1

Methods for Measuring Seepage and Hydraulic Conductivity at Lake Jackson, Leon County, Florida

W.L. Evans, 1996, "Modeling the Hydrodynamics of Closed Lake Basins: A Treatise on Lake Seepage" (master's thesis, Department of Geology, Florida State University).

When investigating the lake mass balance of Lake Jackson, Evans (1996) used a combination of methods for investigating lake seepage. Lake sediment cores were collected for falling-head analysis to determine initial hydraulic conductivity values. Grain size and organic content analyses were completed to determine the occurrence of trending lithology or patterns of sediment distribution. Patterns of plant distribution were also examined to see if specific wetland plants had an affinity for lithologies such as clays or carbonates.

Based on the patterns of hydraulic conductivity (see section 10.6), the bottom of Lake Jackson was contoured and divided into ten sections (Figure 10.8). At least one seepage meter was placed in each section.

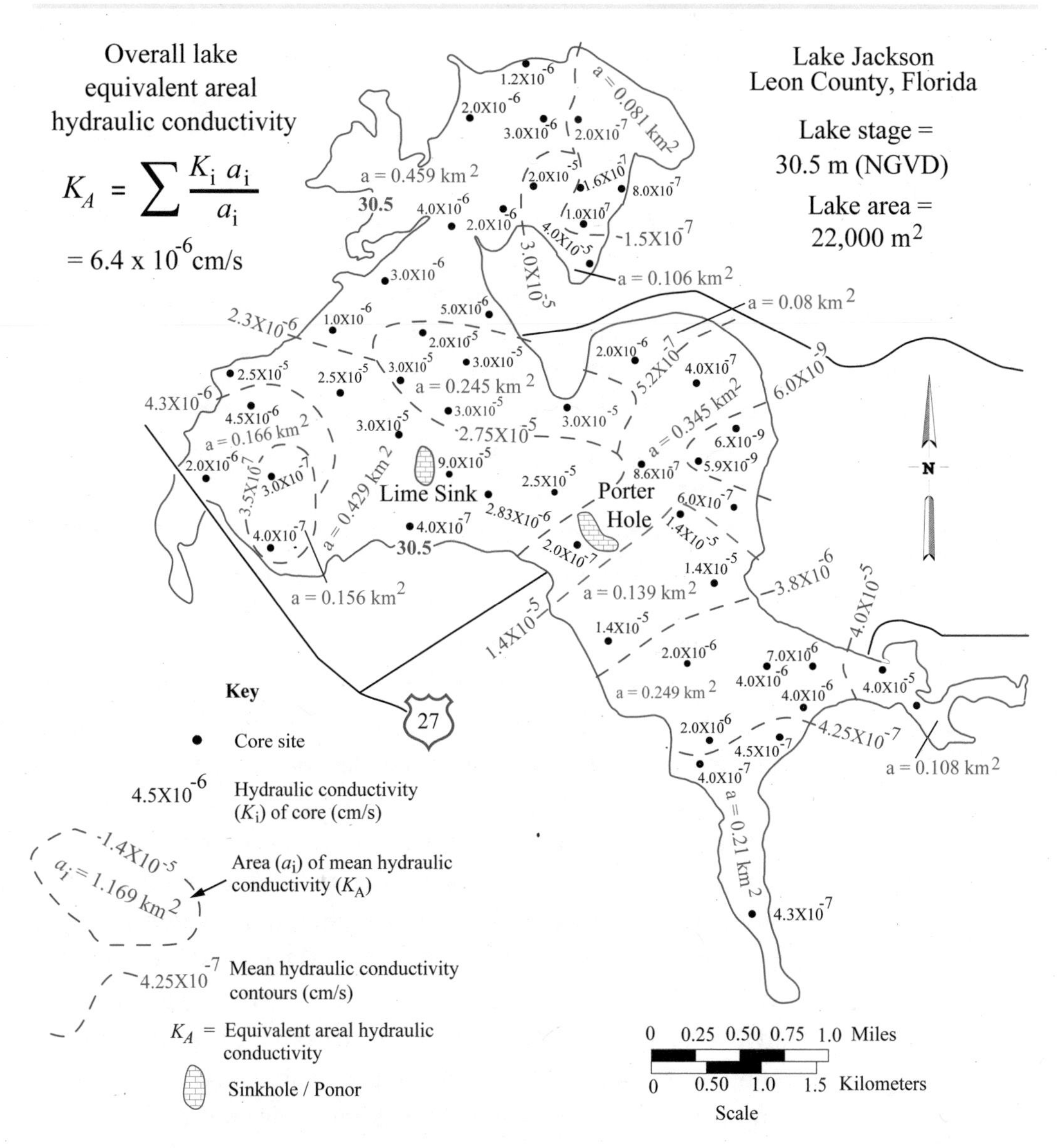

Figure 10.8. Hydraulic conductivity contour map for Lake Jackson, Leon County, Florida, based on falling-head permeameter analysis of lake cores. Modified from Evans (1996).

10.6. Lake Bottom and Hydraulic Conductivity Heterogeneities

As eluded to in Figure 10.8, lake bottom hydrogeology and subsequent hydraulic conductivity (K) can vary both horizontally and vertically in lake systems (see section 9.7). Shoreline erosion from precipitation and runoff events or wave action can result in trending deposition with the coarser sediments near the shoreline and finer sediments deposited away from the edge of the lake.

Fluvial processes, such as streams into or out of a lake system, can result in areas of trending deposition of coarse sediments near the lakeshore and finer silts and clays along the diminishing stream channel into the interior of the lake system. Biological interactions, such as vegetation root systems, bioturbation by benthic invertebrates or feeding birds, and organic deposition, can also impact K as can such anthropomorphic activities as agricultural, dams, urbanization, and aquifer use.

Using a hypothetical example to examine typical heterogeneities of lake hydrogeology that are influenced by stream flow (Figure 10.9), the cross section from A–A′ illustrates that heterogeneities exist both vertically and horizontally. Vertical heterogeneity is known as *layered heterogeneity* and is common in unconsolidated lacustrine and marine deposits.

In this hypothetical lake system, K is greatest within the area of proximal higher-energy coarse stream sediments and becomes progressively smaller toward the finer lower-energy sands, silts, and clays deposited toward the center of the lake. A similar process can occur in closed lake systems during precipitation events and subsequent overland flow into the lake. This is known as *trending heterogeneity*, which is common in sedimentary processes that occur in deltas, alluvial fans, glacial outwash plains, or other depositional environments where there is a trending energy gradient.

Studies by Lee (1977), Erickson (1981), and Attanayake and Waller (1988) have reported exponential (trending) decreases in hydraulic conductivity K with distance from the lake shoreline. In contrast, research by Woessner and Sullivan (1984), Krabbenhoft and Anderson (1986), and Evans (1996) have revealed that lake bottom hydraulic conductivity can vary both vertically and horizontally throughout the lake bottom area without any apparent trends. Thus, understanding of lake bottom sedimentation and hydrogeology is crucial in determining lake behavior and mass balance and must be investigated on a lake-by-lake basis.

As seen in Figure 10.8, one method for estimating lake bottom K is to collect cores to examine the lake bed geology (e.g., grain size analysis) and the nature of the system heterogeneity (Figure 10.9). The cores can be tested for K by using a Darcy-like falling-head permeameter (section 10.3.1). Seepage can be estimated directly

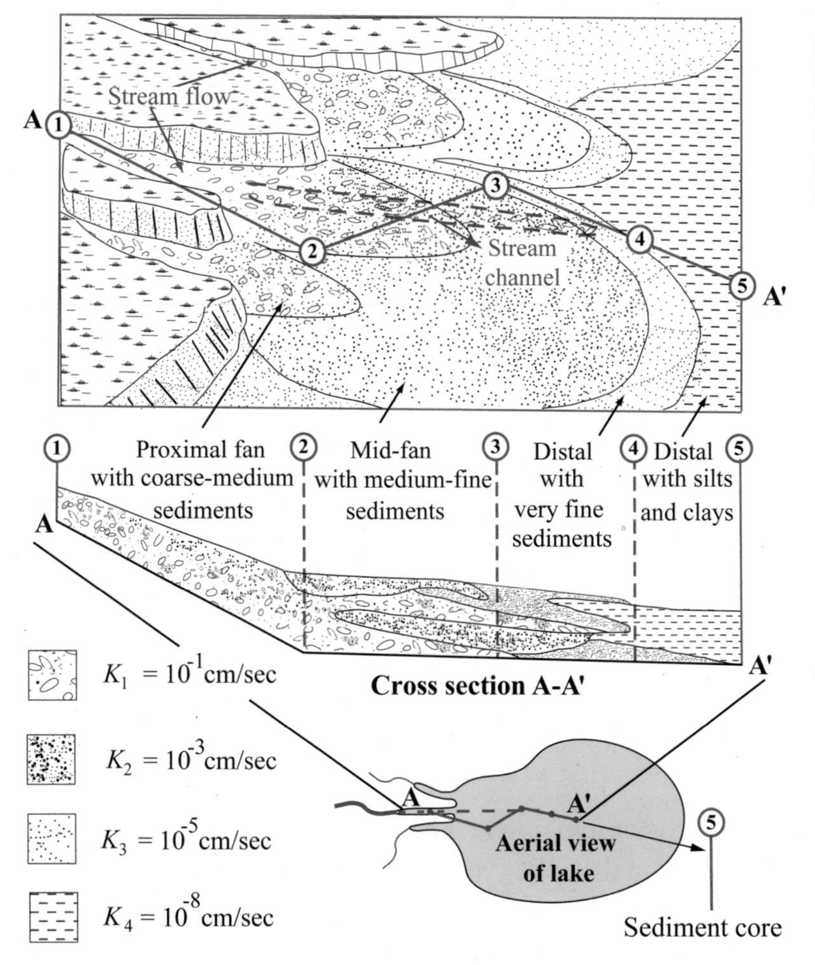

Figure 10.9. Hypothetical stream-fed lake system demonstrating vertical and horizontal heterogeneity in deposition of sediments.

by employing seepage meters that, based on the core data, can be strategically placed on the lake bed.

Equations (10.5) and (10.6) take a heterogeneous and anisotropic system and provide K_z and K_x values for a system analogous to a homogeneous but anisotropic formation (Freeze and Cherry 1979). From the K values a statistical distribution can be established and an overall hydraulic conductivity average (e.g., effective hydraulic conductivity) can be assessed for the lake bed. By taking core 3 as an example (Figure 10.10), one observes that the vertical and horizontal lithology varies over three depositional environments. Each depositional sequence (i) has its own K_{i} value that varies over a range of four orders of magnitude (between K^{-1} to K^{-5}). Using a statistical average approach, Freeze and Cherry

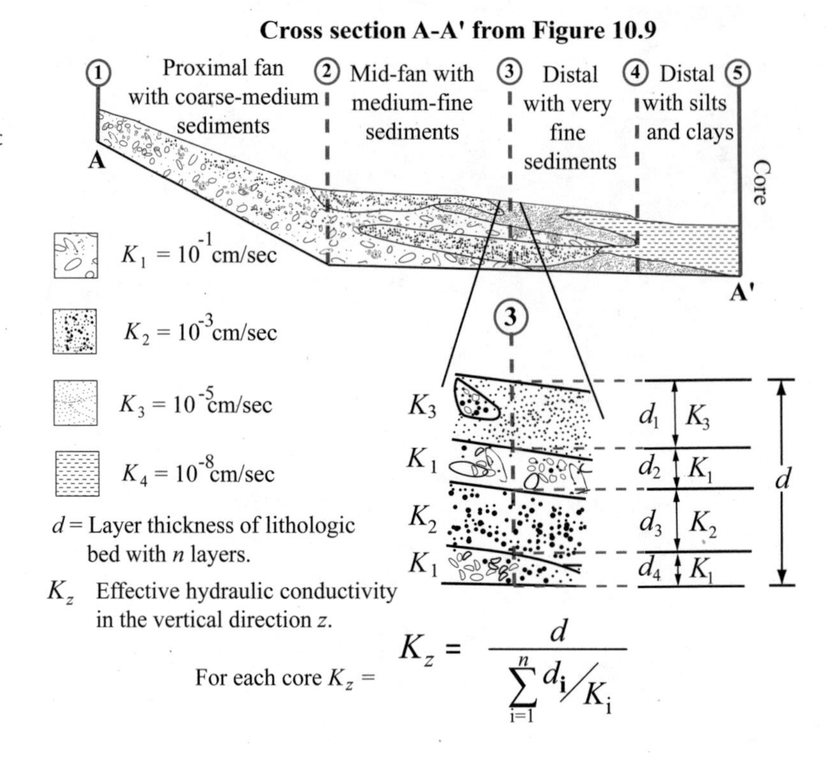

Figure 10.10. Determining effective vertical (z-axis) hydraulic conductivity for a heterogeneous anisotropic hydrogeologic system.

(1979) provide a solution for determining effective K in both the x,y (horizontal) and z (vertical) direction by examining the K values for each lithologic layer (K_i) with layer thickness d_i and n number of layers within the total hydrogeologic unit. For vertical hydraulic conductivity

$$K_z = \frac{d}{\sum_{i=1}^{n} \frac{d_i}{K_i}}. \tag{10.5}$$

Horizontal conductivity, on the other hand, is a measure of the areal distribution of K over the lake bed for the overall lake area and can be estimated as

$$K_A = \sum_{i=1}^{n} \frac{K_i a_i}{a_i} \tag{10.6}$$

where a_i is the contoured areas of K as determined by cores and K_z values from equation 10.1 (Figure 10.11).

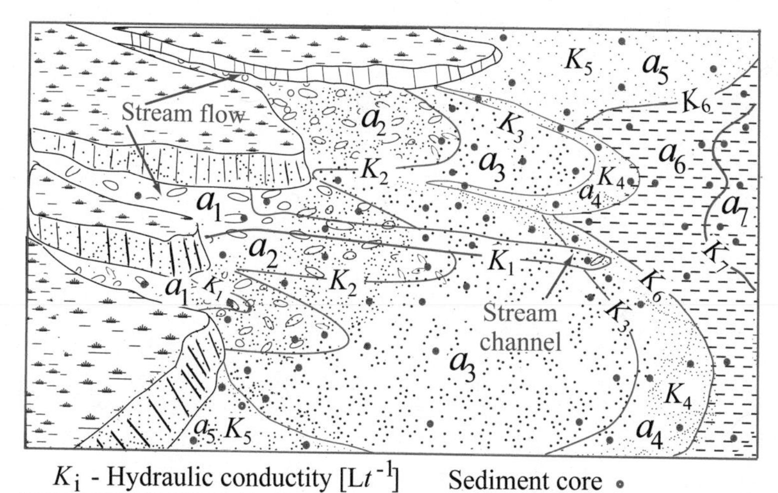

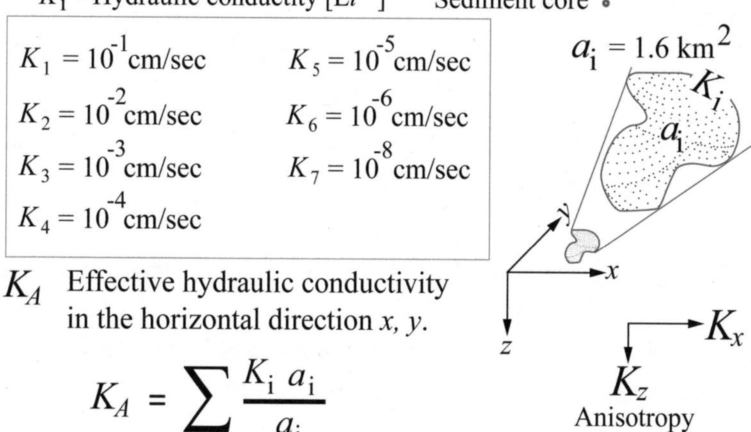

Figure 10.11. Determining effective horizontal (x-axis) or areal hydraulic conductivity for a heterogeneous anisotropic hydrogeologic system.

Case Study 10.2

Hypsometric Effects and Lake Bottom Hydraulic Conductivity Modeling of Lake Jackson, Leon County, Florida

W.L. Evans, 1996, "Modeling the Hydrodynamics of Closed Lake Basins: A Treatise on Lake Seepage" (master's thesis, Department of Geology, Florida State University).

Lakes in Florida display a range of water-level fluctuations in association with wet and dry periods. Many of these lakes originated from karstic processes; that is, they were formed by the dissolution of limestone with a subsequent collapse of overlying sediments forming solution basins.

Between periods of high and low rainfall, the range of fluc-

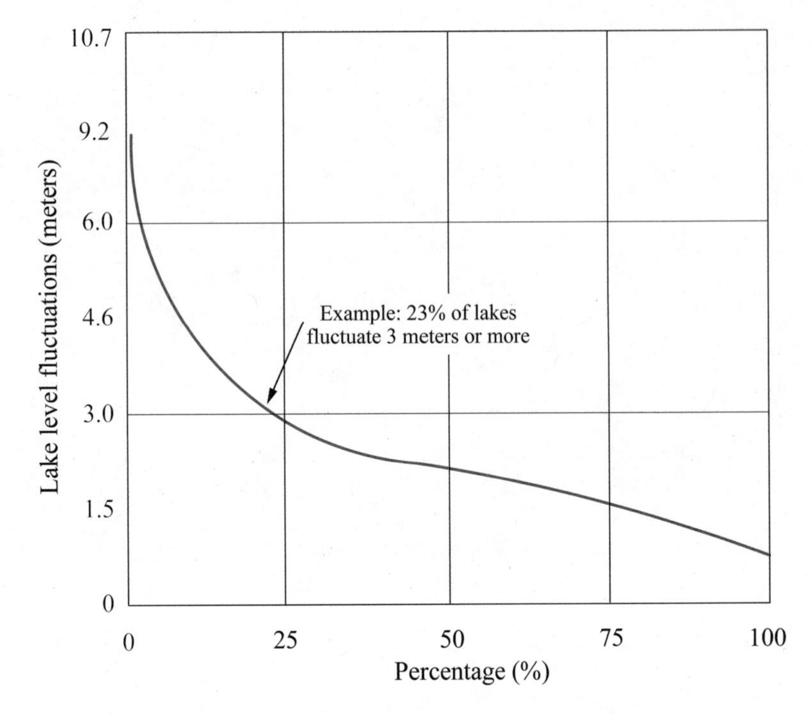

Figure 10.12. Percentage of lakes in Florida for which maximum fluctuation in lake stage equals or exceeds a given magnitude. Modified from Evans (1996).

tuation in stage (lake level) varies greatly among lakes in Florida (Hughes 1974a). Stages can vary as little as 0.6 meters for some lakes to more than 9 meters for others, with 80% of Florida lakes fluctuating greater than 1.5 meters (Figure 10.12).

One of the primary causes of differences in the magnitude of lake-level fluctuation is the variability in permeability of lake bottom sediments (Hughes 1974a). In general, the rate of seepage varies directly with the hydraulic conductivity of the lake bottom, and the range of lake-level fluctuation increases with increasing seepage (Hughes, 1974b). Lakes with large stage fluctuations are often associated with karst formation, where at certain lake stages karst seepage dominates.

Lake systems formed in conjunction with karst processes can contain remnant sinkhole features that become dormant over time as they fill with sediments or collapse. These features can periodically become active due to continued dissolution of the associated carbonate lithology. Depending on the potentiometric surface of underlying aquifers, lake drainage through ponors or dramatic stage rises due to spring-flow can occur (Figure 10.13).

A good example of this is Lake Jackson in Leon County, Florida, which has a history of periodic drainage due to the reactivation of

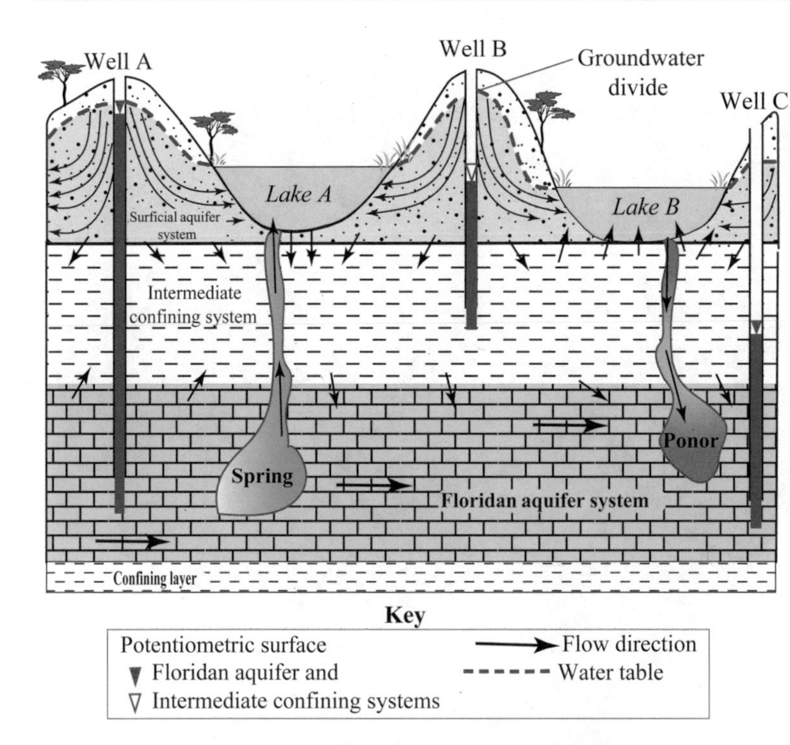

Figure 10.13. Seepage effects of confined and unconfined aquifers on karstic lake basins containing active karst features, such as springs or ponors. Modified from Evans (1996).

Figure 10.14. Southeast view of Porter Hole and dry lake bed at Lake Jackson, Leon County, Florida. Photo courtesy of Florida Geological Survey.

one or two existing sinkholes/ponors (Porter Hole and Lime Sink). The latest drainage event was well documented and occurred in September 1999 when Porter Hole opened and drained the entire lake within days (Figure 10.14).

Porter Hole was large enough for Florida Geological Survey

Figure 10.15. Block diagram of Porter Hole sink. Courtesy of the Florida Geological Survey.

(FGS) geologists to access and map it (Figure 10.15). The upper portion of the ponor is approximately 2.44 meters (8 feet) wide and 2.74 meters (9 feet) deep. At a depth of 2.74 meters, the sink narrows to about 0.46 meters (1.5 feet) wide and 1.72 meters (4 feet) deep, developed along fractures or joints in the Torreya Formation carbonates. Below the restriction, a lower cavity opens into a linear cave honeycombed with fractures and conduits extending 6.1 meters (20 feet) laterally. This ponor becomes filled with sediments and eventually gets plugged over time.

After Lake Jackson stage recovered in 2001, flow through this system was periodically monitored using a flow meter and SCUBA by this author and other FGS staff, and flow varied between 68,191 cubic meters per day (15 million gallons per day) to 59,000 cubic meters per day (13 million gallons per day) (Balsillie 2003).

The distribution of hydraulic conductivity of the lake bottom varies due to the variation of lake bottom sedimentary parameters, such as grain-size distribution, organic content, and porosity. This condition is analogous to the concept of layered heterogeneity

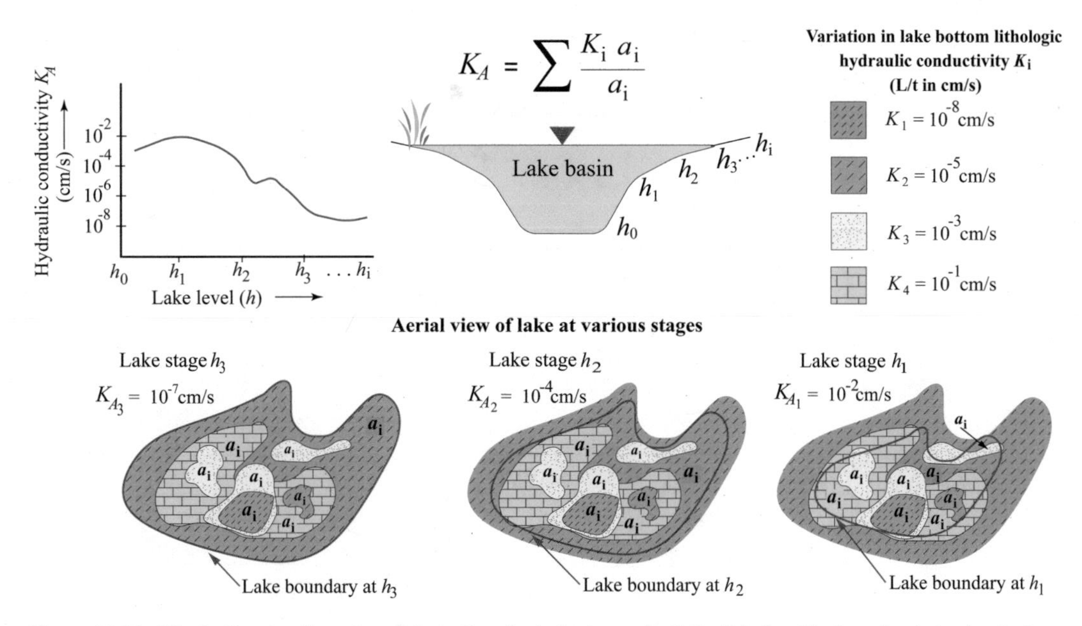

Figure 10.16. Effect of karst sediments on lake bottom hydrologic conductivity K. As karst features begin to dominate at lower lake stages, K increases.

developed by Freeze and Cherry (1979) and discussed in section 10.3 (see Figures 10.6, 10.7, and 10.8); however, in general, for most Florida and lake systems influenced by karst, lake bottom hydraulic conductivity can vary widely with no apparent trends (Evans 1996). This variation is directly observed in the lake bottom conductivity map of Lake Jackson in Leon County, Florida (see Figure 10.11).

Assuming homogeneity and isotropy for each area, the equation from Leonard (1962) for vertical equivalent hydraulic conductivity (equation 10.6 in section 10.6) can be modified and expressed as

$$K_A = \frac{\Sigma K_i a_i}{\Sigma a_i} \tag{10.7}$$

where K_A is the equivalent areal conductivity for the lake bottom, K_i is the homogeneous conductivity of the individual area, and a_i is the size of the individual area. K_A is a measure of overall lake bottom hydraulic conductivity, and as the lake level changes, the individual areas change and K_A changes (Figure 10.16). Hence, K_A is a function of lake hypsometry.

Table 10.1, Figure 10.17, and Figure 10.18 illustrate the hyp-

Table 10.1. Effects of hypsometry on Lake Jackson

Lake stage (meters NGVD)	Lake area (m^2)	Hydraulic conductivity (K_A) cm/sec
A (30.5)	22,000	6.4×10^{-6}
Mean (26.5)	16,160	8.4×10^{-6}
B (25.6)[a]	13,880	8.6×10^{-6}
C (23.8)	7,109	9.8×10^{-5}

Source: Modified from Evans (1996).
[a]At this lake stage, one or more ponors has become active where flow has been estimated at 17 cm/sec, or a volume of 0.57 m^3 sec (13 million gallons per day) in Porter Hole (see Figure 10.15).

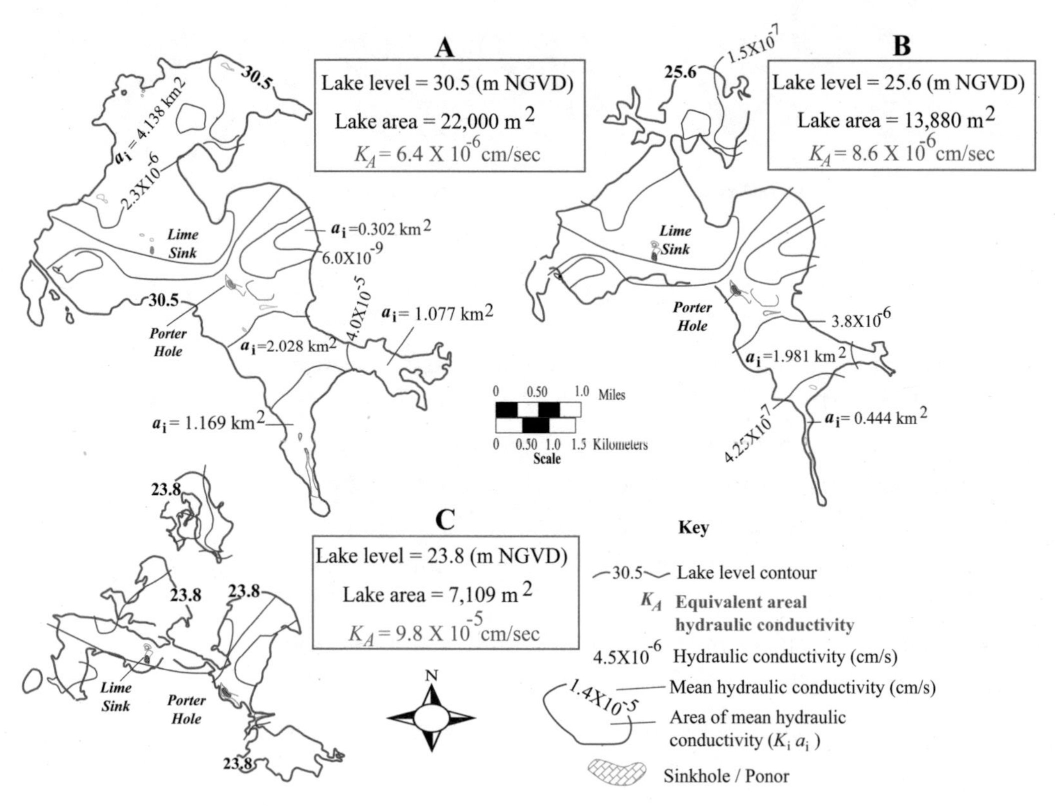

Figure 10.17. Changes in overall equivalent areal hydraulic conductivity (K_A) of lake bottom with changes in lake stage at Lake Jackson, Leon County, Florida. Modified from Evans (1996).

sometric effect on Lake Jackson where the overall K_A varies with lake level and lake area. As seen in Figure 10.17, the lake stage and lake area can vary dramatically due to the periodic activation of either Lime Sink or Porter Hole ponors, resulting in drainage of this lake.

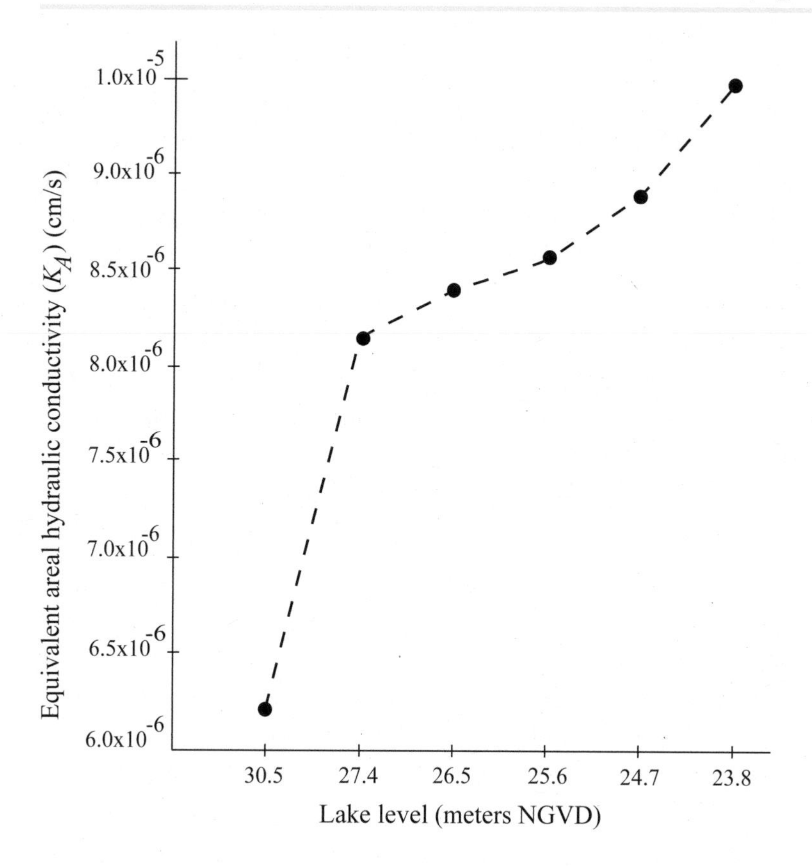

Figure 10.18. Equivalent areal hydraulic conductivity (K_A) versus declining lake levels, Lake Jackson, Leon County, Florida. Modified from Evans (1996).

10.7. Ecological Indicators of Lake Seepage

Plant communities associated with lakes can fall into two general categories: submergent marsh and floating-leaved aquatic, and both communities can be found within a single lake. Submergent aquatic macrophytes tend to occur in deeper water than beds of floating-leaved or emergent species, however, there is often significant overlap between these categories.

Studies in rivers and lakes have revealed strong associations between groundwater seepage and concentrated accumulations of aquatic plants. Groundwater and surface water differ in water chemistry due to different environmental influences and flow regimes. These differences can affect plant species distribution where it is possible to use vegetation as an indicator of the dominant water source at site (Goslee, Brooks, and Cole 1997). For example, Rosenberry, Striegel, and Hudson (2000) found that the distribution of aquatic plants is useful for locating springs on a Minnesota lake by the occurrence of marsh marigold (*Caltha palustris L.*) to locate springs that discharge on land near the shoreline of the lake. Marsh marigold produces large (2–4 cm in diameter) yellow flowers that provide a ready indicator for locating groundwater springs. A study of Pennsylvania lakes by Goslee, Brooks, and Cole (1997) found that some species, such as *Nyssa sylvatica* (commonly known as tupelo, black gum, or sour gum), were strongly associated with the presence of groundwater. Others, such as *Symplocarpus foetidus* (skunk cabbage or swamp cabbage), were strongly associated with the presence of seasonal surface water.

Some plant species or ecological associations may be indicative of a certain sediment or lithology at or near the subsurface. Evans (1996) noted a correlation between distinct concentrations of *Sagittaria latifolia* (broadleaf arrowhead or duck-potato) and clayey sedimentation at Lake Jackson. Other ecosystems, such as karst wetlands and lakes, can have a distinctive ecologic floral and faunal distribution and may be indicative of carbonates sediments at or near the subsurface (Cartwright and Wolfe 2016).

Observation of floral and faunal distribution patterns can be a useful tool for locating and investigating sites that may indicate a variation in lake hydrogeology. A specific plant or invertebrate species may be indicative of seepage sources, such as groundwater versus surface water, or may be associated to a lithology, such as clay or carbonate, which could have a characteristic range of hydraulic conductivity.

10.8. Summary

Seepage in lake systems is an essential component of lake mass balance and management. However, because evapotranspiration is usually the most significant factor in removing water from the lake system, seepage is often underestimated or assigned one value when incorporating it into a mass balance equation.

Studies have revealed that lake bottom hydraulic conductivity can vary both vertically and horizontally throughout the lake bottom area with or without any apparent trends. Thus, understanding of lake bottom sedimentation and hydrogeology is crucial in determining lake behavior and mass balance and must be investigated for each individual lake system under study. In addition, seepage into and out of a lake can vary seasonally as well as during specific events such as a heavy rainfall occurrence, drought, or activation of remnant ponor features where karst geology is present.

For most lakes, seepage and groundwater interaction is under unconfined conditions and mainly influenced by gravity and topography. Domenico and Schwartz (1990) provide an excellent synopsis on surface and groundwater interaction in their section on basin hydrology. For more complex systems of seepage and lake groundwater interactions whose solution require a numerical modeling approach, the reader is referred to Winters (1976) and Toth (1963) as well as the references found in chapter 11 on lake modeling.

Many of the methods used for calculating evapotranspiration (ET) do not adequately account for the influence of emergent aquatic plants or for local meteorology and are a source of error when computing a lake mass balance. Evans (1996) estimated ET by incorporating a model that used lake stage and seepage calculations. When compared with the local pan evaporation data, the seepage-based model results were more consistent with the vegetative effects and more representative of the lake system ET under study. Hence, a combination of seepage modeling with available ET methodology and vegetative cover will result in better understanding of the lake system.

CHAPTER 11

An Overview of Hydrologic Modeling of Lakes, Lake Mass Balance, and Hypsometry

11.1. Systems

When trying to solve such complex problems involving natural environments as lake mass balance, groundwater flow, and evaporation or such anthropomorphic processes as economics and engineering, mathematical models are often incorporated into the descriptions and solutions. Mathematical models are an attempt to represent a process by mathematical equations; they are a tool designed to describe or represent a simplified version of a system. The precise language of mathematics provides a powerful mechanism for describing a system in a compact, simple, and efficient way.

Consider an example of the seepage of a local lake to predict certain outcomes for land management purposes (e.g., a lake as a water resource reservoir). The validity of the predictions about this lake system is dependent upon the data collected or already known about the system's components. However, an attempt to model the system with inadequate data can be instructive in identifying areas where more accurate data is needed to increase the success of a model.

Most hydrologic systems, including lake seepage, can be described by differential equations derived from the basic principles of physics. Processes such as flow of an electric current through a resistive medium or the flow of heat through a solid have been well studied and have many quantitative properties analogous to groundwater flow. The mathematical similarity between Darcy's law of groundwater flow and Ohm's law for the flow of electricity is one such analogy. For example, changes in head in a groundwater model are analogous to changes in voltage in an electrical analog model.

The starting point in modeling is a clear and concise understanding of the processes involved. In the case of a lake system, this entails knowledge of all the variables affecting flow into and out of the lake. The hydrodynamics of a lake may be thought of as a microcosm of the global hydrologic cycle, as discussed in chapter 2, with all its components, including precipitation, surface and subsurface flow, evaporation, and transpiration. In fact, various and multifaceted phenomenon that describe the hydrology of a lake may never be fully understood.

However, in the absence of detailed comprehensive understanding, lakes, like the hydrologic cycle, can be represented in a simplified way by the *systems* concept. A system is a set of connected parts or components that creates the whole (Figure 11.1). As such, the components of the system can be grouped into simpler subsystems (e.g., input can include rainfall, runoff, and streamflow and output can include transpiration and evaporation, lake seepage, or ponor flow and streamflow) to be analyzed separately, or in interaction with other subsystems, and then reincorporated back into the larger system that is being defined.

When modeling hydrologic systems, the objective is to study the system operation and predict some outcome, such as lake stor-

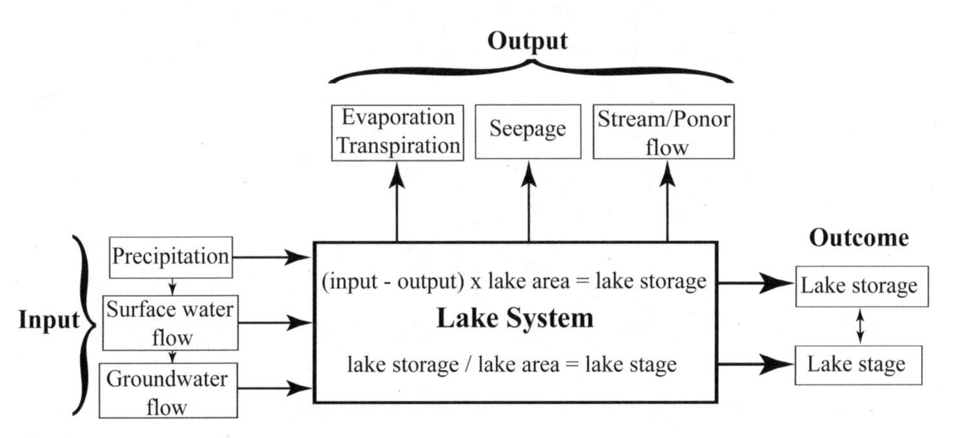

Figure 11.1. Schematic representation of the mass balance components of a lake system that determine lake storage and stage.

age (S) or lake level (h) (also known as lake stage). The model approximates the actual system. The inputs and outputs of the model are measurable hydrologic variables affecting the system.

Crucial to the model are a set of governing equations that describe the system and incorporate the range of variables that influence the system. As seen in the previous chapters, each input to the lake is comprised of its own complex systems of input and output, especially as it applies to storm events and subsequent impacts on surface and baseflow (e.g., Figure 7.16).

11.2. Model Process

Once a system has been identified, the next step is the construction of a *conceptual model* of the specific area of investigation. The conceptual model consists of a set of assumptions that reduces the real problem and the real domain to a simplified version acceptable to the objectives of the modeling and associated management problem (Bear and Verruijt 1987). As an example, let us examine a groundwater interaction between a lake and underlying aquifer as it relates to mass balance of the lake system. Basic assumptions that relate to this problem may include

- the geometry of the boundaries of the lake basin–aquifer system domain;
- such aquifer characteristics as *isotropic* versus *anisotropic*, *homogeneity* versus *heterogeneity*, and *confined* versus *unconfined*;
- lake geology and lithology, such as sediment core analysis to estimate *hydraulic conductivity* of the lake bottom (e.g., Figure 10.8);
- seepage and aquifer hydrology interactions, such as head conditions (h), hydraulic conductivity (K), storage (S), and transmissivity (T);
- mode of flow between the lake and aquifer as two versus three dimensions (i.e., horizontal and vertical flow regimes);
- flow regime (*laminar* versus *turbulent*, or *saturated* versus *unsaturated* flow);
- water properties (chemical composition, saline versus fresh, compressibility, temperature, homogeneity throughout the study area, etc.);

- sources and sinks of water (such as springs, streams, wells, or ponors) within the area of study (Figure 10.13);
- definition and effects of regional and local flow; and
- *boundary conditions* of the system, which refers to the lithological, geochemical, and hydraulic conditions at the boundaries of the watershed, aquifer, or lake system in question.

The second phase is to express the conceptual model in the form of a *mathematical model* by incorporating equations that define the behavior and mass balance of the system or systems under investigation based on the set of governing assumptions. These include equations that define the boundary domain of the study area, mass balance equations describing flow, and equations and coefficients that describe the variables and materials (fluids and solids) and their behavior within the study. Boundary condition equations describing the interaction between the study domain and adjacent environment are necessary, as are *initial conditions* that describe the known state of the system at some initial time. In general, the mathematical models are either deterministic in nature or statistical/stochastic.

11.3. Model Types

For the purposes of this text, mathematical models will be broken down into three basic types:

1. *Deterministic*: developed by applying *first principal equations* such as mass balance, energy balances, kinetic rates, Newtonian equations, and so on that incorporate well-known parameters within a well-known system (see appendix, Figure A2.5). Under the deterministic umbrellas are *continuum* or *analytical models* that use partial differential equations to describe the system and *numerical models* that divide the study area into a finite set of cells or points in the space-time domain. In a numerical model, the defining differential equations are converted into a set of discrete algebraic equations, one equation per cell that then uses a computer to generate a solution to the large number of equations.
2. *Stochastic*: a statistical approach that deals with the probabilistic modeling of hydrological processes that have random

components associated with them. *Geostatistics* is a class of statistics used to analyze and predict the values associated with spatial or spatiotemporal phenomena (e.g., rainfall distribution on a watershed). It incorporates the spatial (and in some cases temporal) coordinates of the data within the analyses (see section 5.3.1b).

3. *Mixed*: incorporating a mixture of deterministic and stochastic methods to model a system.

In the continuum or analytical approach, mass balance equations are in the form of partial differential equations expressed in terms of macroscopic state variables, which are specified as an average over the representative elementary volume of the domain under consideration (Bear and Verruijt 1987) (more in section 9.10). In general, the boundary conditions are expressed mathematically as the flux of the quantity normal to the boundary bordering the study domain (see section 11.3.1a, Figure 11.8).

The analytical model is the preferable method because once the solution is derived it can be used for a variety of similar study domains. However, certain simplifying assumptions are often used when deriving analytical equations, such as steady-state conditions—where continuity requires that water flowing into a system (e.g., a representative elemental volume) is equal to the amount flowing out—or characteristics of an aquifer, for instance the homogeneity and or isotropic lithology of an aquifer. In the case of systems with irregular domain boundaries, or heterogeneity, expressed in spatial distribution of aquifer or lake hydraulic characteristics or sink-source flux functions, numerical approaches are better suited for solving the mathematical model.

Stochastic models are often referred to as *black box models* (Figure 11.2) in science and engineering. A *black box* is a device, system, or object that can be viewed in terms of its inputs and outputs (e.g., stimulus and response or transfer characteristics) without any knowledge of its internal workings. In contrast, a deterministic model, where the process and variables are well known, can be referred to as a *white box model.* Mixed models would be a combination of deterministic and stochastic models to describe a system and are sometimes referred to as *grey box models* (Bohlin 2006).

A. Determistic Model

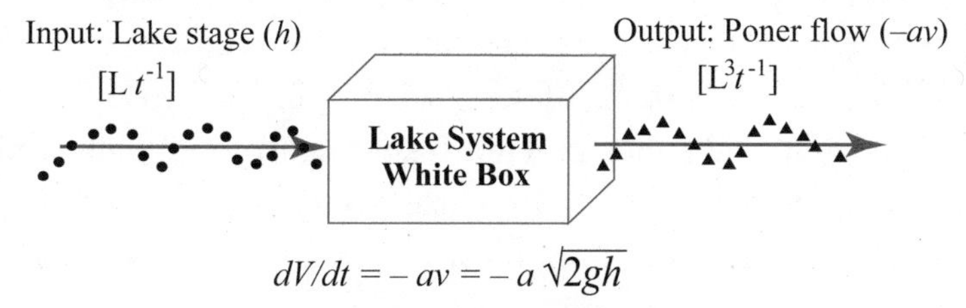

B. Stochastic Model

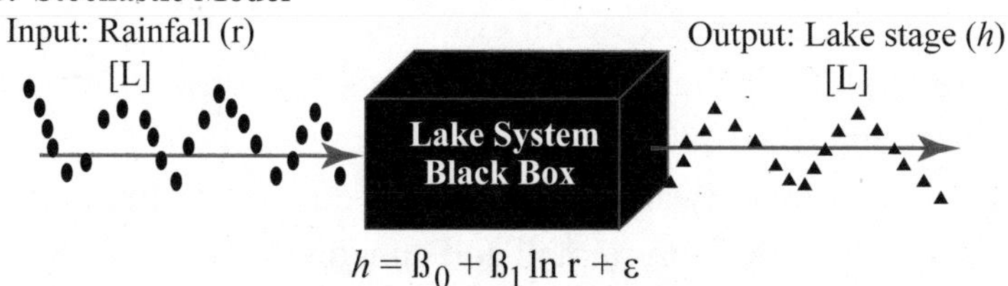

Regression model based on data observations and linear relationships between input and output where $ß_0$ is the y-intercept, $ß_1$ is the fixed coefficient of regression, ln is the natural logarithmic funtion, and ε is the random component or error term.

C. Mixed Model

Input: Rainfall (r) and ponor flow ($-av$) Output: Lake storage (S)

Lake System
Grey Box

$$dV = (ß_0 + ß_1 \ln r + \varepsilon)A \quad (a\sqrt{2gh})dh$$

A = lake basin area

Figure 11.2. Three major types of models for describing hydrologic systems.

11.3.1. Deterministic Analytical Model

Case Study 11.1

Lake Mass Balance and Hypsometry

W.L. Evans, 1996, "Modeling the Hydrodynamics of Closed Lake Basins: A Treatise on Lake Seepage" (master's thesis, Department of Geology, Florida State University).

Phase 1: Conceptual Model

LAKE MASS BALANCE

Lake behavior and stage fluctuation are defined by accurately measuring the quantities of water that move into the lake from various sources and out of the lake to various sinks. For lakes with stream influences, input or output is simply a measure of stream mass balance that can be incorporated into the mass balance equations herein. Observing lakes as a closed system, however, such as karst lakes with no stream influences, inputs include precipitation, condensation, and ephemeral flow resulting from rainfall within the catchment basin (i.e., groundwater seepage and runoff). Outputs consist of evaporation, transpiration, and seepage through lake bottom sediments and ponors. Since it is generally difficult to separate the two processes of evaporation and transpiration, they are often combined and treated as a single process, evapotranspiration (Hillel 1971).

The flow of mass (water) into the lake during the interval of time dt is primarily from rainfall and ephemeral flow (assuming condensation is negligible) and is given as

$$\rho PAdt + \rho Qdt \tag{11.1}$$

where ρ is the density of the water, P is the rate of precipitation (with units of length over time [Lt^{-1}], A is the surface area of the lake [L^2], and Q is the rate of ephemeral discharge [L^3t^{-1}]. The flow of mass out of a lake due to evaporation and transpiration is $\rho ETAdt$, where ET is the rate of evapotranspiration [Lt^{-1}]. Seepage through the lake bed and ponor is described by

$$\rho AK_A \frac{(h + h_b - h_a)}{L} dt \tag{11.2}$$

and

$$\rho aK_p \frac{(h + h_b - h_a)}{L} dt \tag{11.3}$$

where h is the height of the water surface above the lake bed, h_b is the height of the lake bed over the underlying aquifer, h_a is the height of the potentiometric surface of the aquifer, L is the thickness of the sediments separating the lake from the aquifer, a is the effective cross-sectional area of the ponor, and K_A and K_p are the effective hydraulic conductivities of the lake bottom sediments and ponor, respectively (Figure 11.3).

Phase 2: Mathematical Model

LAKE HYPSOMETRY

Lake levels, rates of seepage, and evaporation are influenced by lake hypsometry. When describing lake-level fluctuation, the

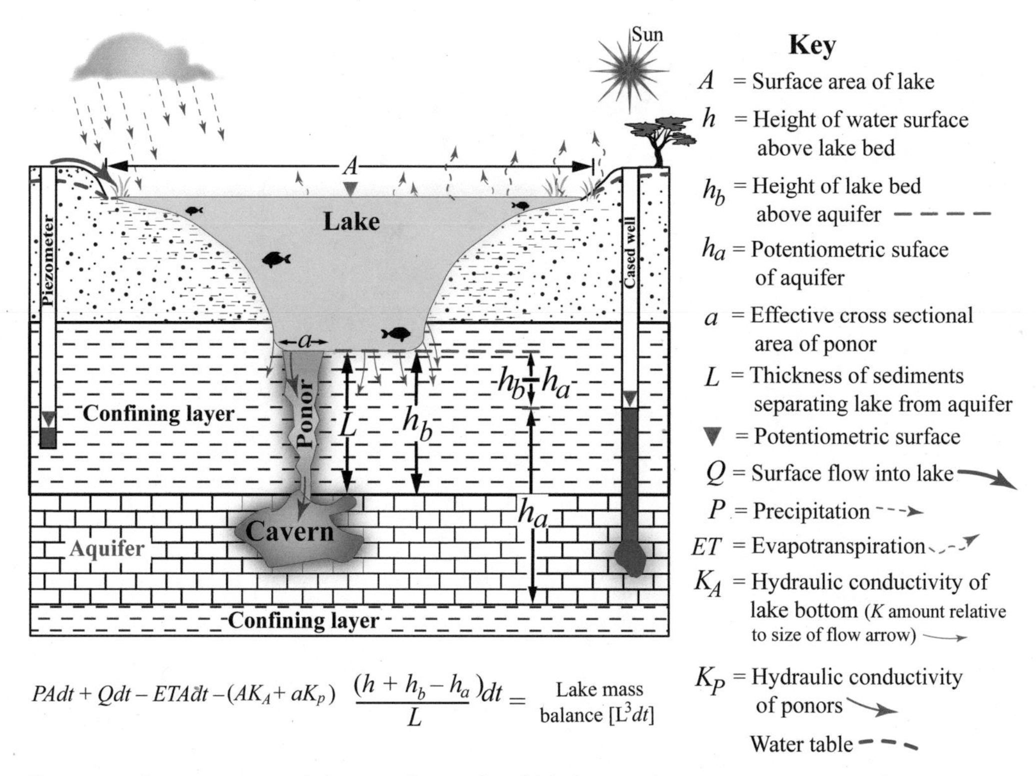

Figure 11.3. Determining mass balance in a karstic closed-lake basin with active ponors.

effects of hypsometry must be considered. The *hypsometric curve* (Figure 11.4) is a graphic representation of the relationship between the surface area of a lake and its depth and subsequent volume. For example, in Figure 11.4, the change in lake volume is greater between lake levels h_0 and h_1 then between h_1 and h_3 whereas the change in lake area is greater between lake levels h_2 and h_3.

The hypsometric function $A(h)$ relates the surface area A of the lake to its level h. The volume V of the lake is then

$$V(h) = \int_a^h A(h)\,dh \tag{11.4}$$

and the mass within it at time t is $\rho V(h)$. The mass within the lake after an interval of time dt is to the first order

$$\rho V + \rho \frac{\partial V}{\partial t} dt \tag{11.5}$$

so that the change in mass is given as

$$\Delta m = \Delta m = \rho \frac{\partial V}{\partial t} dt. \tag{11.6}$$

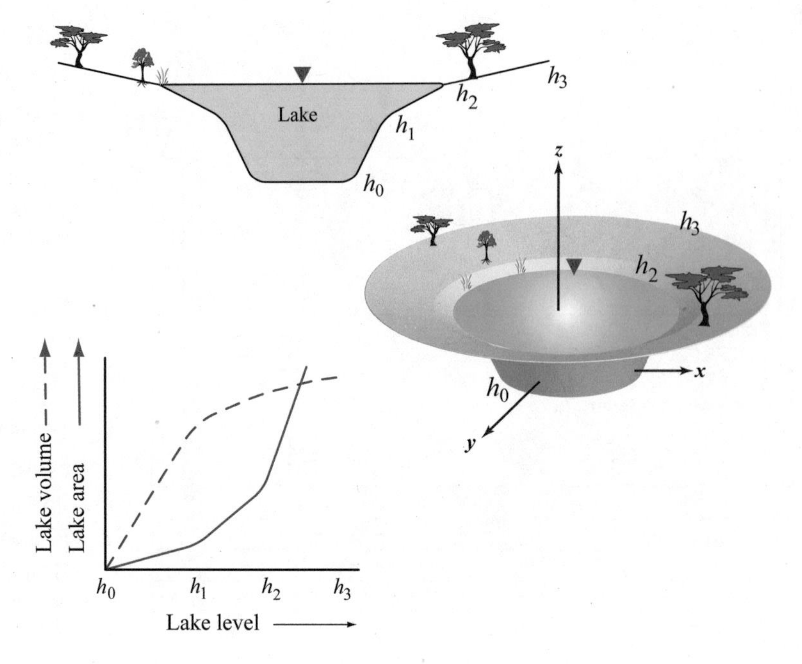

Figure 11.4. Example of a hypsometric curve showing the relationship between lake level (*h*), lake volume, and lake area.

Due to the complexity of lake basin geometry, it is useful to look at lake volume in terms of lake area such that

$$\frac{\partial V}{\partial t} = \frac{\partial V}{\partial h}\frac{\partial h}{\partial t} = A\frac{\partial h}{\partial t}dt \tag{11.7}$$

thus, by the conservation of mass, a lake mass balance can be described by combining equations (11.1), (11.2), and (11.7) to arrive at

$$\rho PAdt + \rho Qdt - \rho ETAdt - \left(\rho AK_A + \rho aK_p\right)\frac{\left(h + h_b - h_a\right)}{L}dt = \rho A\frac{\partial h}{\partial t}dt. \tag{11.8}$$

Assuming constant density, and noting that $h_b = L$ (see Figure 11.1), then

$$PA + Q - ETA - \left[\frac{K_A A}{L} + \frac{K_p a}{L}\right]h + \frac{K_A h_a A}{L} - K_A A + \frac{K_p a h_a}{L} - K_p a = A\frac{\partial h}{\partial t} \tag{11.9}$$

Between rainfall events, after ephemeral flow has subsided, then $PA + Q = 0$ and the mass balance reduces to

$$-ETA - \left[\frac{K_A}{L} + \frac{K_p a}{LA}\right]h + \frac{K_A h_a}{L} - K_A + \frac{K_p a h_a}{LA} - \frac{K_p a}{A} = \frac{\partial h}{\partial t}. \quad (11.10)$$

When the surface area (A) is sufficiently large such that the contribution of ponor seepage to the balance is negligible, then

$$-\frac{K_A}{L}h + \frac{K_A h_a}{L} - K_A - E = \frac{\partial h}{\partial t} \quad (11.11)$$

indicating that the rate of stage decline is linearly related to its level h. After accounting for evapotranspiration by subtracting it from h, the lake-level decline due to seepage through the lake bottom is

$$-\frac{K_A}{L}h + \frac{K_A}{L}h_a - K_A = \frac{\partial h}{\partial t} \quad (11.12)$$

and the total mass removed from the lake is

$$A\left(-\frac{K_A}{L}h + \frac{K_A}{L}h_a - K_A\right) = A\frac{\partial h}{\partial t}. \quad (11.13)$$

Assuming L and K_A are constant, and letting $\alpha = K_A/L$, then lake-level decline due to seepage is

$$\frac{dh}{dt} = \alpha(h - h_a) \quad (11.14)$$

or

$$\alpha dt = \frac{1}{h - h_a}dh. \quad (11.15)$$

This linear differential equation can be expressed as

$$-\int \frac{1}{h - h_a}dh = \alpha \int dt. \quad (11.16)$$

If the water level of the lake (h) is greater than the potentiometric surface of the underlying aquifer, then the effect of the potentiometric surface h_a on lake stage is negligible. If, by convention, the atmospheric pressure is set to gauge, then the flow through the lake bottom will be driven by the lake level. This assumes the flow through the underlying sediments behaves as if draining into an unsaturated lithology that is pneumatically connected with the atmosphere. Therefore,

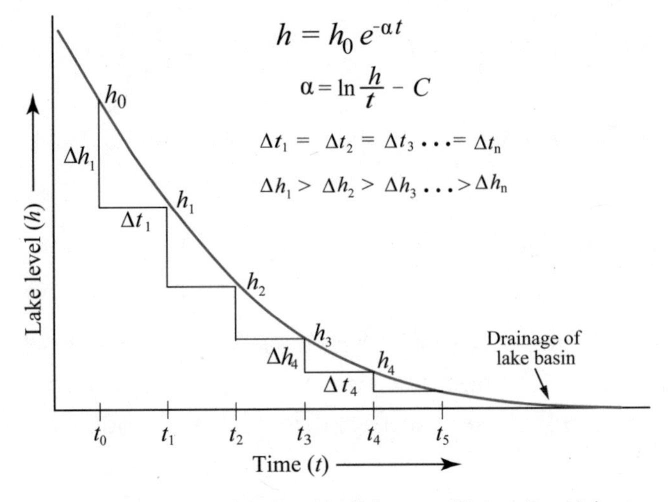

Figure 11.5. Exponential decline (e) of lake stage (h) over time (t) due to seepage; α characterizes seepage through the lake bottom and ponors where C is the constant of integration.

$$-\int \frac{1}{h} dh = \alpha \int dt \tag{11.17}$$

and

$$-\ln h - C = \alpha t \tag{11.18}$$

where C is the constant of integration. Substituting the initial value of h_0 when $t = 0$, $C = \ln h_0$ and

$$h = h_0 e^{-\alpha t} \tag{11.19}$$

which is an exponential decay function. Thus, as lake level decreases, seepage will decrease, and the rate of change in lake level over time will decrease exponentially (Figure 11.5).

After accounting for the effects of evapotranspiration and delayed recharge, fluctuations in lake level should resemble a series of exponential decay curves (Figure 11.6) whose parameters reflect basin hypsometry and lake bottom conductivity. In principle, the parameter α, which reflects seepage through the lake beds and ponors can be estimated from equation (11.18) to become

$$\alpha = \ln \frac{h}{t} - C \tag{11.20}$$

by examining lake-level variations as a function of time t.

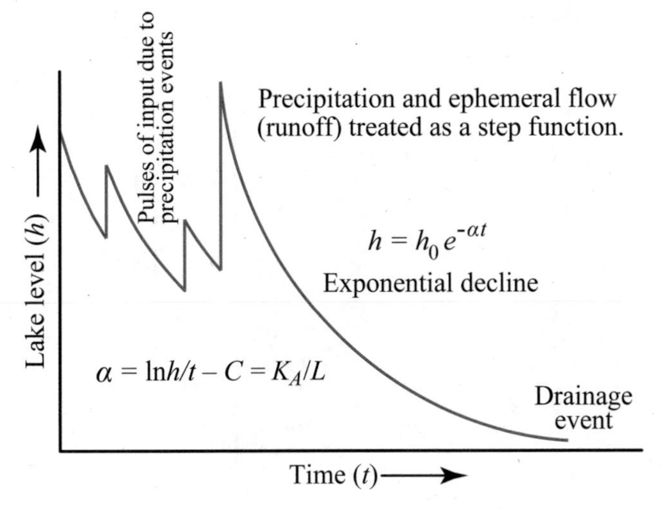

Figure 11.6. Lake level (h) over time (t) due to seepage and ponor flow (α) with inputs of precipitation treated as a step function.

Example: Modeling Ponor Flow

When deriving the mathematical model to explain lake-level behavior in Case Study 11.1, equation (11.11) made the assumption that flow through the ponor was negligible. However, as seen in Figures 10.14 and 10.15, in the case of karstic lakes such as Lake Jackson, the periodic activation of a ponor (e.g., continued dissolution and subsequent collapse of underlying carbonate lithology) can cause rapid water loss resulting in lake drainage. Thus, active ponor flow must be accounted for in the mass balance investigation.

Using the principles of mass balance and continuity, an equation to model flow through a lake ponor can be derived. A simple example of an analytical model homologous to ponor or streamflow from a lake system as it relates to mass balance would be water flowing out of a cylindrical beaker via a circular outlet (i.e., ponor or stream) near the bottom of the beaker (Figure 11.7).

The rate that fluid leaves through the hole must balance with the rate that fluid decreases in the beaker (*Bernoulli's principle*). Assuming that the density of water is constant, one can investigate net flux in terms of changes in volume (V) (rather than mass). The beaker is also open to the atmosphere, thus the flow pressure at both the top of the water level in the beaker and the bottom are equal, or at atmospheric pressure.

If we refer to V as the volume [L^3] of fluid in the beaker, then the rate of change of V is dV/dt, which is equal to the negative rate volume [$-L^3t^{-1}$] lost from the beaker as fluid flows through the ponor (the minus sign indicates that the volume is decreasing). The rate of volume leaving the beaker as fluid flows out of the ponor equals av, where a is the area [L^2] of the ponor and v is the velocity or flow [Lt^{-1}] such that

$$dV/dt = -av \qquad (11.21)$$

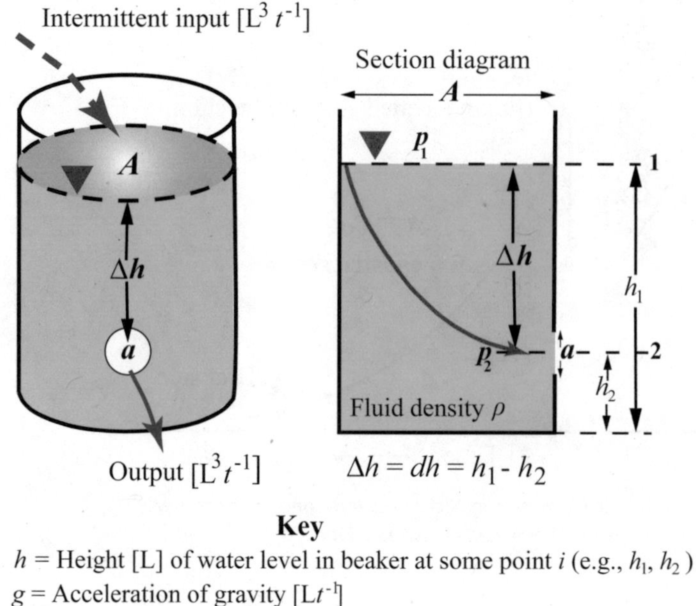

Figure 11.7. Flow through an outlet in a cylindrical beaker representing a first principle equations model homologous to ponor flow through a lake bottom.

From basic physics, work is done by a force when the force moves an object through a distance. Energy (*kinetic* and *potential*) can be defined as the ability to do work. Remember from chapter 9 that *kinetic energy* (KE) is a body of mass m in motion at speed v, or KE = ½ mv^2. And an example of *potential energy* (PE) is associated with the position or configuration of multiple bodies, such as gravitational energy (PEg = mgz) where z is the object above an arbitrary reference point (e.g., z_0), or in the case of our beaker, the height of the beaker water level h. So, the potential energy of a small mass m of water at height h will be mgh, whereas when the water flows out of the hole, its kinetic energy is given by ½ mv^2. Thus, for these to balance (so that total energy is conserved)

$$\frac{1}{2mv^2} = mgh \tag{11.22}$$

and flow through the ponor is related to the height of the water in the beaker (or lake) by

$$v = \sqrt{2gh}. \tag{11.23}$$

Investigation of the overall system in terms of volume of fluid in the beaker over time can be expressed as

$$V(t) = Ah(t) \tag{11.24}$$

where A is the constant (> 0) cross-sectional area of the beaker such that

$$\frac{dV}{dt} = A\frac{d(h(t))}{dt} \tag{11.25}$$

$$\frac{dV}{dt} = -av = -a\sqrt{2gh} \tag{11.26}$$

$$\frac{dh}{dt} = -\frac{a}{A}\sqrt{2gh} = -kh \tag{11.27}$$

where k is a constant representing the size and shape of the cylinder and its opening (e.g., lake basin and ponor conductivity) or

$$k = \frac{a}{A}\sqrt{2g}. \tag{11.28}$$

Using the principles of conservation of mass and energy, it can be shown that the height h of water in the beaker at time t satisfies the following differential equation:

$$\frac{dh}{dt} = -k\sqrt{h}. \tag{11.29}$$

Given a known initial condition of $h(t)$ where $h(0) = h_0$, the equation can be solved by the integration and separation of variables over the interval $0 \leq t \leq t_i$. During this time interval, the value of h changes from its known initial value h_0 to some final value that we will refer to as to $h(t_i)$, or h_i (i.e., $h_0 \leq h \leq h_{ti}$). We do not yet know $h(t_i)$, but part of our goal is to find that value, or, for example, to predict the future behavior of h. Starting with

$$\frac{dh}{\sqrt{h}} = -kdt \tag{11.30}$$

and integrating

$$\int_{h(0)}^{h(t_i)} h^{-1/2}dh = -k\int_0^{t_i} dt \tag{11.31}$$

$$2(\sqrt{h(t_i)} - \sqrt{h_0} = -kT \tag{11.32}$$

$$\sqrt{h(t_i)} = -k\frac{t_i}{2} + \sqrt{h_0} \tag{11.33}$$

and

$$h(t_i) = \left(-k\frac{t_i}{2} + \sqrt{h_0}\right)^2 \tag{11.34}$$

and since this is true for any time t, we can write equation (11.34) as

$$h(t) = \left(\sqrt{h} - k\frac{t}{2}\right)^2. \tag{11.35}$$

This allows us to predict the height of fluid in the beaker at any time t due to flow through the opening in the beaker as well as predicting the time it takes to empty the beaker where $h_t = 0$, or

$$0 = \left(\sqrt{h} - k\frac{t}{2}\right)^2. \tag{11.36}$$

Solving for t yields

$$k\frac{t}{2} = \sqrt{h_0} \tag{11.37}$$

or

$$t = \frac{2\sqrt{h_0}}{k} \tag{11.38}$$

which is the time it takes for the beaker to drain to the water level at the center of the outlet.

The beaker outlet homology with a lake and ponor or streamflow system is a simplistic approach yet it yields the basic equations that can be used for investigating lake or wetland behavior as it applies to outflow due to seepage, ponor flow, or streamflow. This model would be characterized as a deterministic model because it is developed by applying first principal equations that incorporate well-known parameters within a well-known system.

A lake system, however, is a dynamic and changing system, and as such there will be a random component to the beaker model as it is applied to lakes. In the case of stream or ponor flow ($-av$), the flow characteristics will vary over time as it fills with sediments or changes in size or shape due to continued dissolution and erosion processes. The lake surface area (A) will vary with lake stage (h) because of lake hypsometry, hence k in the equations is not a constant and is replaced with

$$\frac{a}{A}\sqrt{2g} \tag{11.39}$$

where a/A would be probabilistic in nature and require nondeterministic methods to account for these random characteristics.

11.3.1a Numerical Simulation Computer Models

AQUIFER-LAKE INTERACTIONS AND BOUNDARY CONDITIONS

Deterministic mathematical models are the most useful types for analyzing systems like aquifers as they are based on recognized governing equations that describe the system to be modeled. The development of the model starts with the conceptual understanding of the aquifer system to be investigated. For example, the response of an aquifer and lake system due to water withdrawal or recharge depends on the transmissivity and storage coefficient of the aquifer, the hydrologic and geologic boundary conditions (section 9.11), the location points of water withdrawal or recharge in relation to the lake (Van der Heijde et al. 1980), and seepage characteristics of the lake.

As discussed in section 9.12.3, simplifying assumptions can be used to solve complex flow equations. Oversimplification, however, can result in reduction of model accuracy. Computer-generated numerical methods can be used for analyzing complex lake systems with underlying multifaceted hydrogeology as well as convoluted surface water inputs from the various influences of urbanization. Numerical models replace the continuous form of the differential equations used to describe the system and turn them into a finite number of algebraic equations. However, the governing equations are still based on first principles, thus these models are deterministic in nature. Use of the computer for solving this series or matrix of algebraic equations makes it possible to simulate the response of extremely complicated systems to changing conditions.

Two basic types of numerical models relevant to lake hydrodynamics are used to describe lake-aquifer interactions and are briefly discussed in this text. These are *finite-difference* models and *finite-element* models. In general, these methods take the differential equations used to describe the system and convert them into a system of algebraic linear equations (Bear and Verruijt 1987). Both models require a system of nodal points superimposed over a system domain with relevant boundary conditions (e.g., Figure 11.8 and Figure 11.9a).

The fourth boundary, plane AD, is the water table. If the water table head is specified as stationary or fixed, this would be a Dirichlet or steady-state condition and could be solved using the Laplace equation (9.50) from section 9.11:

$$\frac{\partial^2 h}{\partial x} + \frac{\partial^2 h}{\partial y} = 0.$$

This solution would estimate the amount of recharge needed to maintain the water table in the observed stationary position. Dirichlet's conditions on a boundary is as if the hydraulic head (h) is independent of the flow conditions of the aquifer. Generally, this occurs where there is contact between the aquifer and an open expanse of water such as a lake, a river, or an ocean (Marsily 1986) (Figure 11.8(a)).

A second solution is to specify the water table under Neumann conditions where flux or recharge is examined to determine the

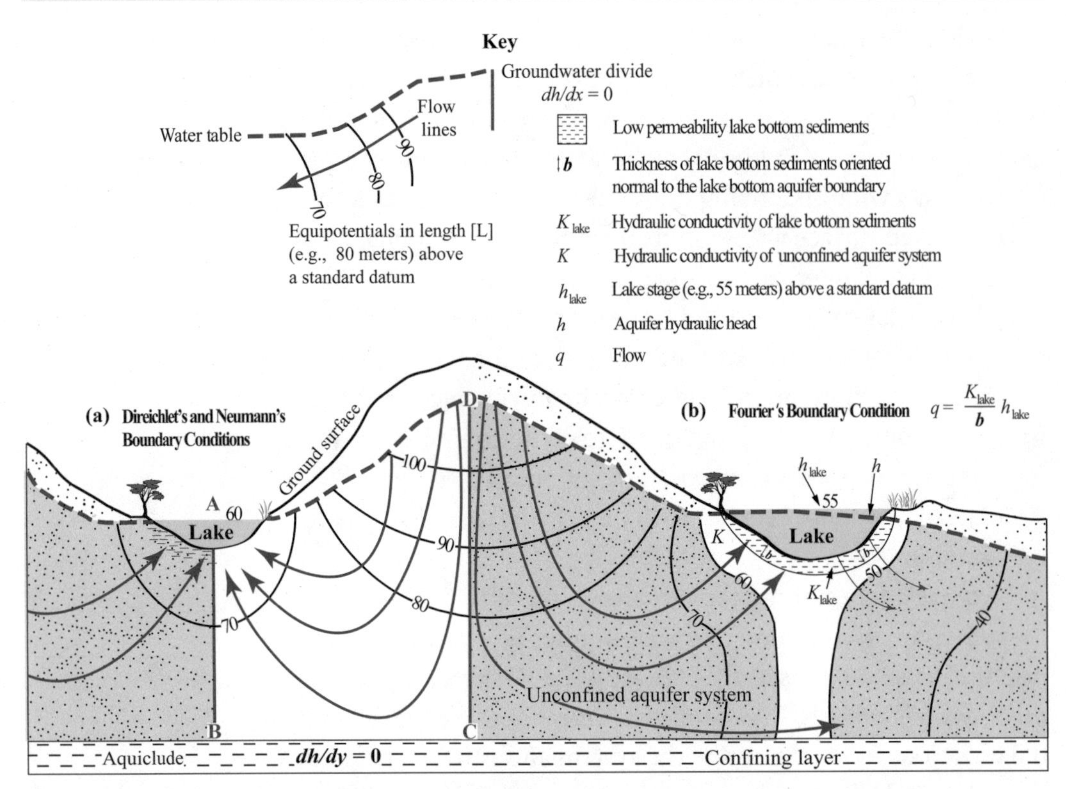

Figure 11.8. Cross section of a regional unconfined aquifer and lake system demonstrating Dirichlet's (A-D) and Neumann's (A-B, D-C and B-C) boundary conditions (a) and Fourier's boundary condition (b).

water table under various recharge or withdrawal conditions. This method would employ the *Poisson equation* where $R(x,y)$ would be the volume of water added per unit time per unit aquifer area (Wang and Anderson 1982) (see discussion on elemental volume in section 9.11). As discussed in section 9.12.2, for an unconfined aquifer, the saturated thickness (b) varies with the water table, and using the *Dupuit assumptions* that (a) flow is horizontal and (b) the hydraulic gradient is equal to the slope of the free surface, then

$$\frac{K}{2}\left(\frac{\partial^2 h^2}{\partial x^2}+\frac{\partial^2 h^2}{\partial y^2}\right)=-R. \tag{11.40}$$

For a confined aquifer system, this solution requires that the aquifer's transmissivity (T)—a product of the aquifer's hydraulic conductivity (K) and thickness (b)—is known (Wang and Anderson 1982), such that

$$\frac{\partial^2 h}{\partial x^2}+\frac{\partial^2 h}{\partial y^2}=-\frac{R(x,y)}{T}. \tag{11.41}$$

Although strictly valid for confined aquifers, equation (11.41) can be used for unconfined aquifers by allowing T to vary as the saturated thickness changes.

Recall from section 9.5 Newton's law of viscosity, equation 9.17, in which the potential is the flow velocity:

$$\tau = \mu \frac{\partial u}{\partial z}.$$

Where τ is the momentum flux, μ is a proportionality coefficient called the dynamic viscosity and du/dz is the gradient of the velocity u as a function of distance z from the boundary. Analogous, to Newton's law of viscosity are Fick's law of diffusion (section 4.41, equation 4.5) and Fourier's law of heat conduction whose governing equations are in the same form as equation 9.17 (Chow, Maidment, and Mays 1988, 43). Fourier's law describes the conduction of heat from regions of high temperature to areas where it is low. Using Newton's equation as an analogy for the Fourier's expression, τ is the heat flux, μ is a proportionality coefficient and du/dz is the gradient of the temperature u as a function of distance z from the boundary.

Therefore, a third boundary known as Fourier's condition can be specified for a lake (or stream) where seepage either discharges to or is recharged by the underlying unconfined aquifer system (Marsily 1986). This condition generally occurs where there is a layer of low permeable sediments (e.g., silt or clay) at the lake bottom (Figure 11.8b). The difference in hydraulic head between the lake and aquifer $\Delta h = h_{\text{lake}} - h$ (aquifer) across lake bottom sediments with a hydraulic conductivity of K_{lake}, and thickness b, creates the gradient necessary for flow q per unit surface area between the lake and aquifer system in agreement with Darcy's law as:

$$q = -K\frac{h}{b} = K_{lake}\frac{h_{lake} - h}{b} \qquad (11.42)$$

The flux through the unconfined aquifer system at the lake bottom-aquifer boundary must also be evaluated and expressed in terms of Darcy's law as

$$q = -K\frac{\partial h}{\partial n} \qquad (11.43)$$

Figure 11.9. Two types of finite-difference grids of an unconfined aquifer system bounded by a lake (see Figure 11.8 for the cross section of boundary conditions).

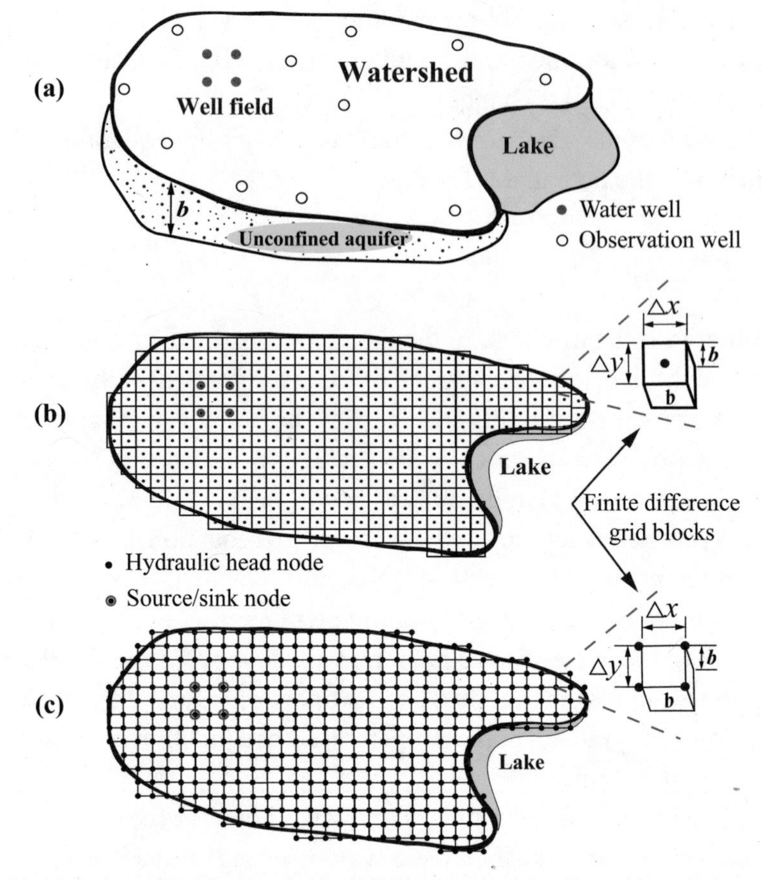

(a) Map view of aquifer showing well field, observation wells, and boundaries.

(b) Block-centered finite-difference grid where Δx is spacing in the x direction, Δy is spacing in the y direction, and b is the aquifer thickness.

(c) Mesh-centered finite-difference grid.

where n is the normal line to the lake bottom aquifer boundary oriented toward the lake. By equating these two equations then

$$q = -K\frac{\partial h}{\partial n} + \frac{K_{lake}}{b} = \frac{K_{lake}}{b}h_{lake} \tag{11.44}$$

which is the definition of a Fourier condition (Marsily 1986, 139)

11.3.1b. Numerical Modeling

FINITE-DIFFERENCE MODEL As discussed in section 9.11, the governing equation for groundwater flow through an isotropic and homogeneous aquifer is the Laplace equation, which combines

Darcy's law and the continuity equation (9.50). When modeling a specific aquifer, boundary conditions are required, and it is assumed that the aquifer properties are known within the boundary domain (Figure 11.9a). In this conceptual model where the aquifer is a continuous flow system, the numerical method breaks the continuum into a finite grid of discrete blocks. Each block in the bounded region has its own set of hydrogeologic properties where the hydraulic head of the block is defined at the nodes (●) either at center (block-centered grid) or corner (mesh-centered grid) of the blocks (Figures 11.9b and 11.9c, respectively). The block-centered grid is most useful where a flow change or flux across a boundary is specified (Neumann condition) versus the mesh-centered grid where the head conditions are specified at the boundary (Dirichlet conditions) (Fetter 1988).

In general, the grid is regular, with the rows and columns perpendicular to each other and the distance in the x direction (Δx) equal to the distance in the y direction (Δy). In the boundary area, if there is a well field or other hydrogeologic phenomenon such as a recharge area, where there are more impacts to the aquifer, nodal points can be increased, and the spatial dimensions of the rows and columns can be varied (Figure 11.10a).

Two types of two-dimensional notation are used to describe the position of the nodes in the grid (Figure 11.10b): a basic (x,y) coordinate designation where adjacent points are located at a distance Δx to the left or right or Δy up or down from the centered (x,y) node, and computer codes using an (i,j) nodal designation, where i designates the column and j the row. The computer notation for i is positive to the right, and for j it is positive downward (Figure 11.10b).

Using these notations in the case of the Dirichlet or steady-state condition (Laplace equation), the finite-difference equation for the system in Figure 11.8 would be

$$h_{i-1,j} + h_{i=1,j} + h_{i,j-1} + h_{i,j+1} - 4h_{i,j} = 0. \tag{11.45}$$

Solving for $h_{i,j}$, equation (11.45) (Wang and Anderson 1982) becomes

$$h_{i,j} = \left(\frac{1}{4}\right)\left(h_{i-1,j} + h_{i=1,j} + h_{i,j-1} + h_{i,j+1}\right). \tag{11.46}$$

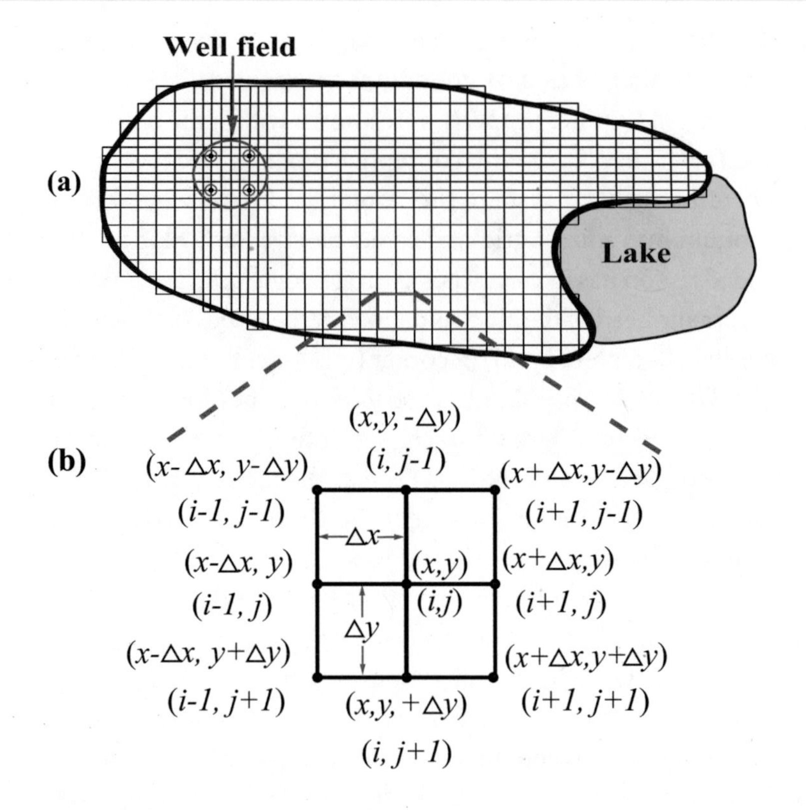

Figure 11.10. Variable spacing finite-difference mesh-centered grid (a) with (*x,y*) and computer node (*i,j*) notation (b).

The value of $h_{i,j}$ is the average of the heads at the four closest nodes in the grid.

For recharge to the aquifer (meeting the Neumann condition for the Poisson equation (11.41), the finite-difference equation (Wang and Anderson 1982) becomes

$$\frac{\left(h_{i-1,j}-2h_{i,j}+h_{i+1,j}\right)}{\left(\Delta x\right)^2}+\frac{\left(h_{i,j-1}-2h_{i,j}+h_{i,j+1}\right)}{\left(\Delta y\right)^2}=-RT \quad (11.47)$$

where

Δx and Δy = the distance between the nodes in the x and y direction,

R = recharge, and

T = aquifer transmissivity.

Each node in the finite-difference grid requires the solution in the form of either equation (11.46) or (11.47). Depending upon the magnitude of the investigation, there can be tens to hundreds

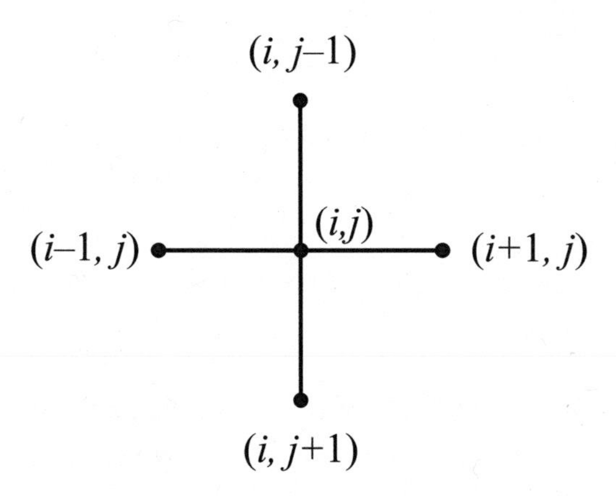

Figure 11.11. Five-point star representing the finite-difference approximation from equation (11.42) for the Laplace equation. The value $h_{i,j}$ at any point in the grid is the average value of the head computed from its four nearest neighbors in the nodal array.

to thousands of nodes in the grid, however, ten thousand is the practical limit (Driscoll 1986). The system of algebraic equations generated at the nodes is solved by generating a computer program or code run on a digital computer by *iterative methods.*

For example, for Dirichlet or steady-state condition (section 9.11, equation 9.50) the hydraulic head of $h_{i,j}$ is the average of the four closet nodes surrounding it. This is known as the *five-point operator* because the algebraic equation (which approximates Laplace's equation) are created one after another at each node by moving the star of five points (Figure 11.11) throughout the grid domain. A computer code or program sweeps the solution path through the finite-difference grid (except for the first and last node) beginning with the best guess initial conditions. Once the head value at each node has been recomputed, the difference between the initial guess and the recomputed head is determined. This sweep process is repeated from one iteration to the next until it becomes less than a preset value known as a *convergence criterion.* When the solution has converged, the equation has been solved. The smaller the value, the more iterations are required and the longer it takes to arrive at the solution.

The *Gauss-Seidel method* and the *successive over relaxation* (SOR) *method* are two common iteration methods used in groundwater numerical models. Detailed discussion of these methods can be found in Wang and Anderson (1982).

Case Study 11.2

Numerical Simulation Analyses of Lake-Groundwater Interaction

T.C. Winter, 1976, "Numerical Simulation Analysis of the Interaction of Lakes and Ground Water," US Geological Survey Professional Paper 1001, 45 pp.

In 1976, Thomas C. Winter was one of the first researchers to use finite-difference numerical modeling to investigate lake and groundwater interaction. Winter assumed a steady-state upper water table with boundary conditions and governing equations similar to those stated for Figure 11.8 and generated a finite-difference grid with a network of uniform rectangles. Approximately nine hundred nodes were used for the one-lake simulations and about eighteen hundred nodes for the three-lake simulations.

To determine the main factors that influence lake-groundwater interaction, Winter ran numerous simulations varying hydrogeologic parameters and investigated a wide range of hydrogeological settings (single versus multiple lakes, underlying aquifers versus no aquifers, changes in hydraulic conductivity, etc.). This approach generated numerous detailed two-dimensional groundwater and lake flow-net diagrams (e.g., Figure 11.12).

Factors that have the greatest influence on lake-groundwater interaction were the height of the water table on the downslope side of the lake relative to lake level, position and relative hydraulic conductivity of aquifers within the groundwater reservoir, ratio of horizontal to vertical hydraulic conductivity, regional slope, and lake depth (Winter 1976).

One of the interesting features from this research is simulation of the flow divides (red dotted lines in Figure 11.12) that separate several types of flow systems (e.g., local versus regional) (see also section 10.2 and Figures 10.10 and 10.11). When following the line dividing local flow and higher magnitude regional flow, there is a point in the divide at lakes 1 and 2 where the head is a minimum compared to the other points along the divide. This point of least head occurs beneath the lake shoreline on the downslope side of the lake and is called the *stagnation point* or *stagnation zone* (Toth 1963). At lakes 1 and 2, these points are where the heads are 0.7

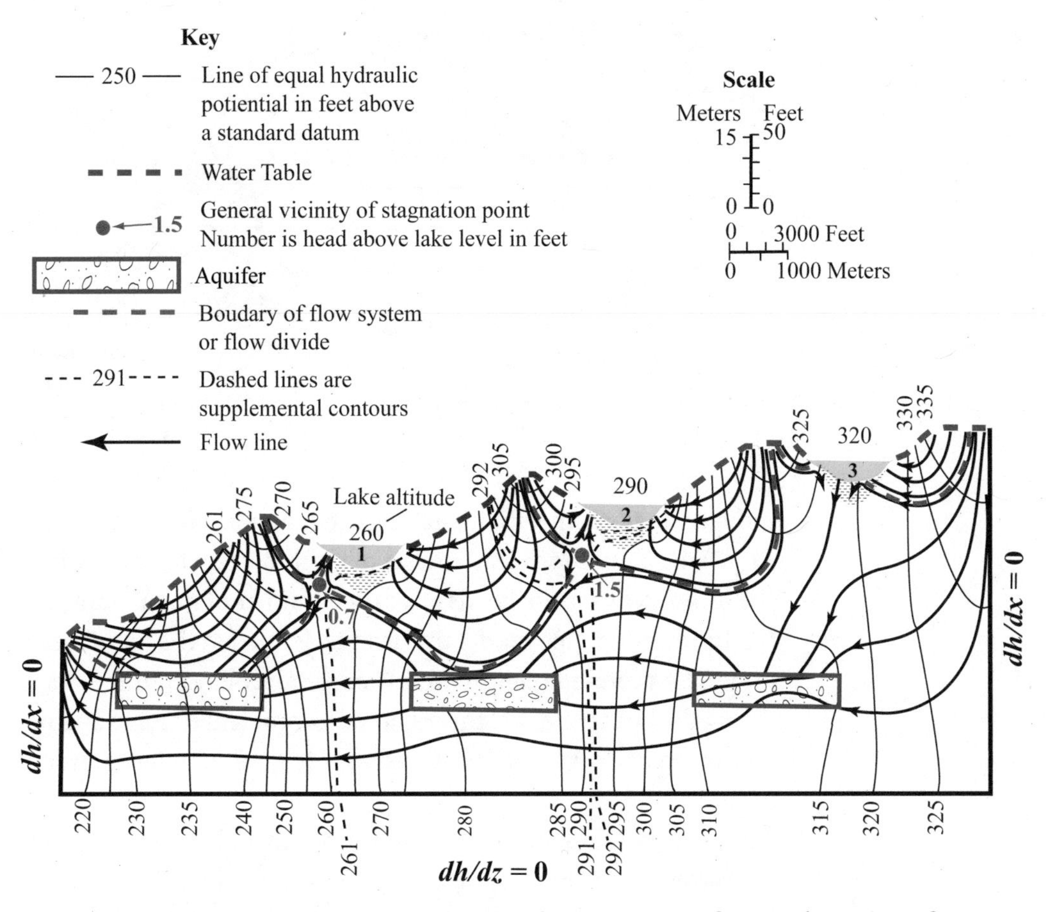

Figure 11.12. Hydrologic section showing a two-dimensional quasi-quantitative flow-net of groundwater flow near lakes in a multiple-lake system that contains aquifers. Modified with permission from Winter (1976).

and 1.5 feet greater than the lake-level altitudes, respectively. Because of the higher head than lake stage conditions at these points, seepage from the lakes to the groundwater does not occur.

Toth (1963) recognized the occurrence of these stagnation zones at the juncture of flow systems when conducting numerical simulations of regional groundwater flow. This point is determined by the head distribution within the groundwater system and ensues where vectors of flow are equal in opposite directions and, thus, cancel each other out.

FINITE-ELEMENT MODELS The finite-difference method is usually implemented with rectangular cells versus the more recently developed finite-element method that uses a variety of

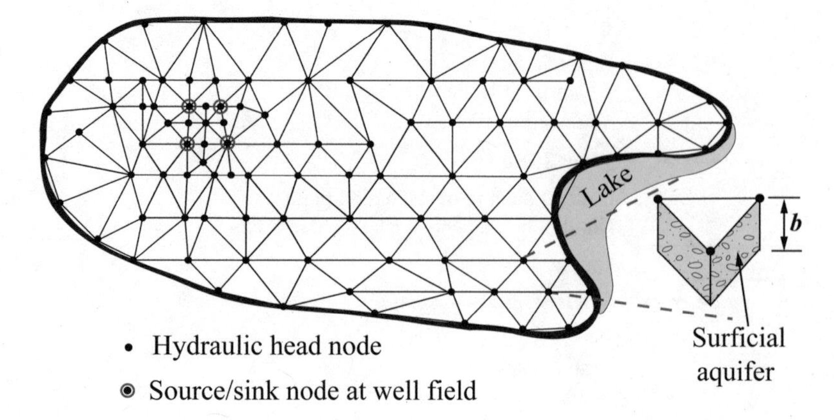

Figure 11.13. Hypothetical finite-element grid for an aquifer system.

polygonal element types, such as a triangular element (Figure 11.13). Both methods lead to a set of algebraic equations in which the unknowns are the heads at a finite number of nodal points. In the finite-difference model the head is defined at the nodal point only. The finite-element methods use interpolation functions to define the head potential throughout the problem domain and not just at the node, which distinguishes it from the finite-difference method.

The finite-element method tends to have more flexibility for problems where boundaries are irregular or in aquifers that are heterogeneous or anisotropic, or in problems of solute transport as well as unconfined aquifers with a variable water table. Both methods have advantages and disadvantages and should be chosen based on the characteristics of the problem domain (Bear and Verruijt 1987). For a discussion of the advantages and disadvantages of each method see Faust and Mercer (1980).

The mathematical theory and methodology for the finite-element method is much more complex than the finite-difference method. The development of the finite-element equations and solutions are based on the *Galerkin method* and require mathematical proficiency that is beyond the scope for an introductory text. In numerical analysis, Galerkin methods are a class of techniques for converting a continuous function, such as a differential equation, into a discrete equation. For further discussion on the development of the mathematical theory as it applies to groundwater solutions, see Trescott, Pinder, and Larson (1976), Pinder and Gray (1977), and Wang and Anderson (1982).

11.3.2. Stochastic Models and Surface Water Applications

Although closely associated, stochastic modeling is perhaps more general than geostatistics, but other differences exist. Stochastic differential equations are models to the extent that they are distinct from geostatistics because of stronger mathematical ties. The principal outcome of stochastic modeling research has been to replace deterministic differential equations with stochastic differential equations, which are especially important when considering transport problems in the subsurface (e.g., contaminate transport). In contrast, *kriging* in its various forms is not really modeling. Although kriging is intricately linked to the modeling of variograms or covariance function, the kriging process itself is not quite the same as modeling in the traditional sense (Myers 2006).

Stochastic hydrology is a statistical approach to modeling that deals with the probabilistic modeling of these hydrological processes that have random components associated with them.

These models are based on data that use mathematical and statistical concepts to link a certain input (e.g., rainfall) to the model output (e.g., lake levels) where simplifications must be carried out to facilitate its integration into different models of analogous systems. Models able to simulate overly complex systems should consider uncertainty due to the lack of data (or data affected by noise) that drive the model itself. These kinds of models are applied to multifaceted natural, ecological, and environmental processes to better understand how intricate phenomena work.

Commonly used techniques are *regression*, *transfer functions*, and *system identification*. Regression analysis is a statistical technique used to investigate and model relationships between variables, such as a linear relationship between rainfall and lake level (see appendix, section 2.2). A transfer function is a mathematical representation or model of a system in that it is an operational method of expressing differential equations that relates the input variables to the output variables. The transfer function is a property or result of the system itself and may be unrelated to the actual physical functions that are driving the system, such as precipitation or transpiration for lake systems as they relate to resulting factors like lake stage or storage.

System identification (Nielsen and Madsen 2000) uses various statistical methods based on the relationship between data input and output to build mathematical models for describing a system. This includes the continued incorporation of the experimental use of input and output data over time to better fit the model.

Models that deal with uncertainty are known as *stochastic hydrology models.* Data based on these types of models have been used within hydrology to simulate the rainfall-runoff relationship, represent the impacts of precipitation and subsequent run-off, and perform real-time control on systems.

Case Study 11.3

Polynomial Regression Seepage Model of Lake Jackson, Leon County, Florida

W. L. Evans, 1996, "Modeling the Hydrodynamics of Closed Lake Basins: A Treatise on Lake Seepage" (master's thesis, Department of Geology, Florida State University).

Investigation of lake hypsometry and its effects on the numerous variables of a lake system involved in mass balance is intricate and nonlinear in behavior. One way to examine these relationships is to use a polynomial regression, which is a special form of multiple linear regression analysis using a least squares approach (see appendix 2.2) to fit a mathematical model to nonlinear data. These models are in the form of $y = \beta_0 + \beta_1 x + \beta_2 x^2 + \beta_3 x^3 + \ldots + \beta_n x^n + \varepsilon$, where β are the coefficients from the regression and ε is a statistical error term.

As stated previously, the hypsometric function $A(h)$ relates the surface area A of the lake to its level h and the change in mass within the lake after interval of time dt (equation (11.7)).

Area is a function of lake level and can be written in the form of a nth ordered polynomial analogous to a polynomial multiple regression, such that

$$A(h) = A_0 + A_1 h + A_2 h^2 + \ldots + A_n h^n \tag{11.48}$$

and, subsequently, the change in lake volume due to evaporation and seepage from the lake bottom and ponors as a second order polynomial is

$$\frac{\partial h}{\partial t}(A(h)) = E(A(h)) - \left[\frac{K_A}{L}(A(h))\right] - \frac{K_p ah}{L} + \left[\frac{K_A h}{L} - K_A\right] + \left[\frac{K_A h_0}{L}(A(h))\right] + \frac{K_p ah_0}{L} - K_p a. \tag{11.49}$$

After removing evaporation and assuming that flow through ponors is negligible, the equation for seepage is reduced to

$$\frac{\partial h}{\partial t}\left(A_0 + A_1 h + A_2 h^2\right) = -\frac{K_A}{L}\left(A_0 h + A_1 h^2 + A_2 h^3\right) + \left[\frac{K_A h_0}{L} - K_A\right]\left(A_0 + A_1 h + A_2 h^2\right). \tag{11.50}$$

Letting $\beta_0 = K_A/L$ and $\beta_1 = (K_A h_0/L - K_A)$, then

$$\frac{\partial h}{\partial t}\left(A_0 + A_1 h + A_2 h^2\right) = -\beta_0\left(A_0 h + A_1 h^2 + A_2 h^3\right) + \beta_1\left(A_0 + A_1 h + A_2 h^2\right). \tag{11.51}$$

More generally, for a polynomial order n, $\partial h/\partial t$ is equal to

$$\frac{-\beta_0\left(A_0 h + A_1 h^2 + A_2 h^3 + \ldots + A_n h^{n+1}\right) + \beta_1\left(A_0 + A_1 h + A_2 h^2 + \ldots + A_n h^n\right)}{\left(A_0 + A_1 h + A_2 h^2 + \ldots + A_n h^n\right)} \tag{11.52}$$

Combining like terms, $\partial h/\partial t$ becomes

$$\frac{\beta_1 A_0 + \left(-\beta_0 A_0 + \beta_1 A_1\right)h + \left(-\beta_0 A_1 + \beta_1 A_2\right)h^2 - \left(\beta_0 A_2 h^3\right) - \beta_0\left(\ldots A_n h^{n+1}\right) + \beta_1\left(\ldots A_n h^n\right)}{\left(A_0 + A_1 h + A_2 h^2 + \ldots + A_n h^n\right)} \tag{11.53}$$

which can be solved numerically.

11.3.2a. Geostatistical Methods

Deterministic methods use mathematical formulas or other relationships to interpolate values. As discussed in section 5.3, common "traditional" or "conventional" approaches include the *nearest neighbor, arithmetic mean, isohyetal method, weighted mean*

(*Thiessen polygon*), and *inverse distance weighting*. These methods range from assigning the value of a measured point to neighboring unmeasured locations (nearest neighbor approach) to estimating ungauged values by use of a weighted average of nearby points or locations (inverse distance weighting). The latter uses distance as the only factor influencing the calculation such that the closer the measured location, the more significant the weight.

Geostatistical methods were first devised in the mining industry and have since spread to the oil industry and most aspects of the earth sciences. Danie G. Krige, a South African mining engineer, and Herbert S. Sichel, a statistician, developed a new innovative estimation method for evaluating disseminated ore reserves after "classical" statistical techniques were found inadequate (Krige 1951; Sichel 1952). These concepts were further developed and formalized by George Matheron (1962) who coined the term "kriging" in recognition of Krige's work.

The main difference between classical statistics and geostatistics is the assumption of spatial dependency. Or specifically, the location of the data elements with respect to one another is essential in the analysis, modeling, and estimation procedures.

The advantage of geostatistical models is the ability to provide estimates of uncertainty for the spatial interpolation estimates. Kriging, which was developed in the mining industry for estimating disseminated ore reserves (Krige 1951; Sichel 1952), is a geostatistical technique because it assigns weights both on distance between surrounding points and on the spatial autocorrelation among the measured points by modeling the variability between the points as a function of distance.

Geostatistical methods are based on the theory of a *regionalized variable*, a concept devised by Matheron (1962, 1963) to describe the significance of the spatial aspect of geostatistical data to represent variables distributed in space. Such variables possess both a random aspect, which accounts for local irregularities, and a structured aspect, which reflects large-scale tendencies (Armstrong 1998). Yet, despite the randomness and spatial aspects, geostatistical methods still possess the usual distribution statistics, such as *mean* and *variance*. Thus, regionalized variables are ideally suited to describe almost all variables encountered in the earth sciences (Journel and Huijbregts 1978).

Geostatistics provides a set of tools for incorporating the spatial correlation of observations in data processing (Goovaerts 1997) and is becoming a preferred method due to the ability to better estimate values at ungauged locations. Several investigations indicate that geostatistics produce better estimates of rainfall than traditional methods (Bacchi and Kottegoda 1995; Christel and Reed 1999; Goovaerts 2000; Campling, Gobin, and Feyen 2001; Drogue et al. 2002; Buytaert et al. 2006; Mair and Fares 2011). This appears especially relevant as it applies to areas of investigation with mountainous terrain or wide-ranging topographic elevations.

Geostatistics can also include secondary variables or attributes, such as weather radar and elevation, in conjunction with a primary element, such as rainfall, to improve precipitation estimates (Mair and Fares 2011). This ability to include additional variables, or *covariates*, for interpolation—such as topography or surface elevation with distance between points of rainfall as an example—can lead to a better predicted surface because the additional information helps identify highs and lows associated with the spatial process that otherwise would be missed. Another potential benefit is the ability to reduce the uncertainty of the interpolation model's output if, in fact, the covariates explain some of the spatial variability in the process of interest (Eberly et al. 2004).

Kriging is one of the most well-known representatives of geostatistical or stochastic methods. Kriging methods are gradual, local, and maybe not exact (i.e., may or may not perfectly reproduce the measured data). Also, they are not by definition set to constrain the predicted values to the range of the measured values. Like the inverse distance weighting method, kriging calculates weights for measured points in deriving predicted values for unmeasured locations. With kriging, however, those weights are based not only on distance between points but also the variation between measured points as a function of distance (see Figure 5.13). The kriging process is composed of two parts: (1) analysis of this spatial variation and (2) calculation of predicted values (Eberly et al. 2004).

Spatial variation is analyzed using *variograms*, which plot the variance of paired sample measurements as a function of distance between samples. An appropriate parametric model (equation) is then typically fitted to the empirical variogram and used to calculate distance weights for interpolation. Kriging selects weights so that the estimates are unbiased, and the estimation variance is minimized. This process is like regression analysis in that a continuous curve is being fitted to the data points in the variogram. Identifying the best model may involve running and evaluating a large number of computer-generated of models, a process made simpler by the geostatistical software packages. There are many software products available for geostatistical analysis, both free and for purchase. (As one example of information and links to these and many other software packages, see Free Geography Tools at https://freegeographytools.com/2010/the-big-list-of-geostatistical-geospatial-analysis-software-iii-r-to-z).

After a suitable variogram model has been selected, kriging creates a continuous surface for the entire study area using weights calculated based on the variogram model and the values and location of the measured points. The analyst can adjust the distance or number of measured points that are considered in making predictions for each point. A fixed search radius method will consider all measured points within a specified distance of each point being predicted, while a variable search radius method will use a specified number of measured points within varying distances for each prediction.

Because kriging employs a statistical model, there are certain assumptions that must be met. First, it is assumed that the spatial variation is homogenous across the area of investigation and depends only on the distance between measured sites. There are different kriging methods, and each has other assumptions that must be met. *Simple kriging* assumes that there is a known constant mean, that there is no underlying trend, and that all variation is statistical. *Ordinary kriging* is similar except it assumes that there is an unknown constant mean that must be estimated based on the data. *Universal kriging* differs from the other two methods in that it assumes that there is a trend in the surface that partly explains the data's variations. This should only be used when it is known that there is a trend in the data (Eberly et al. 2004).

11.3.2b. Summary

Despite the advantages, geostatistical analysis models mathematical objects, not geological or environmental objects. Thus, emphasis on understanding the data, via exploratory analysis, trend analysis error identification, and dealing with sampling issues and nonstationarity is required to successfully use this methodology. André Journel (Isaaks and Srivastava 1989, xi) stated that "geostatistics is an art, and as such, is neither completely automatable nor purely objective," emphasizing that "there are no accepted universal algorithms for determining a variogram/covariance model, that cross-validation is no guarantee that an estimation procedure will produce good estimates at unsampled locations, that kriging needs not be the most appropriate estimation method, and that the most consequential decisions of any geostatistical study are made early in the exploratory data analysis."

For example, Eberly et al. (2004) conducted an experiment where twelve independent geostatisticians were given the same data set and asked to perform the same kriging. The twelve results were vastly different due to widely different data analysis conclusions, variogram models, choices of kriging type, and search strategy.

11.3.3 Mixed Models

A mixed model is a combination of deterministic and stochastic models to describe a system. Nielsen and Madsen (2005) coupled theoretical relations known from the theory of heat transfer with data on heat consumption and climate (temperature, wind speed, and global radiation) and used statistical correlation to establish a mathematical model on heat consumption. (Such an approach is an example of the grey box model in Figure 11.2.)

Case Study 11.4

Seepage Modeling of Lake Jackson, Leon County, Florida

W.L. Evans, 1996, "Modeling the Hydrodynamics of Closed Lake Basins: A Treatise on Lake Seepage" (master's thesis, Department of Geology, Florida State University).

Estimating lake seepage and flow ($[L^3]$/day) through lake bottoms sediments can be done using a multiple regression model (see appendix, sections 2.2-2.5) in conjunction with Darcy's equation. First, use the multiple regression to estimate the equivalent areal hydraulic conductivity (K_A) over an entire range of lake level based on empirical estimates at discrete lake levels (Figure 11.14). The result is an equation in the form of

$$K_A = \theta_0 - \theta_1 h + \theta_2 h^2 + \varepsilon \tag{11.54}$$

where θ_0 is the intercept, θ_1 and θ_2 are the fixed coefficients of regression, and ε is the random component or error term (Table 11.1). The coefficient of determination (R^2) was 0.90. Therefore, 90% of the variation in K_A is explained by changes in lake level (h) (see also appendix, section 2.1). For Lake Jackson, as the lake level decreased so did K_A.

From the regression estimates of K_A, seepage flow (Q_s) (m^3/day) can be modeled by incorporating an alternative form of Darcy's equation

$$Q_s = -K_a \frac{dh}{L} A \tag{11.55}$$

where L is the thickness of the sediments beneath the lake bottom and A is the area of the lake (see Figure 11.4). Seepage (q_s) ([L]/day) can be calculated from Q_{qs} by

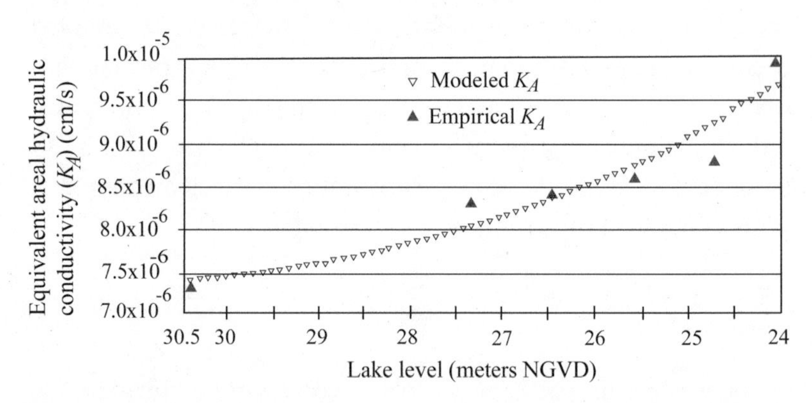

Figure 11.14. Comparison of multiple linear regression model estimates of areal hydraulic conductivity (K_A) with empirical estimated K_A versus declining lake levels at Lake Jackson, Leon County, Florida.

Table 11.1. Multiple regression analysis of K_A

R^2	Error ε	*y*-intercept	θ_1	θ_2	F-test[a]	*t*-test of *y*-intercept[b]	$t-\theta_1$	$t-\theta_2$
0.90	3.45×10^{-9}	4.55×10^{-7}	-2.4×10^{-8}	3.83×10^{-10}	13.6	2.2	1.68[c]	2.2

Source: Modified from Evans (1996). [a]See appendix, sections 2.2–2.4 for H_0, F-test, and t-test discussion. Reject h_0 if F is > 3.84, where h_0 is the null hypothesis.
[b]Reject h_0 if *t* is > 1.96.
[c]θ_1 was retained in the equation because it resulted in better estimates of K_A.

$$q_s = \frac{Q_{qs}}{A}. \tag{11.56}$$

The models demonstrate that there is a direct correlation between the modeled seepage through the lake bottom and decreasing lake levels (Figure 11.15); as lake level declines, the amount of seepage decreases (Figures 11.15 and 11.16). As predicted by the theoretical model, the rate at which seepage decreases is linear (Figure 11.16).

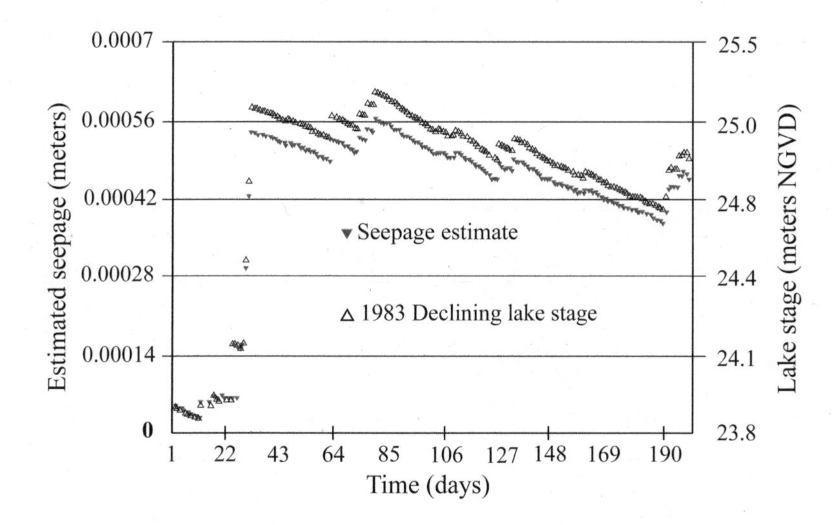

Figure 11.15. Declining lake stage and seepage, Lake Jackson, Leon County Florida, 1983. Modified from Evans (1996).

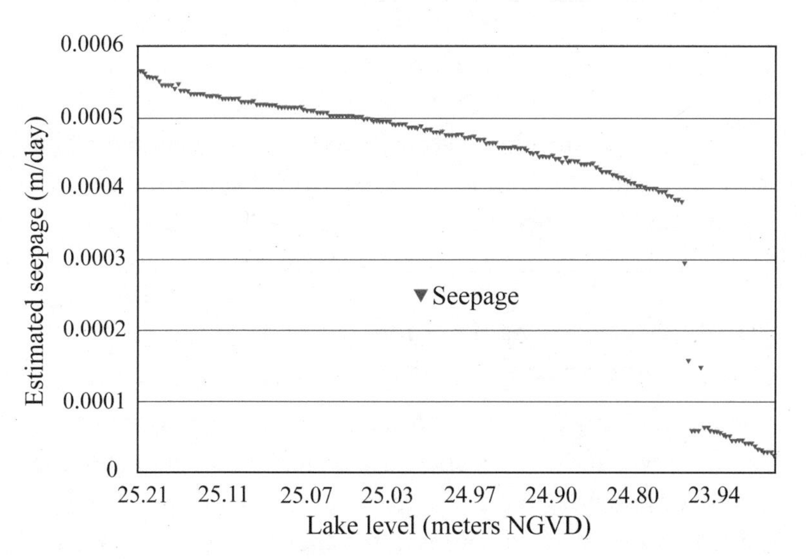

Figure 11.16. Seepage versus declining lake levels, Lake Jackson, Leon County Florida, 1983. Modified from Evans (1996).

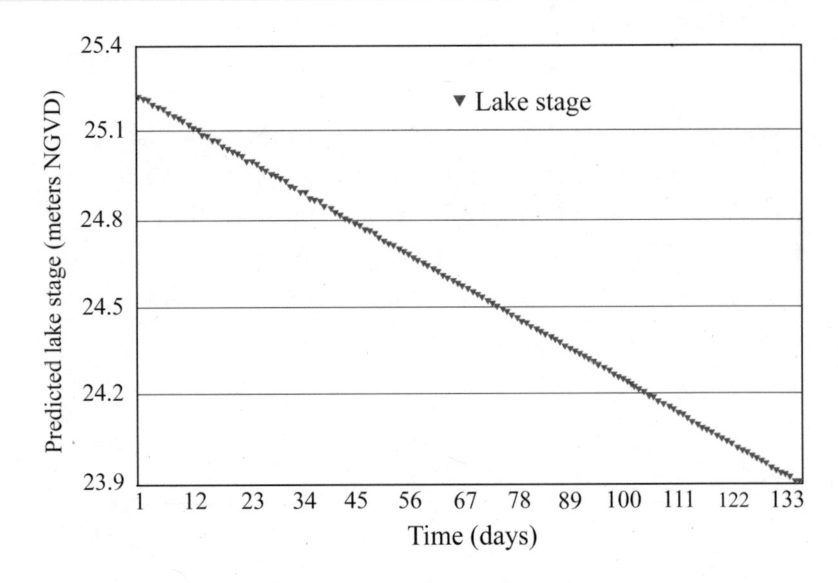

Figure 11.17. Modeled lake level decline due to seepage, Lake Jackson, Leon County, Florida. Modified from Evans (1996).

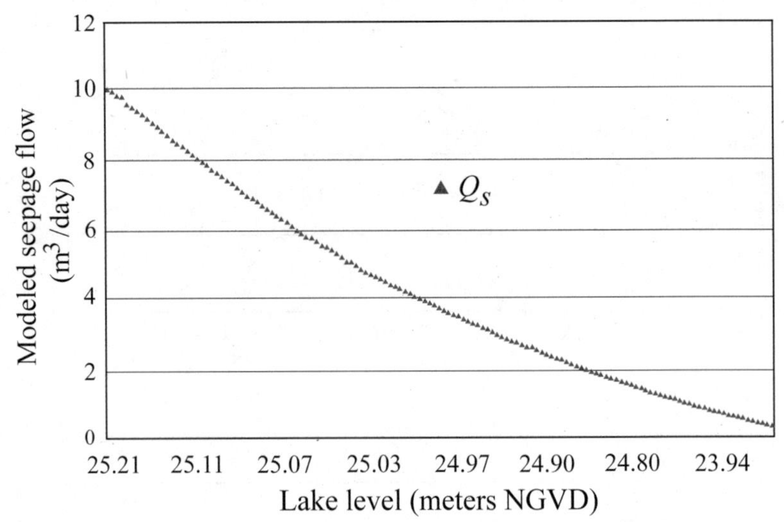

Figure 11.18. Modeled seepage flow (Q_s) (m^3/day) through lake bottom sediments, Lake Jackson, Leon County, Florida. Modified from Evans (1996).

Lake-level decline due to seepage also has a corresponding linear trend (Figure 11.17) and seepage flow (Q_s) appears to decrease exponentially with declining lake stage (Figure 11.18).

11.4. Development of a Model

11.4.1. Introduction

When investigating water budgets or mass balance flows to lakes and wetlands, researchers are typically trying to ascertain current or potential negative impacts to a wetland system as a result of existing or future causes in the area affecting the lake (e.g., watershed). Negative impacts to any environmental system are defined in regulatory terms as a result of *stressors* that, generally, originate from either *point sources* or *nonpoint sources*. The US Environmental Protection Agency (EPA 1998, A-3) defines stressors as "any chemical (e.g., toxics or nutrients), physical entity (e.g., dams, fishing nets, or suspended sediments), or biological entity (e.g., non-native or genetically engineered organisms) that can induce an adverse response. The term 'stressor' is used broadly to encompass entities that cause primary effects, and those primary effects can cause secondary (i.e., indirect) effects."

The EPA (2017) defines nonpoint sources of pollution in terms of its "legal" definition of point source pollution: "The term 'nonpoint source' is defined to mean any source of water pollution that does not meet the legal definition of 'point source' in section 502(14) of the Clean Water Act. . . . The term 'point source' means any discernible, confined and discrete conveyance, including but not limited to any pipe, ditch, channel, tunnel, conduit, well, discrete fissure, container, rolling stock, concentrated animal feeding operation, or vessel or other floating craft, from which pollutants are or may be discharged. This term does not include agricultural storm water discharges and return flows from irrigated agriculture."

Factories and sewage treatment plants are two common types of point sources. "[Nonpoint source pollution] generally results from land runoff, precipitation, atmospheric deposition, drainage, seepage, or hydrologic modification . . . unlike pollution from industrial and sewage treatment plants, [it] comes from many diffuse sources . . . [and] is caused by rainfall or snowmelt moving over and through the ground. As the runoff moves, it picks up and carries away natural and human-made pollutants, finally depositing them into lakes, rivers, wetlands, coastal waters, and groundwaters. Nonpoint source pollution can include:

- Excess fertilizers, herbicides, and insecticides from agricultural lands and residential areas
- Oil, grease, and toxic chemicals from urban runoff and energy production
- Sediment from improperly managed construction sites, crop and forest lands, and eroding streambanks

- Salt from irrigation practices and acid drainage from abandoned mines
- Bacteria and nutrients from livestock, pet wastes, and faulty septic systems
- Atmospheric deposition and hydromodification

States report that nonpoint source pollution is the leading remaining cause of water quality problems. The effects of nonpoint source pollutants on specific waters vary and may not always be fully assessed. However, we know that these pollutants have harmful effects on drinking water supplies, recreation, fisheries, and wildlife" (see "Polluted Runoff: Nonpoint Source Pollution" on the EPA's website at www.epa.gov/nps/basic-information-about-nonpoint-source-nps-pollution).

11.4.2. Awareness of a Problem

Water quality or environmental degradation problems can come to the proper management authority's attention in various ways, including as an *informal* complaint or observation from the public or by experts; a *regulatory issue* where monitoring of a certain resource or environmentally sensitive area (e.g., water supply reservoir, critical ecosystem habitat, etc.) is producing results that may be in violation of regulatory standards (i.e., water quality) or approaching critical levels; a *potential need*, such as a proposal to construct a factory or residential development where point source or nonpoint source pollution are known to be problematic and need to be addressed; or a *proactive need* to protect existing resources and prevent future degradation. For example, a local community may want to ensure protection of its drinking water supply for the next fifty to one hundred years, or a wildlife agency may wish to protect critical habitat.

Once the initial problem is identified, an initial assessment known as *scoping* should be instigated. Scoping is the gathering and analysis of information that an agency or government entity will use to establish the breadth, or scope, of environmental review of a proposed project. This includes the magnitude, source, and causes of the problem, which should be ascertained with inquiries (US Environmental Protection Agency 2019) such as:

What is the extent of the problem or threat (spatial scale)?
What is the existing and projected persistence of the problem (time scale)?
What is the severity of the problem or threat (level of risk)?
Is there a threat to human health?
Is it causing irreversible ecological damage?
Is it repairable or restorable?
Do we know the cause(s) of the problem or threat?
Is it due to point and/or nonpoint sources of pollution?
Are there multiple sources of the problem?
Is the problem exacerbated by interaction with other stressors, including chemical stressors, physical stressors, or the alteration or loss of habitat?

Once these questions are answered, if further assessment required, determine what type of assessment. Typically, the choice is to monitor or model, though both may be required (e.g., modeling quality is dependent on the quality and quantity of data available to it). Monitoring actual data is always preferable to model predictions. Modeling is useful for many purposes, but it may not always be the best tool for a given situation.

Once identified, if the level of understanding of the severity and source of the problem is still inadequate, then simple modeling can initially be used to better understand the scope and nature of the environmental problem. For example, does the sudden atypical eutrophication of a lake, impairing its use, indicate an increase in nutrient loads? Both monitoring and modeling can be used to help answer this question. Monitoring actual lake responses would be more reliable, but there may not be the available data or resources necessary (e.g., funding, personnel, equipment, time, etc.) to collect the data. Lakes often exhibit a high degree of variability in algal response both within a season and year to year, hence short-term collection of chlorophyll (a measurements of algal concentrations) may not provide a definitive answer.

In some cases, a simple model (e.g., some available from the EPA's Center for Exposure Assessment Modeling) can quickly be applied to give a general answer or estimate of the risk posed to water quality. For instance, you could use an empirical model, which predicts trophic state based on a statistical relationship between phosphorus load and lake retention time, and algal concentration.

Simple models can also help to estimate runoff flows or contaminant loads for the purpose of assessing relative magnitudes, thus targeting areas of greatest risk. This is another area where modeling can be more cost-effective than monitoring if a high degree of accuracy is not required for initial estimates. Simple models, such as loading coefficients, can aid in identifying areas where runoff is greatest and areas that are likely to generate the largest loads of a given pollutant. Such modeling is particularly useful for obtaining initial estimates of nonpoint loads; it is difficult to gather monitoring data on nonpoint runoff flow and pollutant loads. This makes modeling of runoff-generated loading an attractive option (US Environmental Protection Agency 2019). For more information, see Mills et al. (1985).

11.4.3. Watershed Modeling

For lake systems and water resource studies in general, drainage basins are a suitable unit of analysis for water resource management because the basin concept of water passing through a stream cross section at the catchment outlet, such as at a lake, originates as upstream precipitation on the basin (see section 3.1). Dependent upon the size and scope of the problem, modeling the entire drainage basin may or may not be necessary. For example, the area of concern may be the proposed construction of a sewage treatment plant or other point source factory. Then all that may be required is to model and monitor the sub-basin factory location and areas impacted downstream from the facility.

Conversely, if the drainage basin is undergoing a population increase and fast-growing economic development, urbanization, or agricultural growth, then there are potential other nonpoint and point sources of eutrophic nutrients and contaminants throughout the watershed that can impact the lake. These include runoff from a number of small and large animal operations as well as urban runoff containing leaf litter, yard fertilizers, and other organic wastes (both nonpoint sources); combined sewer overflows during storm events (an episodic point source); and increased consumptive use of water resources. If the management objective is to protect water quality and to allow economic development, then the modeling strategy is to include the entire catchment impacting the lake, including management strategies for both point and nonpoint sources. For instance, it would be unwise to impose expensive new treatment costs on the sewage treatment plant, resulting in significant increases in local utility rates, if the same level of water quality protection could be obtained with less expense through nonpoint source controls.

The complexity of calculating and modeling drainage basin surface and subsurface flow, especially as it applies to lake systems, has been discussed in previous chapters. As summarized in section 7.6, catchment hydrological methods of analysis and modeling can be adequately estimated with the conceptual methods described throughout this textbook and with the various models available online. It is therefore essential to start by asking how detailed an assessment is required and to develop a modeling strategy that meets these requirements. In general, the level of detail that is required will be closely related to the types of decisions that need to be made, the level of uncertainty that is acceptable in results, and the financial and ecological implications of these decisions.

11.5 Model Selection, Validation, Calibration, and Documentation

11.5.1. Introduction

Models are most beneficial for extrapolating from current conditions to potential future conditions. For example, given the expected increase in the rate of population growth and development within a watershed, and the accompanying conversion of land use and land cover, what impacts are expected to environmental resources under existing management measures and regulatory programs? Are these protections adequate, or are additional management strategies needed?

To best explain the processes and principles of *model selection*, *validation*, and *calibration*, we can use a hypothetical example of a watershed with various types of land use that drain to a large lake reservoir (Figure 11.19). The lake has a dam at the lake stream outlet for controlling flow leaving the lake to provide water for drinking, agricultural irrigation, and industrial uses. In the past, the watershed was mostly agricultural cattle ranches, but the town in the watershed is growing rapidly and land use is

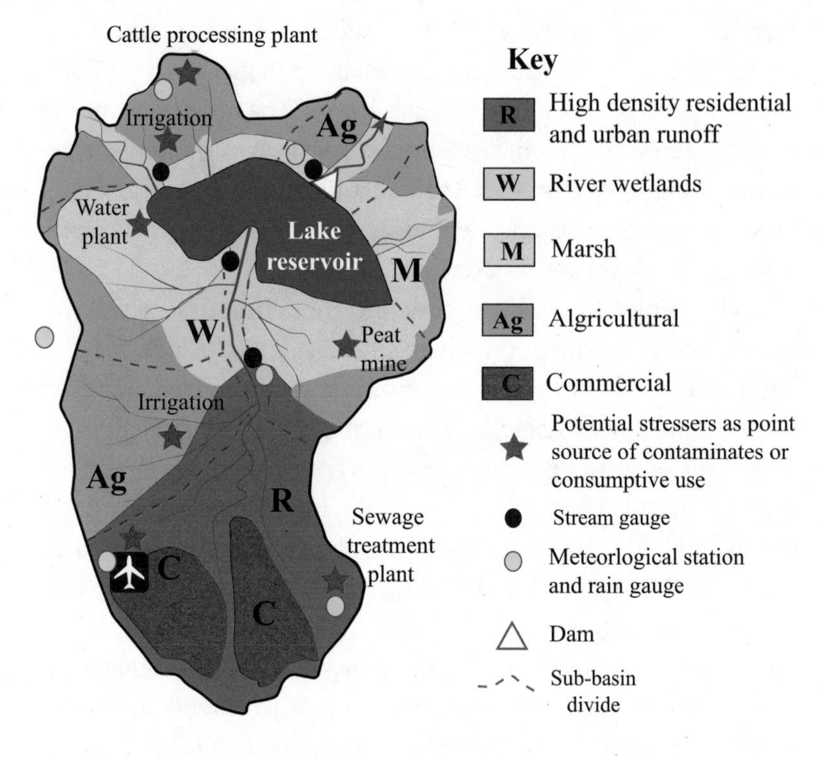

Figure 11.19. Hypothetical land use map with projected increase of population and urbanization. The lake is a water supply reservoir.

changing. Based on information obtained from the local planning department, the area is undergoing rapid population growth and urbanization as agricultural land is converted to suburban, commercial, and industrial uses. Future distribution of land uses in the watershed can be estimated by current forest to agricultural land conversion rates conducted by the department. As such, there is concern for potential impacts, including amplified flooding, increased contaminant loading to the wetlands, and potential water quality degradation and lake decline due to nutrient inputs and increased consumptive use.

11.5.2. Phase I: Early Assessment/Scoping

In the early assessment phase, the purpose of modeling is to provide an initial scoping-level estimate rather than to make final management decisions based on model results. Therefore, it is appropriate to undertake the initial modeling with a simple but appropriate model. Early model results will be subject to considerable uncertainty, but this is not a great concern for initial scoping. After examining the initial results, decisions can be made to determine how much additional thoroughness or complexity should be introduced (US Environmental Protection Agency 2019).

In modeling, *uncertainty analysis* draws upon a number of techniques for determining the reliability of model predictions accounting for various sources of uncertainty in model input and design. At the scoping level of an environmental assessment, a detailed quantitative uncertainty analysis is not necessary (US Environmental Protection Agency 2019). However, it is preferable that some information is available on the quality of predictions to interpret results and make an informed decision about the need for additional modeling. For instance, if the scoping model indicates that no water quality impacts are expected, what is the confidence that no problems will occur, or is there a good chance that the model prediction of no adverse impacts is wrong? To address the initial assessment of no adverse outcome, the model could be run with the worst-case estimates of nutrient loading to see if the outcome still holds. A second test would be to conduct a *sensitivity analysis* of the model. Sensitivity analysis is performed to describe how sensitive the outcome variables are to variation of individual input parameters. Since there may be multiple input parameters, sensitivity analysis can help determine which factors drive most of the variation in the outcome (US Environmental Protection Agency 2019).

11.5.2a. Spatial Considerations

The purpose of the scoping application is to predict nutrient loads to the lake. Therefore, it is not necessary to try to accurately predict nutrient concentrations in each tributary stream if the general loading pattern is adequately represented. Thus, sample points can be located downstream from various tributaries or near the point of entry into the lake. In addition, since this is a scoping-level study, it is not critical that we separate the impact of each individual source of nutrient pollution, as it might be if the modeling was to be used to impose pollutant limitations on various stakeholders (i.e., sewage treatment plant).

11.5.2b. Time Considerations

The receiving water of interest is a lake, with a reasonably long residence time due to dam controls. Hence the initial intention of scoping is to predict an approximate average nutrient concentration in the lake in response to land use changes. Thus, the focus should be during the growing-season or yearly average concentrations (discrete parameters) as opposed to transient changes in nutrient concentrations (continuous model). This would require a simpler *steady-state model* (variables remain constant) rather than a more complex *continuous simulation model* for the initial scoping assessment.

A similar process of scoping is undertaken for the initial assessment of the impacts of consumptive use on the lake system as population and land use changes. Increased consumptive use of lake and groundwater can result in decreasing reservoir capacity and decline of the overall health of the lake system. The initial focus should be examining historical consumptive use with storm patterns (input) and variation with consumptive use. This should include seasonal or temporal patterns of use (e.g., does use increase or decrease during certain times of the year).

11.5.3. Phase II: Model Selection

Once the scoping assessment is complete, the next step is to develop a more-detailed modeling strategy based on the following four principles of model selection:

1. *Have a clear definition of the purpose of the investigation, including defining the problem, objectives, area of investigation and boundary conditions.* The problem in our hypothetical example (Figure 11.19) is defined as a projected dramatic rise in human population causing increased urbanization, consumptive use of water (e.g., decreased agricultural use with projected increased drinking and industrial uses), and sewage treatment. Urbanization has the potential for increased nonpoint source runoff and flooding (episodic sources) with increased nutrient and contaminant loads to the lake. Point source pollution of concern are the peat mine, airport, sewage treatment facility, and cattle processing plant.

Nutrient and contaminant increases can cause lake eutrophication and water quality degradation. Increased consumptive use of lake and groundwater can result in decline in lake stage, area, and volume, all decreasing reservoir capacity and impacting, lake ecology and recreational use. Thus, the purpose of the investigation will be a multifaceted approach for looking at the effects of *urbanization* and increased consumptive use impacts on the lake due to population growth.

Due to this definition of the problem, the model's objectives should be to (1) estimate rainfall and subsequent runoff and baseflow to the lake based on design storms selected from historical rain and stream gauge records; (2) use storm flow models to determine existing and predict future contaminant and nutrient loads to the lake and their impact in terms of trophic states (e.g., eutrophication and oxygen depletion) from phosphorous and nitrogen concentrations; and (3) examine past and present consumptive use amounts and patterns using historical water use and meteorological data (precipitation and evapotranspiration) and to see if there are any trends associated with lake-level behavior and water use.

Well-defined boundary conditions include the study area (e.g., drainage basin, sub-basins, streams, lake, or areas of point source and nonpoint source pollution) and surface flow versus groundwater flow, or both. This should include existing sampling sites and determination of new sites locations (e.g., downstream from sewage treatment facility, stream gauges locations etc.). For this example, primary focus should be on surface water evaluation.

2. *Verify the necessity for modeling by asking questions.*

How can a model help address the questions and problems relevant to decisions?

What models are available that can aid in gathering information used for predictive and forecasting for the various problems and objectives stated in step 1?

How can a model be used to link stressors?

Are there patterns and correlations between meteorological and flooding impacts as well as contaminant loads from an array of design storms?

Are there patterns between stressors such as contaminant concentration and trophic states and water quality changes in the lake?

Can lake behavior (changes in lake stage) be predicted based on past consumptive use patterns and projected use?

Can these predictions help management in deciding actions to acquire quantitative measure endpoints of water-body conditions?
Is modeling appropriate for examination of the stressors of concern in this situation?

3. *Use the simplest model(s) that will satisfy the project objectives.* Assumptions that more complex models always produce better accuracy and results is not always true. The various formulations in a complex model may contain unseen parameters that may not be relevant to a specific investigation and thus increase the probability of generating false model predictions and observations. In general, utilization of complex modeling systems increases the expertise, time, data requirements, and cost associated to run these programs. Therefore, careful examination of model uncertainty and inherent parameters is recommended before choosing a complex model. When selecting a model, a good rule is to not include any more spatial or temporal data than is necessary or required for meeting the specified objectives.

4. *Define carefully the parameter(s) of interest.* As part of the model selection process, it is essential to define the parameter types and how they are quantified. Determining the adequate level of accuracy for model prediction and forecasting is also necessary as part of the calibration process. Modeled parameters should be selected to coincide with an appropriate endpoint for the analysis, which can serve as an indicator of water quality prediction. This ties in closely with model complexity.

In the examples presented so far for management of a lake, we saw that we might choose to predict stressor loads, water quality, and lake trophic status. Choosing to model effects on the water body or biota will add to the complexity of the modeling effort but may be worthwhile in order to produce results that are more useful for making informed management choices. For flooding and mass balance or for a water budget analysis of lakes, select models that can forecast basin flow with rainfall, runoff, and baseflow. Stream inputs such as routing models, discussed in section 7.5.1, should also be considered.

For this contaminant/nutrient increase scenario, investigation of existing and past nutrient and contaminant concentrations should be considered by selecting specific types of nutrients that are known to accelerate *eutrophic states*, resulting in degradation of lake ecology and water quality.

Look for contaminants associated with the appropriate point source facility, such as nitrogen with sewage treatment, cattle manure or phosphorous with fertilizer, and petroleum toxins associated with areas of high automotive use. (The EPA lists typical stressors associated with specific types of land use, such as sewage plants and meat processing plants, on their page "EPA Eco-Box Tools by Stressors" at www.epa.gov/ecobox/epa-ecobox-tools-stressors.) To the extent possible, use a prediction method consistent with available data. Data availability should be evaluated before beginning the model selection process. Models selected for analysis should have input requirements that match up well with data already available or that can be collected during and after the investigation. This is also essential in model checks and calibration by checking model results with past data trends.

It will often be necessary to use multiple models or link models to address watershed assessment problems (Figure 11.20). The multimodel approach is used to ascertain the various impacts to the system based on a specific problem, such as growth in population and related land use change. Beginning with rainfall models, there is one model to predict loading to a water body from

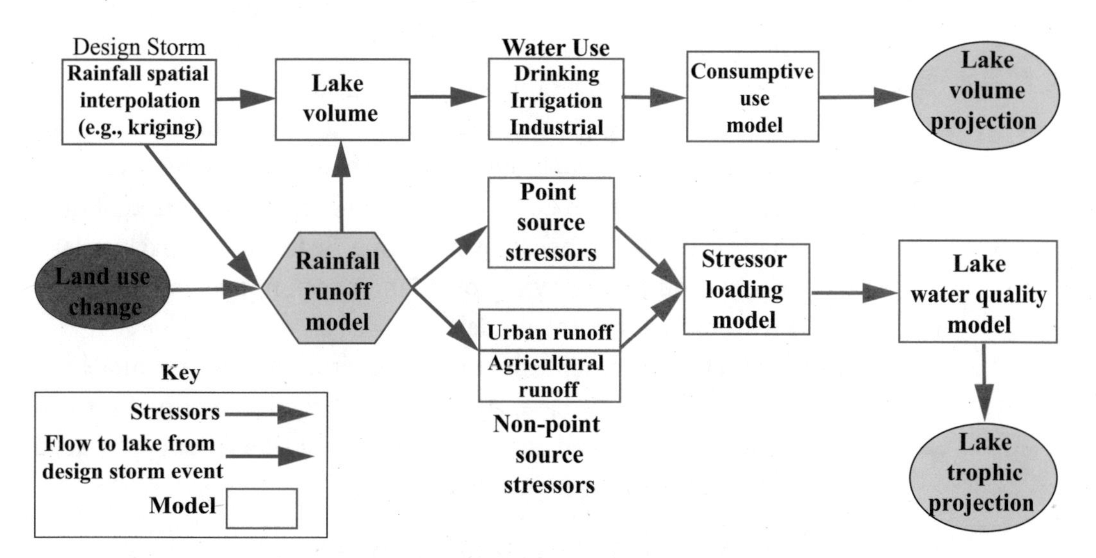

Figure 11.20. Flow chart of the modeling process for investigating projected population increase and impact to the reservoir lake due to increase stressor loading and consumptive use.

nonpoint sources, a second model to predict fate and transport of point source pollutants in the water body, and a third and fourth model to predict effects of the pollutant on water quality and trophic states. Then there is a second set of models to predict consumptive use and its impact and prediction of lake volume. It is important to choose models with compatible data input/output. For instance, if the fate and transport model require daily load values, then the loading model should provide daily output of the appropriate parameters.

11.5.3a. Overview of Hypothetical Model Strategy

In summary, there are two requirements for this hypothetical situation of population growth and land use change as it relates to affecting the lake reservoir system: (1) estimate and predict the runoff of water and loading of stressors from land areas in the watershed (nonpoint sources) and from sites with the potential for point sources (sewage, treatment, cattle processing, airport, peat mine, etc.) and how it affects the lake (water quality and eutrophication); and (2) estimate current and predict future impacts of consumptive use from irrigation, drinking and industrial use to predict changes in lake volume and behavior.

To represent spatial components in the lake system model, the entire basin was required to account for all inputs and outputs to the lake. Initially, surface water is modeled, but groundwater will be impacted as well and should be considered for modeling as part of the investigation as soon as feasible.

Representing time in the model will vary from duration of storms (for initial loading of contaminants and increases in lake volume) to more frequent monitoring over a long period of time to observe system behavior over time. Seasonal patterns for lake macro- and microphyte growth and consumptive use should be incorporated into the models.

Finally, due complexity of basin hydrology, multiple modeling is required in this case. At some point the interaction between stressor loading and consumptive use and lake volume will need to be addressed and modeled.

11.5.4. Phase III: Calibration and Validation

Most environmental models include parameters that must be tuned or adjusted to obtain realistic matches between model predictions and observed conditions. All models require checking and testing to evaluate performance. The first activity is referred to as *model calibration,* and the second is *model validation.* Without calibration and validation, a modeling application is only an educated guess. This may be enough for some scoping applications but generally not for management decisions.

The US Environmental Protection Agency (2019, 29) defines model calibration as "minimization of the deviation between measured field conditions and model output by adjusting parameters of the model. Data required for this step are a set of known input values along with corresponding field observation results. Calibration typically includes a sensitivity analysis, which provides information as to which parameters have the greatest effect on output. Careful consideration should be given to adjustment of model parameters to ensure that values are within the range of reasonable possibility" (Figure 11.21). (The EPA also provides guidance on reasonable values of parameters for water quality models.)

"Model validation," they continue, "involves the use of a second, independent set of information to check the model calibration. The data used for validation should consist of field measurements of the same type as the data output from the model. Specific features such as mean values, variability, extreme values, or all predicted values may be of interest to the modeler and require testing" (29).

There is a variety of statistical testing available for evaluating a model's goodness of fit during calibration and validation. Reckhow, Clements, and Dodd (1990) offer a good introductory summary on statistical evaluation of water quality models. Also, see basic statistical analysis methods in the appendix.

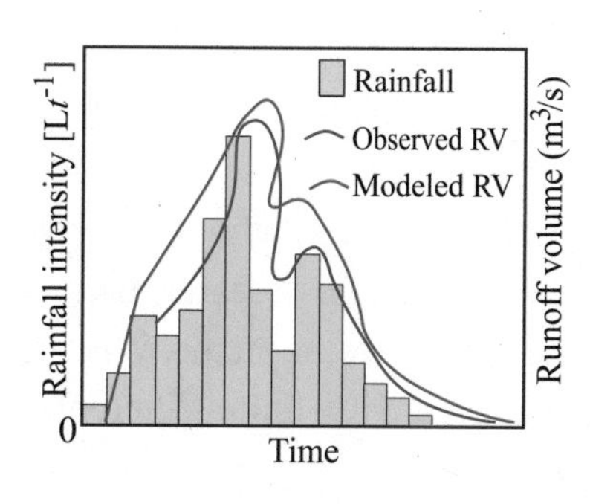

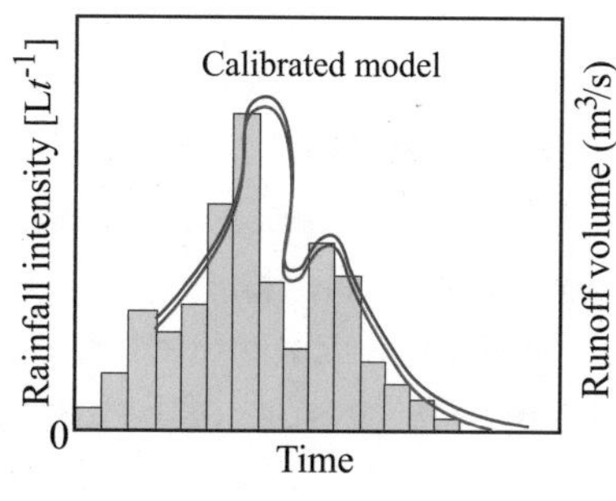

Figure 11.21. Observed runoff volume (RV) versus initial model and calibrated model runoff from hypothetical basin study in Figures 11.19 and 11.20.

11.5.5. Phase IV: Documentation

All models must be documented so that the derivation of the model can be understood and the results reproduced and verified by anyone seeking to use this model. Per the National Research Council (1990), documentation should include:

- a description of the problem;
- a description of the fundamental equations that conceptualize the solution to the problem;
- a list of all assumptions used in the model and the rationale for their use;
- a description of codes (symbolic arrangement of data or instructions in a computer program or the set of such instructions) used in the model;
- verification of the model codes against other solutions to the problem to verify the accuracy;
- an application of the model to a problem with a known solution, even if a simpler problem, for comparison with known results;
- sensitivity analysis;
- quality assurance results;
- model validation;
- a list of prior uses, if any;
- a clear identification of the site-specific data used in the application of the model;
- a characterization of the level of precision, accuracy, and degree of uncertainty in the model results;
- a description of the statutory/policy criteria, if any, used to shape and select the assumptions and the acceptable level of precision, accuracy, and uncertainty; and
- any other information that is essential to understanding or being able to replicate the results.

11.6 Summary

Because of the complex nature of lake-basin hydrogeology, most large-scale lake, wetland, and river basin research incorporates numerical modeling to investigate these systems. Winter (1976, 1978) was one of the first researchers to use finite-difference models to generate complex flow patterns under a variety of hydrogeologic settings to investigate groundwater flow interactions between lakes and underlying aquifers.

The overall theory of lake hydrodynamics (derived Case Study 11.1) is deterministic in nature and, as such, entails integration, partial differentiation, and exponential as well as other linear and nonlinear functions. The final governing equation is in the form of an nth-order polynomial (equation 11.50) resembling a convergent power series. In theory, this mathematical model can be solved numerically. However, while investigating lake seepage, Evans (1996) found that the theoretical models led to a more simplified mixed model. This model used a stochastic solution in the form of a second-order polynomial (equation 11.51) for modeling lake hydraulic conductivity that was derived by applying a multiple linear regression and a modification of Darcy's equation for seepage and overall flow through the lake bottom. From the modeled seepage, more accurate estimates of evapotranspiration can be determined (see section 4.5.1).

When modeling surface flow to lakes from storm events, the process becomes more complex as each component or subsystem of a greater lake system (illustrated in Figure 11.1, 11.20, and 7.16) are systems unto themselves with the components of each subsystem also a subsystem *ad infinitum*. In addition to the governing equations for the lake system, specific numerical values and equations may be needed to characterize all the other processes involved.

Development of an effective modeling implementation requires an initial scoping assessment followed by a more rigorous modeling approach. Basic modeling strategy represents the principles discussed so far and should, in summary, incorporate the following factors:

1. A clear definition of potential problems and types of stressors to be examined, an estimation of their impacts to the lake (eutrophication, water quality), a comparison of levels of stress caused, and identification of prioritized areas of stressor sources (point and nonpoint)
2. Spatial scale and resolution of the application
3. Time scale and resolution of the application

4. Level of complexity and detail required for extrapolation of monitoring data
5. Evaluation of the system through validation and calibration for both current assessments and the prediction of future outcomes/endpoints
6. Documentation of the model
7. Linkage between model output and management decision-making

More advanced computer programs such, as the US Geological Survey (USGS) MODFLOW program, have been developed to stimulate a variety of flow types in both surface and groundwater scenarios (Harbaugh 2005). Numerous MODFLOW finite-difference models, as well as other model packages, have been developed to quantify flow and estimate a range of fluxes between groundwater and surface water interactions in both one and three dimensions. For example, to simulate lake-aquifer interaction, Merritt and Konikow (2000) used a MODFLOW groundwater model couple called MOC3D Solute-Transport Model to estimate lake-aquifer flux. (MODFLOW and related programs information can be found online at USGS, www.usgs.gov/mission-areas/water-resources/science/modflow-and-related-programs?qt-science_center_objects=0#qt-science_center_objects.)

As discussed throughout the text, the area of review for assessing lake mass balance is often at the drainage basin or sub-basin level. The EPA's BASINS model combines spatially distributed land use data, EPA's Reach File 1, and the Hydrological Simulation Program-FORTRAN simulation model for watersheds and rivers into a convenient package with an ArcView Geographic Information System (GIS) interface. Only by using the power of a GIS interface is it convenient to undertake a continuous simulation model of such a large watershed. (It should be noted that the EPA's BASINS version 2 will have the ability to address routing of streamflows, and transformations of pollutant loads during stream transport is included as well. BASINS is free and available to anyone to download and use for any purpose; "BASINS User Information and Guidance" is available at www.epa.gov/ceam/basins-user-information-and-guidance.)

Due to the complexity of lake systems, many questions can

arise as to the validity or adequacy of a chosen model. The natural reaction is for investigators to devise and conduct long-term experiments, which in the scientific tradition will gradually improve the understanding of the process. Hydrogeological and surface/stormwater computer modeling is a separate and complex multidiscipline. Despite the advances in computer modeling, a model is only as good as the accuracy and precision of hydrological data input.

APPENDIX

Section 1. Calculus

1.1. Introduction

Two of the greatest minds of the seventeenth century, Gottfried Wilhelm Leibniz and Isaac Newton, formulated the principles of calculus independently as a means to analyze physical quantities. It was developed to study four classes of scientific and mathematical challenges of the time (Anton 1980):

1. Find the tangent to a curve at a point.
2. Find the length of a curve, the area of a region, and the volume of a solid.
3. Find the maximum or minimum value of a quantity—such as the maximum and minimum distance of a planet from the sun or the maximum range attainable for a projectile by varying the angle of fire.
4. Find the velocity and acceleration of a body at any instant given a formula for the distance traveled by the body in any specified amount of time. Conversely, given the formula that specifies the acceleration or velocity at any instant, find the distance traveled by the body in a specified interval of time.

The most fundamental concept in mathematics is the concept of a function. First defined by Leibniz in 1673, a function is a relation of two terms called variables (i.e., x and y) because their values vary. If every value of x is associated with one value of y, then y is a function of x.

Traditionally x is designated the independent variable and y is the dependent variable because its value is dependent on the value of x.

Calculus consists of two major components: the derivative and the integral. The derivative measures the rate of change at any given point on a curve defined by a function. Or simply put how the variable y is changing as a function of x at a single point along the curve. The integral or antiderivative is the reverse process of the derivative. For example, the instantaneous velocity (y) at which an object is traveling at a certain point in time (x) is its derivative at this position. Conversely, the integral of the velocity over time is the object's position.

1.2. Derivative

When y is a direct linear function of x, the graph of this function is a straight line. This is written mathematically as $y = f(x) = mx + c$. Where c is the y intercept and the slope m of the line is the rate of change between two points on the line where

$$m = \frac{\text{change in } y}{\text{change in } x} = \frac{\Delta y}{\Delta x} \tag{A1.1}$$

and the symbol Δ (Greek letter Delta) is an abbreviation for "change in." This formula is true because

$$y + \Delta y = f(x+ \Delta x) = m\,(x + \Delta x) + c = mx + c + m\Delta x$$
$$= y + m\Delta x$$

or

$$\Delta y = m\Delta x. \tag{A1.2}$$

This gives an exact value for the slope of a straight line (i.e, the slope is the same along all points on the line). However, if the function f is not linear (i.e., its graph is not a straight line), then the change in y divided by the change in x differs, thus differentiation is a method to find an exact value for this rate of change at any given value of x. This is done by determining the slope of the tangent to a curve at a point (Figure A1.1). Thus, the slope of the secant lines converges to the slope of the tangent line as Δx approaches zero.

Differential notation, as defined by Leibniz, is where such an infinitesimal change in x is denoted by dx, and the derivative of y with respect to x is written dy/dx. Applications of differentiation include related rates where one tries to find the rate at which some quantity is changing by relating it to other quantities whose rates of change are known. Lake levels with lake volume would be such

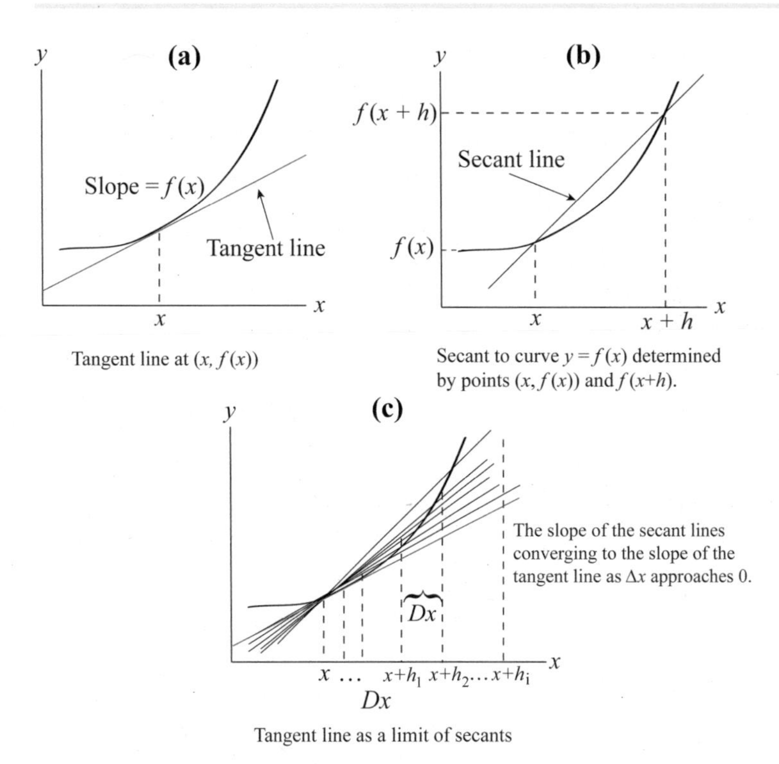

Figure A1.1. Derivation of the slope at *x* by computing the rate of change as the limiting value for the ratio of differences between $\Delta y/\Delta x$ as Δx becomes infinitely small.

an example. Determining maximum and minimum values of a function, known as optimization problems, is another common use important in economic or engineering problems, such as conditions where cost is a minimum or efficiency is at a maximum.

1.3. Integral

The integral measures area under a curve within a specified interval of the curve. Figure A1.2 illustrates this concept, which was defined mathematically by Bernhard Riemann. The integral is based on a limiting procedure that approximates the area of a curvilinear region by dividing the region into rectangles. However, as the number of rectangles increases to infinity, the sum of the rectangle areas do not exactly measure the area under the curve as there is always an "error" on the right half of the interval. To correct for the error, the widths of *all* rectangles must decrease to zero as the number of rectangles approaches infinity.

In summary, integration is the process of taking the curve defined by a function and determining the area under the curve by adding up all the segments (dx) as they approach zero. The symbol for integration ∫ is merely an elongated S for "the sum of." Therefore ∫ dx equals the sum of all the rectangles as defined by the intervals of x. A definite interval defines the area under the curve of interest, such as the area bound between a and b in Figure A1.2, and is designated as $\int_a^b dx = x$. Hence, the definite interval is not a function but a number that is the limit sum of all the rectangles under the curve as their widths approach zero and their number becomes infinite.

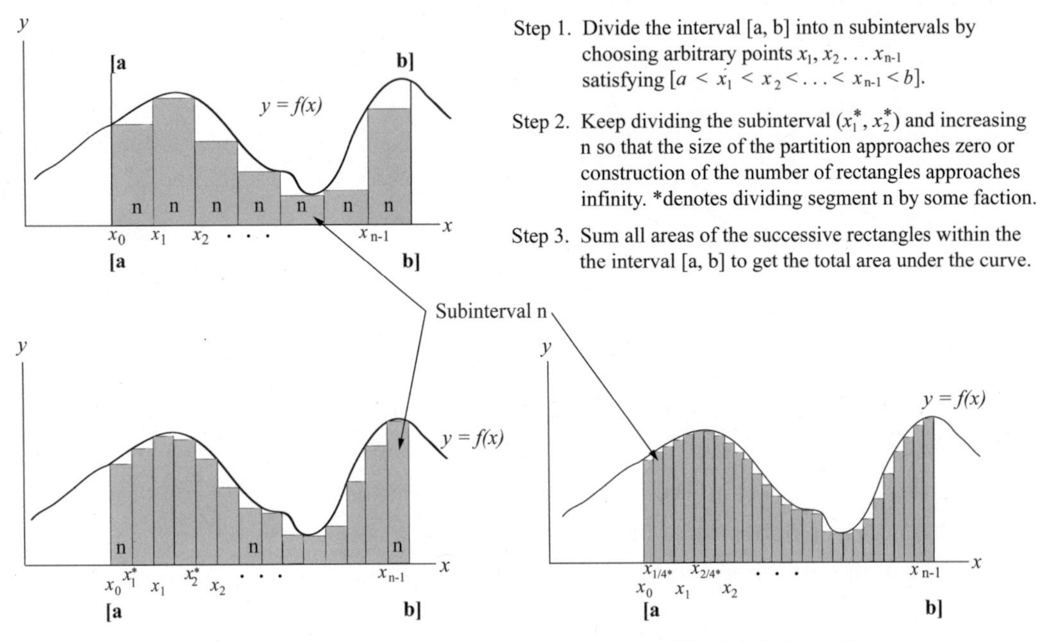

Figure A1.2. Determining the area under a curve using Riemann's concept of the definite integral.

1.3.1. Applications of Intervals

Not only can integrals be used to determine the area under a curve bounded by an axis, they can also find the area between two curves. This application has numerous uses when trying to determine the area of an irregular surface, such as a lake system. Integration is used to calculate volumes by rotating the area under the curve around an axis as defined by

$$V = \int_a^b A(x)\,dx \tag{A1.3}$$

With the definition of volume, the function of the area of a given cross section (Figure A1.3) is determined similarly to finding the area between two curves. Most volume problems are resolved by calculating a *solid of revolution*. These are solids generated by rotating the region bounded by curve(s) about some specified line where the solid generated will have a disk-shaped cross section and be defined by the area of a circle in the form of $A(x) = \pi[f(x)]^2dx$. For each cross-sectional disk, the radius is determined by the bounded region of the curves, which is defined by the function of the curve (Figure A1.3).

This method of volume determination is known as *volumes by slicing disks.* Variations on this basic principle for calculating volumes relevant to understanding lake systems—such as a change of the axis of rotation (i.e., rotation about the y-axis) and volumes of washers (see Figure A1.4) and cylindrical shells—are discussed in basic calculus texts.

Let f be nonnegative and continuous on $[a,b]$, and let R be the region bounded from above by the graph of f, below by the x-axis, and on the sides by the lines $x = a$ and $x = b$ (see shaded area in graph I). When this region is revolved about the x-axis, it generates a solid with circular cross sections (II). The cross section at x has a radius $f(x)$; cross section area is given in I and volume is given in III.

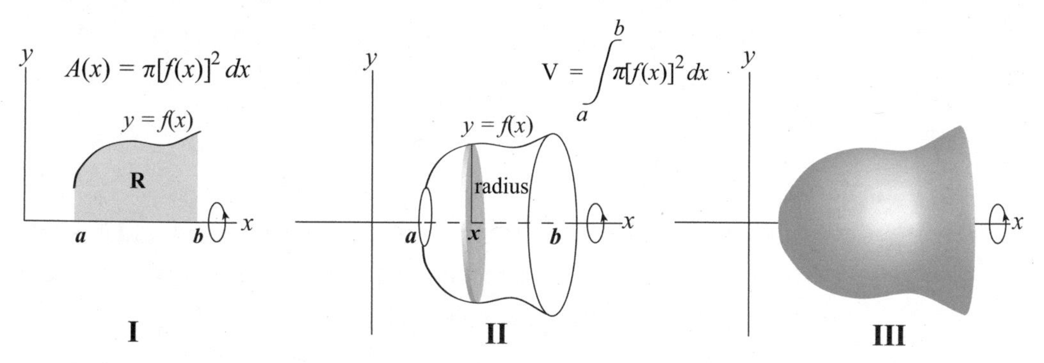

Figure A1.3. Determining volume with the disk method of integration.

Now consider more general solids of revolution where f and g are nonnegative continuous functions such that $g(x) < f(x)$ for $a < x < b$ and let R be the region enclosed between the graphs of these functions and the lines $x = a$ and $x = b$ (IV). When this region is revolved about the x-axis, it generates a solid having annular or washer-shaped cross sections (V). Since the cross section at x has an inner radius of $g(x)$ and outer radius of $f(x)$, its area is seen in IV and volume in VI.

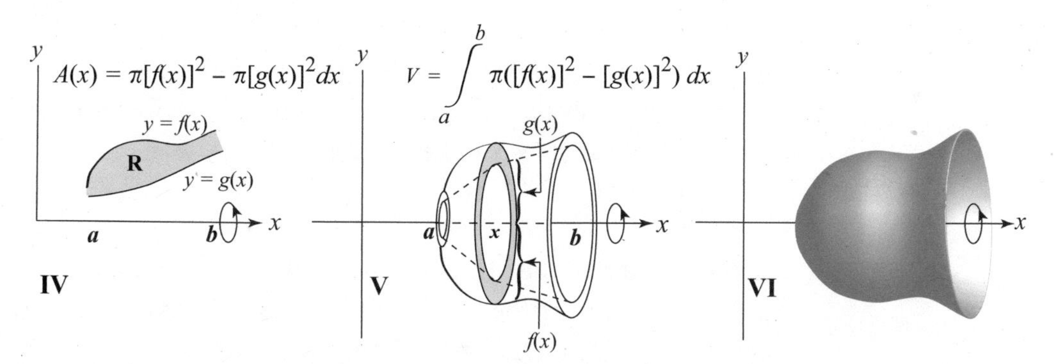

Figure A1.4. Determining volume with the washer method of integration.

1.4. Partial Derivatives

Most variables in natural systems, such as limnologic systems, are functions of several related variables. For example, the volume of water in lake systems over time is dependent upon precipitation, runoff, evapotranspiration, groundwater interactions, lake hypsometry, and so on. Thus, understanding the concept of partial derivatives when describing such a system is essential.

A partial derivative of a function of several variables is its derivative with respect to one of those variables with the others held constant (as opposed to the total derivative, in which all variables are allowed to vary). The partial derivative of a function f with respect to the variable x is written as $f'x$, ∂xf, or $\partial f/\partial x$. The partial derivative symbol ∂ is a rounded letter, distinguished from the straight d of total derivative notation.

There are numerous familiar formulas where a given variable is dependent upon two or more variables. For example the area A of a triangle depends upon the base length b and height h by the formula

$$A = \tfrac{1}{2}bh. \tag{A1.4}$$

This is stated as A is a function of the two variables b and h. Similarly, the volume V of a rectangular box is dependent upon length l, width w, and height h such that

$$V = lwh. \tag{A1.5}$$

V as a function of l, w, and h is expressed mathematically similar to the terminology used for functions of one variable ($y = f(x)$) such that the expression

$$z = f(x,y) \tag{A1.6}$$

states that z is a function of x and y (i.e., the unique value of the dependent variable z is determined by the specific values of the independent variables x and y. Correspondingly

$$w = f(x,y,z) \tag{A1.7}$$

signifies that that the independent variable w is uniquely determined by specifying the values for the independent variables x, y, and z depending on the system being described mathematically.

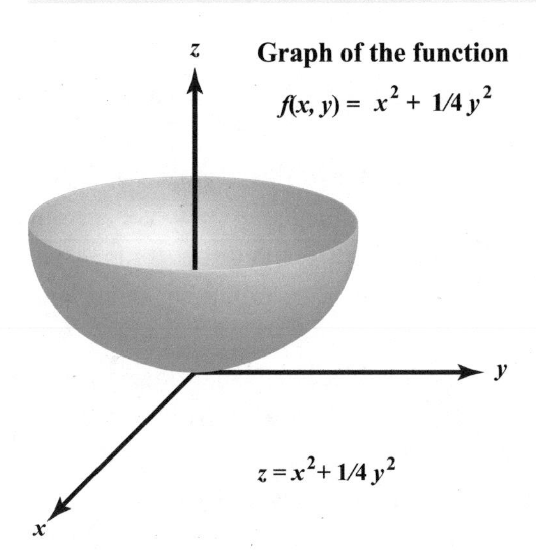

Figure A1.5. Using the partial derivative to determine the volume of an elliptic paraboloid.

For example, the graph of the function

$$f(x,y) = x^2 + \tfrac{1}{4}y^2 = z \qquad \text{(A1.8)}$$

can be used to find the volume of an elliptic paraboloid (Figure A1.5).

Hypsometry is the three-dimensional representation of the lake bottom as represented by dimensional contour lines or curves of constant elevation as it relates to a datum such as sea level. Thus, the physical parameter of a lake's hypsometry such as volume, lake level, and area can be described and modeled utilizing partial derivatives.

Section 2. Statistical Analyses

This section is a basic discussion of linear correlation, simple and multiple regression, and other fundamental concepts for developing a mathematical model used to describe a lake system. A review of the basics of statistical analyses—such as fundamental probability and distribution, mean, median, standard deviation, variance, linear correlation, and hypothesis testing—as well as other essential concepts will be beneficial when reading this segment of the text.

2.1. Linear Correlation

When describing a system such as a lake it useful to compare the interactions of different variables assumed to affect the system (i.e., variables x and y) to determine if there is a relationship or trend. One way to do this is to construct a graph called a scatterplot or scatter diagram between the parameters in question to distinguish if there is visual relationship between the two (Figure A2.1). These correlations can be either positive (graphs a, b, and c), where y increases as x increases, or negative (graphs d, e, and f) where y decreases with increasing x. Graph g shows no correlation and graph h indicates a pattern, although a nonlinear one (Triola 2001).

To better quantify the mostly subjective visual observations of a scatterplot, a more precise and objective measure is needed. The linear correlation coefficient r measures the strength of the paired x and y values in a data sample and is found by determining the relationship of the residuals. Residuals ($\boldsymbol{r}$) are defined for a sample of paired data where the residual is the difference between the observed sample y-value and the value of $\hat{y}$, which is the value of y that is predicted by using the regression equation (see section 2.2) (Triola 2001).

Figure A2.1. Scatterplots with different levels of correlation.

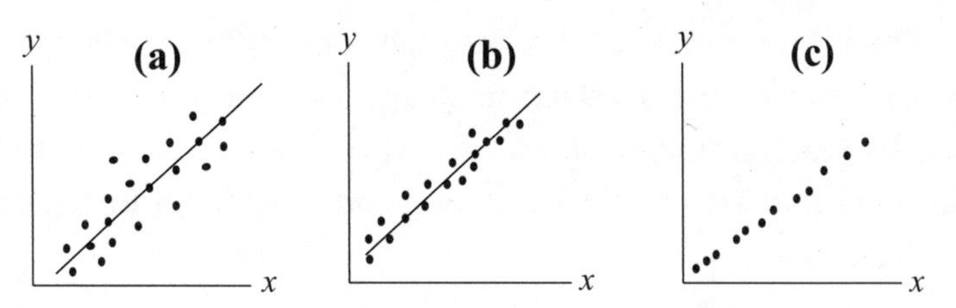

Positive correlation between x and y increasing ($\boldsymbol{r}$ approaching 1) from (a) to (c).

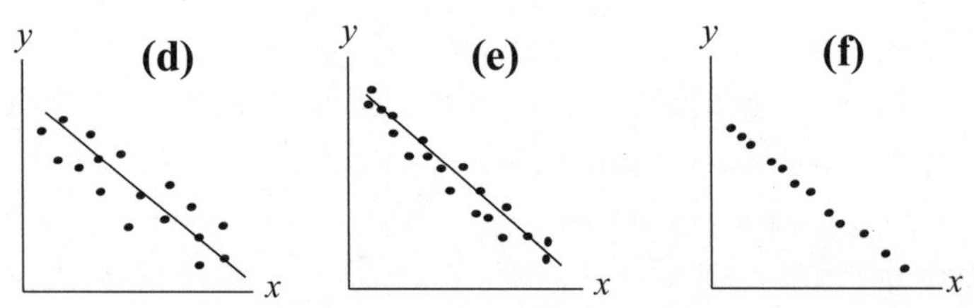

Negative correlation between x and y increasing ($\boldsymbol{r}$ approaching –1) from (d) to (f).

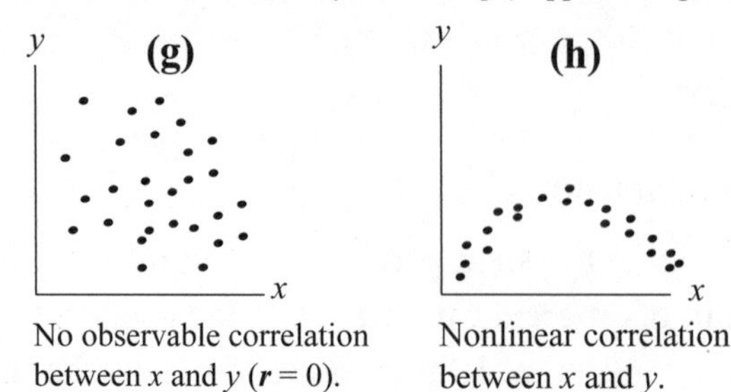

No observable correlation between x and y ($\boldsymbol{r} = 0$).

Nonlinear correlation between x and y.

The value of r is always between the value of -1 and 1 inclusive, or $-1 \leq r \leq 1$. If r is "close" to 0 then there is no correlation between x and y. If $\boldsymbol{r}$ is close to 1 or -1 there is a significant linear correlation. This computation was developed by Karl Pearson and is often referred to as the Pearson product-moment correlation coefficient. The derivation of $\boldsymbol{r}$ is readily referenced in most statistics textbooks as are tables with critical values of the Pearson correlation coefficient for determining if a linear correlation is significant or not.

The linear correlation coefficient measures the strength of a linear relationship (Figure A2.1 a–g) and is not designed to measure the strength of a nonlinear trend (Figure A2.1h). If there is a significant linear correlation between x and y, a linear equation that expresses y in terms of x can be developed. This equation can be applied to predict values of y for given values of x. However, the predicted value of y will not necessarily be the exact result, because of factors other than x affecting the system under study. This might include random variation and other unknown parameters not included in the investigation. The value of $\boldsymbol{r}^2$ therefore is the proportion of the variation in y that is explained by the linear relationship between x and y. For example, if $\boldsymbol{r}$ in Figure A2.1b is equal to 0.855, we get an $\boldsymbol{r}^2 = 0.731$. From this value we conclude that 0.731 (or approximately 73%) of the variation in y can be explained by the relationship between x and y. If 73% of the change in y can be explained by x, then the variation-affecting y from other factors would be approximately 27%.

2.2. Regression Analysis

Regression analysis is a statistical technique used to investigate and model the relationship between variables. Regression applications are numerous and wide ranging and are utilized in such disciplines as economics, business management, social studies, medicine, the physical and natural sciences, and engineering.

A regression analysis determines whether there is a significant linear relationship between two variables. This correlation is determined by finding the graph (scatterplot) and equation of a straight line that represents the relationship. This definition expresses a relationship between x (called the independent variable, or predictor/regressor variable) and y (the independent or response variable). The typical equation is one of a straight line, $y = mx + b$, expressed as

$$y = b_o + b_1x_1 + E \quad \text{(A2.1)}$$

where b_o is the y-intercept, b_1 is the slope of the line, and E is an error term. The error term E can be thought of as a statistical error that accounts for the discrepancies of the model. The error term may be the effects of measurement or sampling inaccuracy or other variables not properly accounted for that could be affecting the system under examination (Montgomery and Peck 1982). The mathematical formulas for deriving b_o, b_1, and E are referenced in most statistical texts and are programmed into many calculators and software spreadsheet programs so these values are easily attained.

To utilize regression, two assumptions are made:

1. The relationship under investigations is linear.
2. For each x-value, y is a random variable having a normal (bell-shaped) distribution.

For a given x-value the distribution of y-values has a mean that lies on the regression line and the y-distributions all have the same variance.

An example of such a linear relation analogous to a lake system would be water level (or the predictor variable x) of an irregular-shaped vessel and its volume (y) (Figure A2.2). For instance, measurements of the volume of water in the vessel at different water levels (h) indicate that as the altitude of h increases so does the vessel volume (V). A graph of this relationship is the regression line (or the line of best fit, or the *least squares line*). The equation of this relationship would be in the form of

$$V = b_o + b_1h_1 + E \quad \text{(A2.2)}$$

Using this equation as a model, prediction of vessel volume V at various levels of h could be estimated.

From the hypothetical data so far, r could be calculated to measure the strength of the relationship between V and h and to explain the variation in the model. For example, if $r = 0.799$ and it is determined based on the tables of Pearson correlation coefficients that this is a significant linear correlation, then the varia-

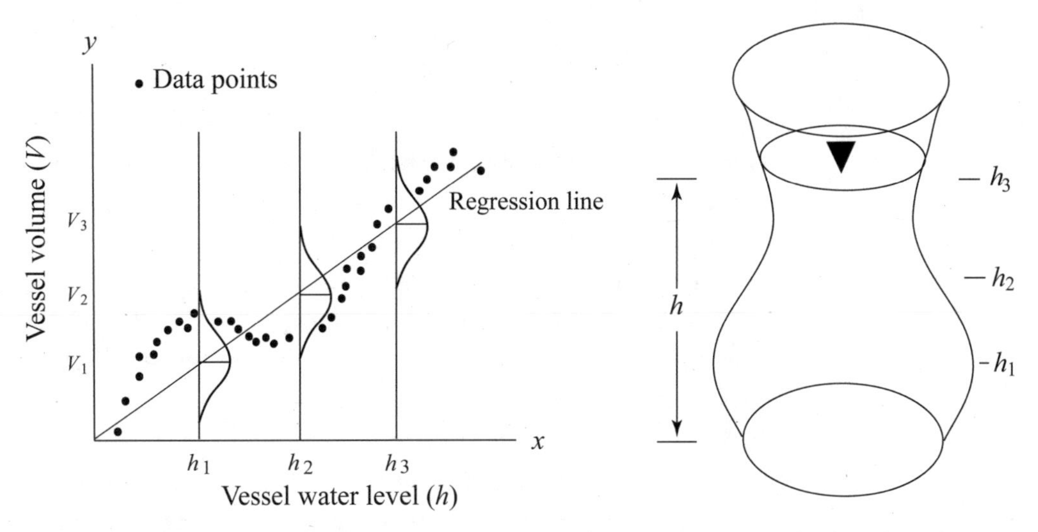

Figure A2.2. Linear regression analysis of water volume (*V*) and water level (*h*) in an irregular-shaped vessel.

tion could be explained by calculating r^2, which would be 0.638. This states that approximately 64% of the variation of volume can be explained by the relationship between the volume and the vessel level. This implies that about 36% of the vessel volume is because of factors other than *h*. These factors could include such things as changes in vessel area with altitude or random sampling procedures.

2.3. Multiple Regression

Multiple regression involves the relationship between more than two variables. Specifically, it expresses a linear relationship between a dependent variable *y* and two or more independent variables ($x_1, x_2, \ldots, x_k$). The general form of a multiple regression equation is

$$y = b_0 + b_1x_1 + b_2x_2 + \ldots + b_kx_k + E \qquad \text{(A2.3)}$$

where *y* is the predictor variable, *k* is the number independent variables, b_0 is the *y*-intercept, and $b_1, b_2, \ldots, b_k$ are the coefficients of the independent variables.

The adjective term "linear" is used to indicate that the model is linear in the parameters of $b_1, b_2, \ldots, b_k$ not because *y* is a linear function of *x*. A model in which *y* is nonlinear to *x* can still be modeled as a linear regression as long as the equation is linear to the *b* (Montgomery and Peck 1982).

A relationship between two variables modeled by regression does not imply a cause and effect between the variables. A strong empirical association between variables cannot be considered evidence for the regressor variable having a direct effect upon the response variable. A source outside the database needs to be established as verification for causality. For example, in the case of the irregular-shaped vessel, the basic volume equation for the frustrum of a right circular cone (Figure A2.3) directly relates changing radii (*R*) and altitude (*h*) with volume and thus establishes the true relationship between altitude, radius, and volume. Therefore, a multiple regression model examining the relationship of radius or area (pR^2) and altitude with volume would be more accurate when modeling volume in an irregular-shaped vessel.

Regression analysis can be helpful in confirming a cause-effect relationship but cannot be the sole basis for such a declaration. The regression equation is only an approximation of the true relationship between variables. In the vessel example, the effect of the changing radius is not directly accounted for, and this relationship could be part of the model discrepancy indicated in the error term E. Also, in this example there exists a straight-line approximation generated by the simple regression model (Figure A2.2 and A2.4a). But based on the r and r^2 values, level h only explains 64%

Figure A2.3. Volume calculation of the frustrum of a right circular cone.

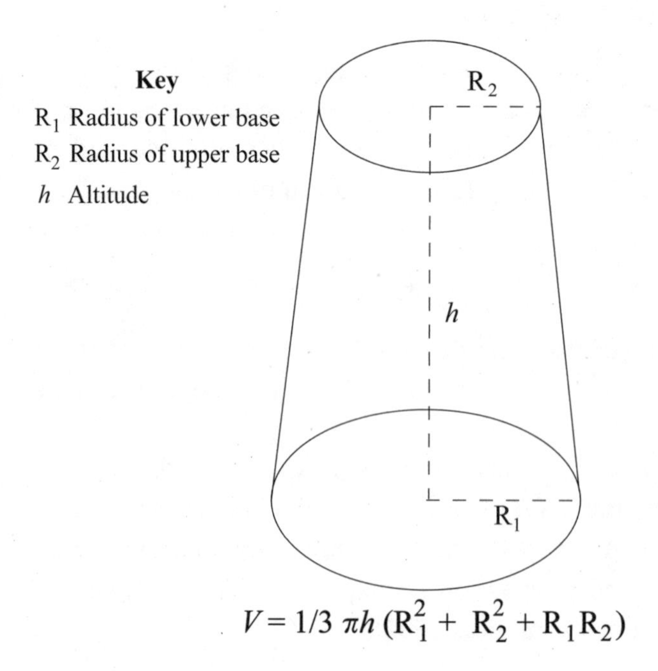

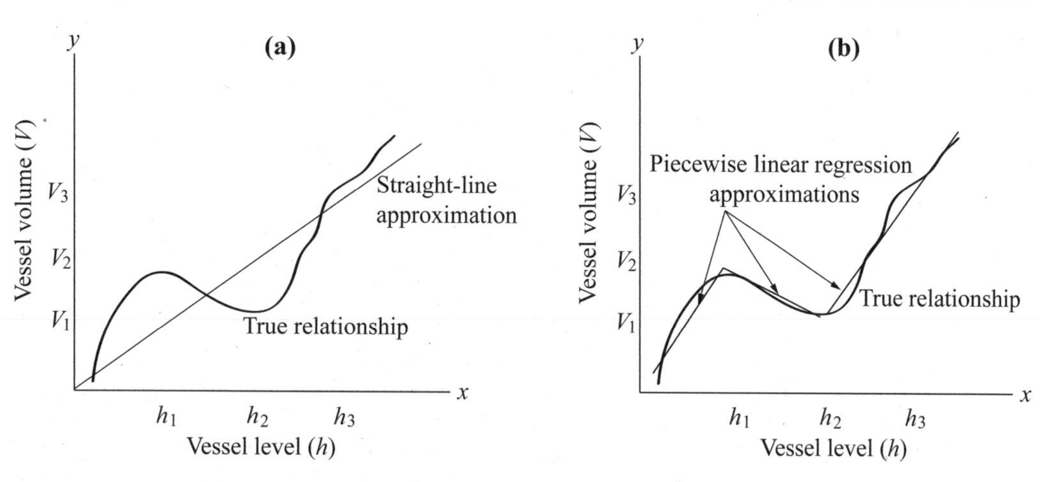

Figure A2.4. Piecewise linear approximation of a complex relationship in an irregular-shaped vessel.

of the variation in volume. The influence of the radius/area with volume is not suitably accounted for. Therefore, a more complex approximation function may be necessary to adequately describe the specific relationship between x and y. One method is to analyze increments of the data producing a *piecewise linear* regression function (Montgomery and Peck 1982), which may reduce the variation within the model (Figure A2.4b).

2.4. Hypothesis Testing

A statistical hypothesis is a statement about the relationship or properties between variables affecting the system in question. For example, when investigating the relationship between the water level h and vessel volume in Figure A2.2, one hypothesis states that as h increases, the volume increases linearly. To validate this statement, statistical hypothesis testing procedures are conducted and described in a very formal way.

The null hypothesis (H_0) formally describes some aspect of the statistical behavior of a set of data; this description is treated as valid unless the actual behavior of the data contradicts this assumption. The adjective "null" is used by statisticians because the basic hypothesis to be tested is often one that asserts the claim of an invalid relationship between x and y. For the vessel system in our example, a formal test for the null hypothesis would be,

H_0: level h has no effect on vessel volume
H_1: level h has an effect on vessel volume

From the water level and volume data, a test statistic t is calculated and if t falls into the "critical region" (the set of values for which H_0 is to be rejected), the null hypothesis is rejected. (Calculation of the t-statistic and tables of its critical values are referenced in most statistic textbooks.)

The null hypothesis is then contrasted against another (alternate) hypothesis (H_1), and statistical hypothesis testing is used to make a decision about whether the data contradicts the null hypothesis: this is called *significance testing*. A null hypothesis is never proven by such methods, as the absence of evidence against the null hypothesis does not establish it. In other words, one may either reject or not reject the null hypothesis; one cannot accept it. Failing to reject it gives no strong reason to change decisions predicated on its truth, but it also allows for the possibility of obtaining further data and then reexamining the same hypothesis.

One example of the null and alternative hypothesis is expressed as

H_0: $P = 0$ (No linear correlation)
H_1: $P \neq 0$ (Linear correlation)

The p-value is the probability of obtaining a result at least as extreme as the one that was actually observed in the sample data, assuming that the null hypothesis is true. The fact that p-values are based on this assumption is crucial to their correct interpretation (Triola 2001). The lower the p-value, the less likely the result, assuming the null hypothesis and the more "significant" the result in the sense of statistical significance (one often uses p-values of 0.05 or 0.01, corresponding to a 5% chance or 1% of an outcome that extreme, given the null hypothesis). Therefore, the p-value is a measure of the overall significance of the multiple regression equation and is usable for predictions.

As discussed, two common methods used to either accept or reject the null hypothesis would be the test statistic t or the test statistic r. Tables with the critical values of these two tests are readily available in statistical textbooks. If the absolute values of the test statistics exceed the critical values, reject H_0; if not then fail to reject H_0. If H_0 is rejected, then conclude that there is a significant linear correlation.

2.5. Regression Analysis for Predictions and Modeling

Once it has been determined that there is a significant linear correlation, then the regression equations can be used to predict a variable given a certain value for the other variable. The best predicted y-value is estimated by substituting the x-value into the regression equation. The system need not be restricted to a linear system to use regression analyses. Observation of the data may indicate patterns that are defined by such well-known functions as exponential ($y = ab^x$), quadratic ($y = ax^2 + bx + C$), logarithmic ($a + b \ln x$), power ($y = ax^b$), logistic ($y = C/1+ae^{-bx}$), or exponential decay ($y = xe^{-bx}$) equations (where C is a constant derived from the data and e in the exponential decay function = 2.71828). Examples of these types of models are illustrated in Figure A2.5. The exponential decay model in Figure A2.5e was utilized when describing lake-level decline over time.

Regression analyses tools such as R^2 values can be used to compare nonlinear or curvilinear models to determine whether the

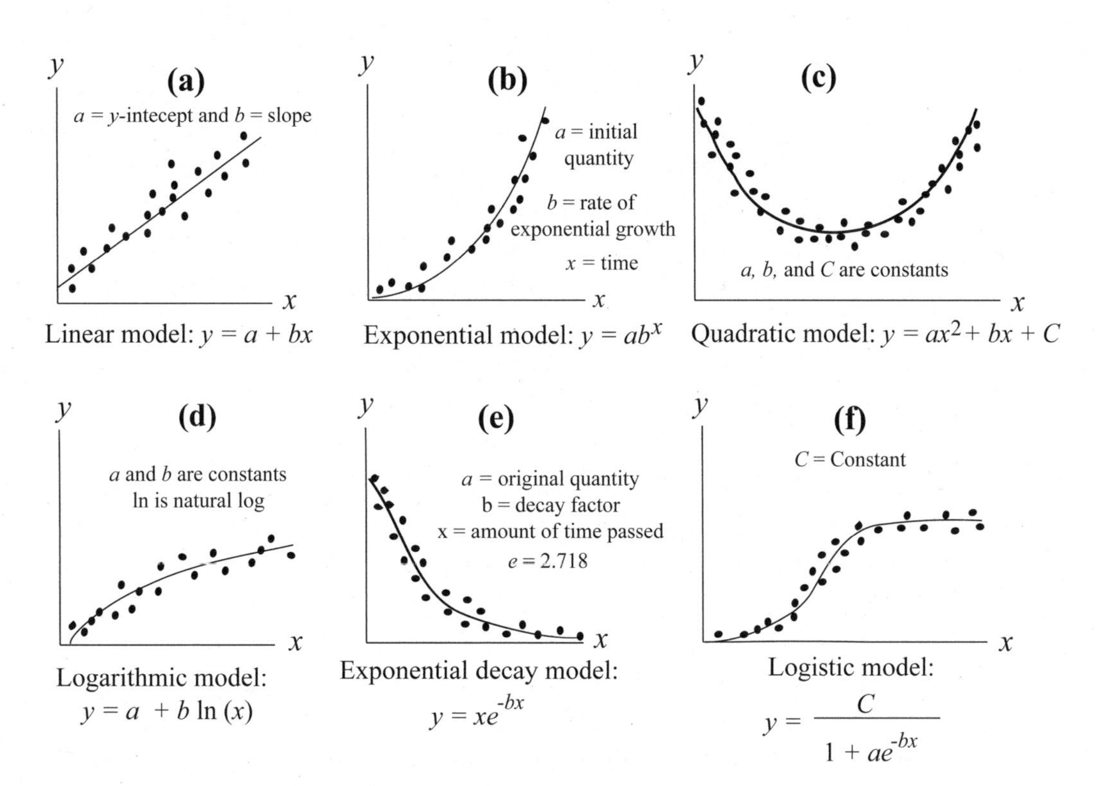

Figure A2.5. Examples of patterns of data that fit known mathematical functions.

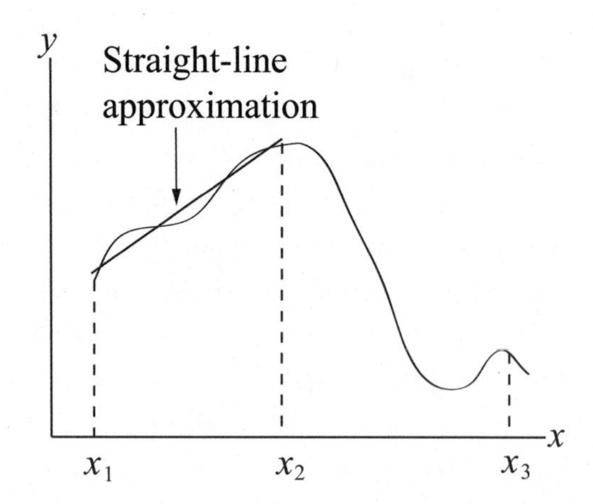

Figure A2.6. The danger of extrapolation in regression.

model has a good or adequate fit. Or the data exhibiting a curvilinear pattern could be analyzed in a piecewise manner, as shown in Figure A2.4.

Use common sense. Do not use a model that leads to predicted values known to be unrealistic. One verification method is to use the model to predict known values, as well as future, past, and missing values to see if the predicted results are reasonable.

Accurate data collection is essential when using regression analysis because the validity of the model is dependent on the quality of the data. The data should correctly reflect the system and thus the better the data, the more applicable the model. Without representative data the regression-model results and conclusions drawn from them will most likely be in error and seriously distorted.

Generally, when there is limited understanding of how the system in question works, the regression equations are valid only over the region of the regressor variables contained in the observed data. In the example depicted in Figure A2.6, suppose that the data on y and x were collected in the interval $x_1 \leq x \leq x_2$. Over this interval the linear regression equation generated from this figure is a good approximation of the relationship. However, if this equation is used to predict values greater than x_2, the model would be inadequate.

Another source of error is outliers where a data point is significantly away from the other data. These data can strongly influence the graph of the regression line as well as other statistical measurements, such as mean and standard deviation. Knowing the source of the outlier can be imperative when evaluating the model. Is this data an error in measurement or is it a true result measured during an unusual event affecting the system in question? If the result of the outlier is human error, then the investigator can justify removing these data. If the data is legitimate, for example generated under atypical circumstances, then the researcher can report it as such and adjust the model accordingly.

When utilizing regression, it is important to realize that the model is a part of the broader data-analytic approach to understanding the system under investigation. The regression equation is only a tool for analysis and should not be the primary objective of the study. It is essential to understand the system generating the data. From this understanding, a better model can be developed.

REFERENCES

Aguado, E., and J.E. Burt. 2001. *Understanding Weather and Climate.* Upper Saddle River, NJ: Prentice Hall.

Aldridge, B.N., and J.M. Garret. 1973. "Roughness Coefficients for Streams in Arizona." US Geological Survey Open-File Report 73-3, 87 pp.

Allen, R.G. 1986. "A Penman for All Seasons." *Journal of Irrigation and Drainage Engineering* 112 (4): 348–68.

AMEC Earth & Environmental Inc. 2008. "Technical Guidance." Chapter 3 in *Stormwater Hydrology, Knox County, Tennessee Stormwater Management Manual*, vol. 2, 3–9. Knox County, TN: Department of Engineering and Public Works.

American Society of Civil Engineers. 1975. "Aspects of Hydrological Effects of Urbanization." *Journal of the Hydraulics Division* 101 (5): 449–68.

Anton, H. 1980. *Calculus with Analytical Geometry*. New York: John Wiley & Sons.

Aracement, G.J., Jr., and V.R., Schneider. 1987. "Roughness Coefficients for Densely Vegetated Flood Plains." US Geological Survey Water-Resources Investigations Report 83-4247, 62 pp.

———. 1989. "Guide for Selecting Manning's Roughness Coefficients for Natural Channels and Flood Plains." US Geological Survey Water-Supply Paper 2339, 38 pp.

Armstrong, M. 1998. *Basic Linear Geostatistics*. New York: Springer.

Arnold, J.G., and P.M. Allen. 1999. "Automated Methods for Estimating Baseflow and Groundwater Recharge from Streamflow Records." *Journal of the American Water Resources Association* 35 (2): 411–24.

Askew, A.J. 1975. "Use of Risk Premium in Chance-Constrained Dynamic Programming." *Water Resources Research* 11 (6): 862–66.

Attanayake, M.P., and D.H. Waller. 1988. "Use of Seepage Meters in a Groundwater-Lake Interaction Study in a Fractured Rock Basin: A Case Study." *Canadian Journal of Civil Engineering* 15 (6): 984–89.

Averett, R.C., and L.J. Schroder. 1994. "A Guide to the Design of Surface-Water-Quality Studies." US Geological Survey Open-File Report 93-105, 39 pp. http://pubs.er.usgs.gov/publication/ofr93105.

Bacchi, B., and N.T. Kottegoda. 1995. "Identification and Calibration of Spatial Correlation Patterns of Rainfall." *Journal of Hydrology* 165

(1–4): 311–48. https://www.sciencedirect.com/science/article/pii/0022169494025908.

Ball, J., M. Babister, R. Nathan, W. Weeks, E. Weinmann, M. Retallick, and I. Testoni, eds. 2016. *Australian Rainfall and Runoff: A Guide to Flood Estimation*. Canberra: Commonwealth of Australia (Geoscience Australia). Accessed January 3, 2017. http://book.arr.org.au.s3-website-ap-southeast-2.amazonaws.com.

Balsillie, J.H. 2003. Personal communication, Florida Department of Environmental Protection, Florida Geological Survey.

Barnes, H.H., Jr. 1967. "Roughness Characteristics of Natural Channels." US Geological Survey Water-Supply Paper 1849, 213 pp.

Barnston, A.G. 1991. "An Empirical Method of Estimating Rain Gauge and Radar Rainfall Measurement Bias and Resolution." *Journal of Applied Meteorology* 30 (3): 282–96.

Bear, J. 1972. *Dynamics of Fluids in Porous Media*. New York: American Elsevier.

Bear, J., and A. Verruijt. 1987. *Modeling Groundwater Flow and Pollution*. Dordrecht, Holland: D. Reidel.

Bedient, P.B., and W.C. Huber. 1992. *Hydrology and Floodplain Analysis*. New York: Addison-Wesley.

Benson, M.A., and T. Dalrymple. 1967. *General Field and Office Procedures for Indirect Discharge Measurements*. US Geological Survey Techniques of Water Resources Investigations, book 3, chap. A1, 30 pp.

Benton, A.R., W.P. James, and J.W. Rouse, Jr. 1978. "Evapotranspiration from Water Hyacinth in Texas Reservoirs." *Water Resources Bulletin* 14 (4): 919–30.

Berner, E.K., and R.A. Berner. 1987. *The Global Water Cycle: Geochemistry and Environment*. Englewood Cliffs, NJ: Prentice Hall.

Beven, K.J. 2004. "Robert E. Horton's Perceptual Model of Infiltration Process." *Hydrological Processes* 18 (17): 3447–60.

Bhaduri, B., M. Grove, C. Lowery, and J. Harbor. 1997. "Assessing the Long-Term Hydrologic Effects of Land Use Change." *Journal of the American Water Resources Association* 89 (11): 94–106.

Biggs, J., P. Williams, M. Whitfield, N. Pascale, and A. Weatherby. 2005. "15 Years of Pond Assessment in Britain: Results and Lessons Learned from the Work of Pond Conservation." *Aquatic Conservation: Marine and Freshwater Ecosystems* 15 (6): 693–714. doi:10.1002/aqc.745.

Bjerklie, D.M., S.L. Dingman, and C.H. Bolster. 2005. "Comparison of Constitutive Flow Resistance Equation Based on Manning and Chézy Equation Applied to Natural Rivers." *Water Resources Research* 41:W11502. doi:10.1029/2004WR003776.

Blake, G.J. 1968. "Infiltration at the Puketurua Experimental Basin." *Journal of Hydrology* 7 (1): 38–46.

Bohlin, T.P. 2006. *Practical Grey-Box Process Identification: Theory and Applications*. Berlin and London: Springer Science & Business Media.

Boussinesq, J. 1877. *Essai sur la théorie des eaux courantes. Mémoires présentés par divers savants a l'Academie des Sciences de l'Institute National de France, Tome XXIII, No. 1* [Essay on the theory of running water, memoirs by various scholars to the Academy of Sciences of the National Institute of France, volume 23, no. 1]. Paris: Academy of Sciences (Mathematics and Physics) of the National Institute of France.

Bouwer, H. 1986. "Intake Rate: Cylinder Infiltrometer." In *Methods of Soil Analysis*, Part I, edited by A. Klute, 825–44. Madison, WI: American Society of Agronomy.

Boyd, C.E. 1985. "Pond Evaporation." *Transactions of the American Fisheries Society* 114 (2): 299–303.

Bradley, S.G., W.R. Gray, L.D. Pigott, A.W. Seed, C.D. Stow, and G.L. Austin. 1997. "Rainfall Redistribution over Low Hills Due to Flow Perturbation." *Journal of Hydrology* 202 (3–4): 33–47.

Bras, R.L., and I. Rodriguez-Iturbe. 1985. *Random Functions and Hydrology*. Reading, MA: Addison-Wesley.

Bray, D.I. 1979. "Estimating Average Velocity in Gravel-Bed Rivers." *Journal of the Hydraulics Division* 105 (9): 1103–22.

Browning, K.A. 1979. "The FRONTIERS Plan: A Strategy for Using Radar and Satellite Imagery for Very-Short-Range Precipitation Forecasting." *Meteorological Magazine* 108:161–84.

Bryan, J., S. Scott, and G. Means. 2008. *Roadside Geology of Florida*. Missoula, MT: Mountain Press.

Buytaert, W., R. Celleri, P. Willems, B. De Bièvre, and G. Wyseure. 2006. "Spatial and Temporal Rainfall Variability in Mountainous Areas: A Case Study from the South Ecuadorian Andes." *Journal of Hydrology* 329 (3–4): 413–21. https://www.sciencedirect.com/science/article/pii/S0022169406001144.

Campling, P., A. Gobin, and J. Feyen. 2001. "Temporal and Spatial Rainfall Analysis across a Humid Tropical Catchment." *Hydrological Processes* 15 (3): 359–75. doi:10.1002/hyp.98.

Carpenter, S.R. 1983. "Lake Geometry: Implications for Production and Sediment Accretion Rates." *Journal of Theoretical Biology* 105 (2): 273–86.

Carr, M.R., and T.C. Winter. 1980. "An Annotated Bibliography of Devices Developed for Direct Measurement of Seepage." US Geological Survey Open-File Report 80-344, 38 pp.

Cartwright, J.M., and W.J. Wolfe. 2016. "Insular Ecosystems of the Southeastern United States—a Regional Synthesis to Support Biodiversity Conservation in a Changing Climate." US Geological Survey Professional Paper 1828, 162 pp.

Case, H.L. III, J. Boles, A. Delgado, T. Nguyen, D. Osugi, D.A. Barnum, D. Decker, et al. 2013. "Salton Sea Ecosystem Monitoring and Assessment Plan." US Geological Survey Open-File Report 2013-1133, 220 pp.

Center for Watershed Protection. 2003. *Impacts of Impervious Cover on Aquatic Systems*. Watershed Protection Research Monograph. Ellicott City, MD: Center for Watershed Protection.

Cess, R.D., G.L. Potter, J.P. Blanchet, G.J. Boer, A.D. Del Genio, M. Déqué, V. Dymnikov, et al. 1990. "Intercomparison and Interpretation of Climate Feedback Processes in 19 Atmospheric General Circulation Models." *Journal of Geophysical Research: Atmosphere* 95 (D10) 16601–15.

Chahine, M.T. 1992. "The Hydrologic Cycle and Its Influence on Climate." *Nature* 359:373–80.

Chapman, T. 1991. "Comment on 'Evaluation of Automated Techniques for Base Flow and Recession Analyses,' by R.J. Nathan and T.A. McMahon." *Water Resources Research* 27 (7): 1783–84.

Chézy, Antoine de. 1775. *Manuscript Report on the Canal de I´Yvette*. In C. Herschel, 1897, "On the Origin of the Chezy Formula," *Journal Association of Engineering Society* 18 (6): 363–69.

Choudhury, B.J., N.E. DiGirolamo, J. Susskind, W.L. Darnell, S.K. Gupta, and G. Asrar. 1998. "A Biophysical Process-Based Estimate of Global Land Surface Evaporation Using Satellite and Ancillary Data II. Regional and Global Patterns of Seasonal and Annual Variations." *Journal of Hydrology* 205 (3–4): 186–204.

Chow, V.T. 1959. *Open Channel Hydraulics*. New York: McGraw-Hill.

———, ed., 1964. *Handbook of Hydrology*. New York: McGraw-Hill.

Chow, V.T., D.R. Maidment, and W. Mays. 1988. *Applied Hydrology*. New York: McGraw-Hill.

Chow-Fraser, P. 1991. "Use of the Morphoedaphic Index to Predict Nutrient Status and Algal Biomass in Some Canadian Lakes." *Canadian Journal of Fisheries and Aquatic Sciences* 48 (10): 1909–18.

Christel, P., and D.W. Reed. 1999. "Mapping Extreme Rainfall in a Mountainous Region Using Geostatistical Techniques: A Case Study in Scotland." *International Journal of Climatology* 19 (12): 1337–56.

Chu, S.T. 1978. "Infiltration during an Unsteady Rain." *Water Resources Research* 14 (3): 461–66.

Clark, C.O. 1945. "Storage and the Unit Hydrograph." *Transactions of the American Society of Civil Engineers* 110 (1): 1419–46.

Collier, C.G. 1996. *Applications of Weather Radar Systems: A Guide to Uses of Radar in Meteorology and Hydrology*. 2nd ed. Chichester, NY: Wiley.

Cooke, G.D., E.B. Welch, S.A. Peterson, and S.A. Nichols. 2005. *Restoration and Management of Lakes and Reservoirs*. New York: Taylor and Francis.

Coon, W.F. 1998. "Estimation of Roughness Coefficients for Natural Stream Channels with Vegetated Banks." US Geological Survey Water-Supply Paper 2441, 133 pp. Prepared with the cooperation of the New York State Department of Transportation. doi:10.3133/wsp2441.

Coordinating Committee on Great Lakes Basic Hydraulic and Hydrologic Data. 1977. *Coordinated Great Lakes Physical Data*. Detroit, MI; Burlington, Ontario: Coordinating Committee on Great Lakes Basic Hydraulic and Hydrologic Data.

Cordery, I. 1976. "Some Effects of Urbanization on Streams." *Civil Engineering Transactions / The Institution of Engineers, Australia* CE18 (1): 7–11.

———. 1987. "The Unit Hydrograph Method of Flood Estimation." Chapter 8 in *Australian Rainfall and Runoff: A Guide to Flood Estimation*, edited by D.H. Pilgrim. Barton: Institution of Engineers, Australia.

Cordery, I., and D.H. Pilgrim. 2000. "The State of the Art of Flood Prediction." In *Floods*, vol. 2, edited by D.J. Parker, 185–97. London: Routledge.

Cordery, I., D.H. Pilgrim, and B.C. Baron. 1981. "Validity of Use of Small Catchment Research Results for Large Basins." *Civil Engineering Transactions / The Institution of Engineers, Australia* CE23:131–37.

Cordery, I., and S.N. Webb. 1974. "Flood Estimation in Eastern New South Wales: A Design Method." *Civil Engineering Transactions / The Institution of Engineers, Australia* CE16:87–93.

Couette, M. 1890. "Etude sur le frottement des liquides, Gauthiers-Villars [Studies on the friction of liquids]." PhD diss., University of Paris.

Cowan, W.L. 1956. "Estimating Hydraulic Roughness Coefficients." *Agricultural Engineering* 37 (7): 473–75.

Critchley, W., K. Siegert, C. Chapman, and M. Finkel. 1991. *Water Harvesting: A Manual for the Design and Construction of Water Harvesting Schemes for Plant Production*. FAO Paper AGL/MISC/17/91. Rome: Food and Agriculture Organization of the United Nations. http://www.fao.org/3/U3160E/u3160e00.htm.

Croley, T.E., II, T.S. Hunter, and S.K. Martin. 2001. Great Lakes Monthly Hydrologic Data. NOAA Technical Report TM-083. Ann Arbor, MI: Great Lakes Environmental Research Laboratory. (First published 1994).

Curtis, O.F., and D.G. Clark. 1950. *An Introduction to Plant Physiology*. New York: McGraw-Hill.

Cutnell, J.D., and K.W. Johnson. 2004. *Physics*. 6th ed. John Wiley & Sons.

Darcy, H. 1856. *Les fontaine publiques de la ville de Dihon*. Paris: Victor Dalmont.

Debler, W.R. 1990. *Fluid Mechanics Fundamentals*. Englewood Cliffs, NJ: Prentice Hall.

Dingman, S.L. 2002. *Physical Hydrology*. Englewood Cliffs, NJ: Prentice Hall.

Dixon, H., and J. Joly. 1894. "On the Ascent of Sap." *Annals of Botany* 8 (4): 468–70.

Domenico, P.A., and F.W. Schwartz. 1990. *Physical and Chemical Hydrogeology*. New York: John Wiley & Sons.

Driscoll, F.G. 1986. *Groundwater and Wells*. St. Paul, MN: Johnson Screens.

Drogue, G., J. Humbert, J. Deraisme, N. Mahr, and N. Freslon. 2002. "A Statistical-Topographic Model Using Omnidirectional Parameterization of the Relief for Mapping Orographic Rainfall." *International Journal of Climatology* 22 (5): 599–613.

Duff, J.H., B. Toner, A.P. Jackman, R.J. Avanzino, and F.J. Triska. 1999. "Determination of Groundwater Discharge into a Sand and Gravel Bottom River: A Comparison of Chloride Dilution and Seepage Meter Techniques." *Verhandlungen: Proceedings of the International Society of Limnology* 27 (1): 406–11.

Duncan, J.G., W.L. Evans, and K. Taylor. 1994. *Geologic Framework of the Lower Floridan Aquifer System, Brevard County, Florida*. FGS Bulletin No. 64. Tallahassee: Florida Geological Survey.

Dunne, T., and R.D. Black. 1970. "Partial Area Contributions to Storm Runoff in a Small New England Watershed." *Water Resources Research* 6 (5): 1296–1311.

Dunne, T., and L.B. Leopold. 1978. *Water in Environmental Planning*. New York: W. H. Freeman.

Dunne, T., T.R. Moore, and C.H. Taylor. 1975. Recognition and Prediction of Runoff-Producing Zones in Humid Regions. *Hydrological Sciences Bulletin* 20 (3): 305–27.

Dupuit, J. 1863. *Estudes theoriqes et pratiques sur le mouvement des eaux dans les canaux decouverts et a travers les terrains permeables*. 2nd ed. Paris: Dunod.

Eberly, S., D. Swall, D. Holland, B. Cox, and E. Baldridge. 2004. *Developing Spatially Interpolated Surfaces and Estimating Uncertainty*. EPA-454/R,h.04-004. Triangle Park, NC: US Environmental Protection Agency, Office of Air Quality Planning and Standards Research.

Ebert, E.E., and M.J. Manton. 1998. "Performance of Satellite Rainfall Estimation Algorithms during TOGA COARE." *Journal of the Atmospheric Sciences* 55 (9): 1537–57.

Eckhardt K. 2005. "How to Construct Recursive Digital Filters for Baseflow Separation." *Hydrological Processes* 19 (2): 507–15.

Eichenlaub, V.L. 1979. *Weather and Climate of the Great Lake Region*. Notre Dame, IN: University of Notre Dame Press.

England, J.F., Jr., T.A. Cohn, B.A. Faber, J.R. Stedinger, W.O. Thomas, Jr., A.G. Veilleux, J.E. Kiang, and R.R. Mason, Jr. 2018. "Guidelines for Determining Flood Flow Frequency—Bulletin17C" (ver. 1.1, May 2019): *US Geological Survey Techniques and Methods*, book 4, chap. B5, 148 pp. doi:10.3133/tm4B5.

Enright, R.V. 1990. "Relating the Effects of Oceanic Tidal Loading of a Confined Aquifer in Sarasota County, Florida, to Fluctuations in Well-

Water Levels." Master's thesis, Department of Geology, Florida State University.
Erickson, D.R. 1981. "A Study of Littoral Groundwater Seepage at Williams Lake, Minnesota, Using Seepage Meters and Wells." Master's thesis, University of Minnesota.
Espey, W.H., and D.E. Winslow. 1974. "Urban Flood Frequency Characteristics." *Journal of the Hydraulics Division* 100 (HY2): 279–93.
Evans, W.L. 1996. "Modeling the Hydrodynamics of Closed Lake Basins: A Treatise on Lake Seepage." Master's thesis, Department of Geology, Florida State University.
———. 2018. *Stormwater Management Report: Analysis of Stormwater Runoff at CASA 12 Property 2410 Monday Road, Tallahassee, Florida.* Tallahassee: EIII Environmental Consulting Inc., 15–24.
Farvolden, R.N. 1963. "Geologic Controls on Ground-Water Storage and Base flow. *Journal of Hydrology* 1 (3): 219–49. doi:10.1016/0022-1694(63)90004-0.
Faust, C.R., and J.W. Mercer. 1980. "Ground-Water Modeling: Numerical Models." *Groundwater* 18 (4): 395–409.
Feldman, A.D., P.B. Ely, and D.M. Goldman. 1981. *The New HEC-1 Flood Hydrograph Package.* Davis, CA: US Army Corps of Engineers Hydrologic Engineering Center. Accessed January 3, 2016. https://www.hec.usace.army.mil/publications/TechnicalPapers/TP-82.pdf.
Fellows, C.R., and P.L. Brezonik. 1980. "Seepage into Florida Lakes." *Water Resources Bulletin* 16 (4): 635–41.
Ferguson, B.K., and P.W. Suckling. 1990. "Changing Rainfall-Runoff Relationships in the Urbanizing Peachtree Creek Watershed, Atlanta, Georgia." *Water Resources Bulletin* 26 (2): 313–22.
Fernald, E.A., and E.D. Purdum, eds. 1998. *Water Resources Atlas of Florida.* Tallahassee: Institute of Science and Public Affairs, Florida State University.
Fetter, C.W. 1988. *Applied Hydrogeology.* 2nd ed. Columbus, Ohio: Merrill.
Fick, A. 1855. "On Liquid Diffusion." In *Annalen der Physik und Chemie,* vol. 94, edited by J.C. Poggendorff, 59. Reprinted 1995 in *Journal of Membrane Science* 100 (1): 33–38.
Florida LAKEWATCH. 2001. *A Beginner's Guide to Water Management—Lake Morphometry.* Information Circular 104. Gainesville: Department of Fisheries and Aquatic Sciences, University of Florida/Institute of Food and Agricultural Sciences (UF/IFAS).
Forel, F.A. 1901. *Handbuch der Seenkunde. Allgemeine Limnologie.* Stuttgart: J. Engelhorn.
Foster, S.S.D., B.L. Morris, and A.R. Lawrence. 1994. "Effects of Urbanization on Groundwater Recharge." In *Groundwater Problems in Urban Areas,* edited by W.B. Wilkinson, 43–63. London: Tomas Telford.
Fredrick, R.H., V.A. Myers, and E. P. Auciello. 1977. Five- to 60-Minute Precipitation Frequency for the Eastern and Central United States.

Technical Memorandum NWS HYDRO-35. Silver Spring, MD: National Oceanic and Atmospheric Administration.

Freeze, R.A. 1972. "Role of Subsurface Flow in Generating Surface Runoff: 2. Upstream Source Areas." *Water Resources Research* 8 (5): 1272–83. doi:10.1029/WR008i005p01272.

Freeze, R.A., and J.A. Cherry. 1979. *Groundwater*. New York: Prentice Hall.

Gams, I. 1993. "Origin of the Term 'Karst,' and the Transformation of the Classical Karst (Kras). *Environmental Geology* 21:110–14. doi:10.1007/BF00775293.

Giancoli, D.C. 1988. *Physics for Scientists and Engineers*. London: Prentice Hall.

Giorgi, F., E. Coppola, F. Solmon, L. Mariotti, M.B. Sylla, X. Bi, N. Elguindi, et al. 2012. "RegCM4: Model Description and Preliminary Tests over Multiple CORDEX Domains." *Climate Research* 52:7–29. doi:10.3354/cr01018.

Goovaerts, P. 1997. *Geostatistics for Natural Resources Evaluation*. New York: Oxford University Press.

———. 2000. "Geostatistical Approaches for Incorporating Elevation into Spatial Interpolation of Rainfall." *Journal of Hydrology* 228 (1–2): 113–29. https://www.sciencedirect.com/science/article/pii/S002216940000144X.

Goslee, R.P., C.A. Brooks, and A. Cole. 1997. "Plants as Indicators of Wetland Sources." *Plant Ecology* 131 (2): 199–206.

Grannemann, N.G., and T.L. Weaver. 1998. "An Annotated Bibliography of Selected References on the Estimated Rates of Direct Ground-Water Discharge to the Great Lakes." US Geological Survey Water-Resources Investigations Report 98-4039, 22 pp.

Gray, D.M. 1961. "Synthetic Hydrographs for Small Drainage Areas." *Proceedings of the American Society of Civil Engineers* 87 (HY4): 33–54.

Gray, D.M., and T.D. Prowse. 1993. "Snow and Floating Ice." In Maidment, *Handbook of Hydrology*, 7.1–7.58. New York: McGraw-Hill.

Green, W.H., and G.A. Ampt. 1911. "Studies on Soil Physics I: Flow of Air and Water through Soils." *Journal of Agricultural Science* 4 (1): 1–24.

Green, W.R., D.M. Robertson, and F.D. Wilde. 2015. "Lakes and Reservoirs: Guidelines for Study Design and Sampling." In US Geological Survey Techniques and Methods, book 9, chap. A10, 48 pp. Updated May 31, 2018. doi:10.3133/tm9a10.

Gregory, K.J., and D.E. Walling. 1973. *Drainage Basin Form and Process: A Geomorphological Approach*. New York: John Wiley & Sons.

Haan, C.T., B.J. Barfield, and J.C. Hayes. 1994. *Design Hydrology and Sedimentology for Small Catchments*. San Diego, CA: Academic Press.

Haan, C.T., and R.E. Schulze. 1987. "Return Period Flow Prediction with Uncertain Parameters." *Transactions of the ASAE* 30 (3): 665–69. doi:10.13031/2013.30457.

Hack, J.J., J.M. Caron, S.G. Yeager, K.W. Oleson, M.M. Holland, J.E. Truesdale, and P.J. Rasch. 2006. "Simulation of Global Hydrological Cycle in the CCSM Community Atmosphere Model Version 3 (CAM3): Mean Features." *Journal of Climate* 19 (11): 2199–221.

Håkanson, L. 1981. *A Manual of Lake Morphology*. Berlin: Springer-Verlag.

Håkanson, L., and M. Jansson. 1983. *Principles of Lake Sedimentology*. Berlin: Springer-Verlag.

Harbaugh, A.W. 2005. "MODFLOW-2005: The US Geological Survey Modular Ground-Water Model—the Ground-Water Flow Process." US Geological Survey Techniques and Methods 6-A16, 253 pp.

Haverkamp, R., L. Rendon, and G. Vachaud. 1987. "Infiltration Equations and Their Applicability for Productive Use." In *Infiltration Development and Application: Proceedings of the International Conference on Infiltration Development and Application*, edited by Yu-Si Fok, 142–52. Honolulu: Water Resources Research Center.

Henderson, F.M. 1966. *Open Channel Flow*. New York: Macmillan.

Henningson, Durham, and Richardson Inc. 1979. *Preliminary Engineering Report, Winters Elm Creek Dam and Reservoir.* Winters, TX: Submitted to the Farmers Home Administration, US Department of Agriculture.

Hershfield, D.M. 1963. *Rainfall Frequency Atlas of the United States for Durations from 30 Minutes to 24 Hours and Return Periods from 1 to 100 years*. Technical Paper No. 40. Washington, DC: Engineering Division, Soil Conservation Service, US Dept. of Agriculture, 61 pp.

Hewlet, J.D. 1982. *Principles of Forest Hydrology*. Athens: University of Georgia Press.

Hewlet, J.D., and A.R. Hibbert. 1967. "Factors Affecting the Response of Small Watersheds to Precipitation in Humid Regions." In *Forest Hydrology*, edited by W.E. Sopper and H.W. Lull, 275–90. Oxford: Pergamon Press.

Hill, M.S. 1997. *Understanding Environmental Pollution*. Cambridge: Cambridge University Press.

Hill, P.F., R.G. Mein, and L. Siriwardena. 1998. *How Much Rainfall Becomes Runoff? Loss Modelling for Flood Estimation*. Clayton, Victoria, Australia: Cooperative Research Centre for Catchment Hydrology.

Hillel, D. 1971. *Soil and Water: Physical Principles and Processes*. New York: Academic Press.

———. 1982. *Introduction to Soil Physics*. New York: Academic Press.

———. 1989. *Environmental Soil Physics*. San Diego: Academic Press.

Hinman, C. 2005. *Low Impact Development: Technical Guidance Manual for Puget Sound*. Olympia: Puget Sound Action Team and Washington State University Pierce Country Extension.

Hollis, G.E. 1975. "The Effect of Urbanization on Floods of Different Recurrence Interval." *Water Resources Research* 11 (3): 431–35.

Holman, K.D., A. Gronewold, M. Notaro, and A. Zarrin. 2012. "Improving Historical Precipitation Estimates over the Lake Superior Basin." *Geophysical Research Letters* 39: L03405. doi:10.1029/2011GL050468.

Holtan, H.N. 1961. *A Concept for Infiltration Estimates in Watershed Engineering*. Washington, DC: Agricultural Research Service, US Department of Agriculture.

Holtschlag, D.J., and J.R. Nicholas. 1998. "Indirect Ground Water Discharge to the Great Lakes." US Geological Survey Open-File Report 98-579, 25 pp.

Hornberger, G.M., P.L. Wibert, J.P. Raffensperger, and P. D'Odorico. 2014. *Elements of Physical Hydrology*. 2nd ed. Baltimore: John Hopkins University Press.

Horne, A.J., and C.R. Goldman. 1994. *Limnology*. 2nd ed. New York: McGraw-Hill.

Horton, R.E. 1933. "The Role of Infiltration in the Hydrologic Cycle." *Eos, Transactions American Geophysical Union* 14 (1): 446–60.

———. 1939. "Analysis of Runoff-Plat Experiments with Varying Infiltration Capacity." *Eos, Transactions American Geophysical Union* 20 (4): 693–711.

———. 1940. "An Approach toward a Physical Interpretation of Infiltration Capacity." *Soil Science Society of America Journal* 5 (C): 399–417.

———. 1945. "Erosional Development of Streams and Their Drainage Basins: Hydrophysical Approach to Quantitative Morphology." *GSA Bulletin* 56 (3): 275–370. doi:10.1130/0016-7606(1945)56[275:EDOSAT]2.0.CO;2.

Houghton, J.T., Y. Ding, D.J. Griggs, M. Noguer, P.J. van der Linden, and D. Xiaosu, eds. 2001. *Climate Change 2001: The Scientific Basis*. Cambridge: Cambridge University Press.

Hubbert, M.K. 1940. "The Theory of Ground-Water Motion." *Journal of Geology* 48 (8): 785–944.

———. 1956. "Darcy's Law and the Field Equations of the Flow of Underground Fluids." *Transactions of the AIME*. 207:222–39. doi:10.2118/749-G.

Huff, F.A. 1967. "Time Distribution of Rainfall in Heavy Storms." *Water Resources Research* 3 (4): 1007–19.

Hughes, G.H. 1974a. *Water-Level Fluctuations of Lakes in Florida*. FGS Map Series 62. Tallahassee: Florida Bureau of Geology.

———. 1974b. *Water Balance of Lake Kerr: A Deductive Study of a Landlocked Lake in North-central Florida*. FGS Report of Investigations No. 73. Tallahassee: Florida Bureau of Geology.

Huizinga, R.J., and Heimann, D.C. 2018. "Hydrographic Surveys of Rivers and Lakes Using a Multibeam Echosounder Mapping System." US Geological Survey Fact Sheet 2018-3021, 6 pp. doi:10.3133/fs20183021.

Hutchinson, G.E. 1937. *A Contribution to the Limnology of Arid Regions*.

Transactions of the Connecticut Academy of Arts and Sciences, vol. 33. New Haven: Connecticut Academy of Arts and Sciences.

———. 1938. "Chemical Stratification and Lake Morphology." *Proceedings of the National Academy of Sciences* 24 (2): 63–69. doi:10.1073/pnas.24.2.63.

———. 1957. *A Treatise on Limnology*. Vol. 1: Geography, Physics, and Chemistry. Chichester, NY: Wiley.

Hutchinson, G.E., and H. Löffler. 1956. "The Thermal Classification of Lakes." *Proceedings of the National Academy of Sciences* 42 (2): 84–86.

Isaaks, E.H., and R.M. Srivastava. 1989. *Applied Geostatistics*. New York: Oxford University Press.

Isiorho, S.A., and J.H. Meyer. 1999. "The Effects of Bag Type and Meter Size on Seepage Meter Measurements." *Groundwater* 37 (3): 411–13.

Jarret, R.D. 1985. "Determination of Roughness Coefficients for Streams in Colorado." US Geological Survey Water-Resources Investigations Report 85-4004, 54 pp.

Jefferies, D.S., W.R. Snyder, W.A. Scheider, and M. Kirby. 1978. "Small-Scale Variations in Precipitation Loading near Dorset, Ontario." *Water Quality Research Journal* 13 (1): 73–96. From *Proceedings of the 13th Canadian Symposium on Water Pollution Research*. doi:10.2166/wqrj.1978.007.

Joss, J., and A. Waldvogel. 1990. "Precipitation Measurement and Hydrology." In *Radar in Meteorology*, edited by D. Atlas, 577–606. Boston: American Meteorological Society.

Journel, A.G. 1989. *Fundamentals of Geostatistics in Five Lessons*. Short Course in Geology, vol. 8. Washington, DC: American Geophysical Union.

Journel, A.G., and C.J. Huijbregts. 1978. *Mining Geostatistics*. London: Academic Press.

Kadlec, J.A. 1993. "Effect of Depth of Flooding on Summer Water Budgets for Small Diked Marshes." *Wetlands* 13:1–9.

Kalff, J. 2002. *Limnology*. Englewood Cliffs, NJ: Prentice Hall.

Kessler, E. 1992. *Thunderstorms: A Social, Scientific, and Technological Documentary*. Norman: University of Oklahoma Press.

Khosroshahi, F. 1991. "Evaluation of Unit Hydrograph in Flood Estimation." [*Proceedings of First*] *Iran Hydrology Conference*, 1–18.

King, F.H. 1899. "Principles and Conditions of the Movement of Groundwater." In USGS, *19th Annual Report to the Secretary of the Interior: Part 2. Papers Chiefly of a Theoretic Nature*, 59–294. Washington, DC: Government Printing Office.

King, K.W. 2000. "Response of Green-Ampt Mein-Larsen Simulated Runoff Volumes to Temporally Aggregated Precipitation." *Journal of the American Water Resources Association* 36 (4): 791–97.

Kirpich, Z.P. 1940. "Time of Concentration of Small Agricultural Watersheds." *Civil Engineering* 10 (6): 362.

Klein, R. 1979. "Urbanization and Stream Quality Impairment." *Water Resources Bulletin* 15 (4): 948–63.

Klute, A., and C. Dirksen. 1986. "Hydraulic Conductivity and Diffusivity. Laboratory Methods." In *Methods of Soil Analysis, Part I: Physical and Mineralogical Methods*, edited by A. Klute, 687–732. Madison, WI: America Society of Agronomy.

Konrad, C.P. 2003. "Effects of Urban Development on Floods." US Geological Survey Fact Sheet 076-03, 4 pp.

Kostiakov, A.N. 1932. "On the Dynamics of the Coefficient of Water Percolation in Soils and on the Necessity for Studying It from a Dynamic Point of View for Purposes of Amelioration." *Transactions of 6th Committee International Society of Soil Science*, Russia, Part A: 17–21.

Krabbenhoft, D.P., and M.P. Anderson. 1986. "Use of a Numerical Ground-Water Flow Model for Hypothesis Testing." *Groundwater* 24 (1): 49–55.

Krabbenhoft, D.P., M.P. Anderson, and C.J. Bowser. 1990. "Estimating Groundwater Exchange with Lakes, 2: Calibration of a Three-Dimensional, Solute Transport Model to a Stable Isotope Plume." *Water Resources Research* 26 (10): 2455–62.

Krige, D.G. 1951. "A Statistical Approach to Some Basic Mine Evaluation Problems on the Witwatersrand." *Journal of the Chemical, Metallurgical, and Mining Society of South Africa* 52: 119–39.

Kummerow, C., H. Masunaga, and P. Bauer. 2007. "A Next-Generation Microwave Rainfall Retrieval Algorithm for Use by TRMM and GPM." In *Measuring Precipitation from Space*, edited by V. Levizzani, P. Bauer, and F.J. Turk, 235–52. Advances in Global Change Research 28. Dordrecht, Netherlands: Springer.

Landon, M.K., D.L. Rus, and F.E. Harvey. 2001. "Comparison of Instream Methods for Measuring Hydraulic Conductivity in Sandy Streambeds." *Groundwater* 39 (6): 870–85.

Lane, E. 1981. *Karst in Florida*. Special Publication No. 29. Tallahassee: Florida Geological Survey.

Lane, R.K. 1998. "Lake (Physical Feature)." In *Encyclopedia Britannica Online*. Accessed March 15, 2017. http://www.britannica.com/science/lake.

Larson, L.W. 1971. *Precipitation and Its measurement: A State of the Art*. Water Resources Series No. 24. Laramie: University of Wyoming Water Resources Research Institute.

Laurenson, E.M. 1975. "Streamflow in Catchment Modeling." In *Proceedings of Prediction in Catchment Hydrology*, 25–27. Canberra: Australian Academy of Science.

———. 1987. "Back to Basics on Flood Frequency Analysis." *Civil Engineering Transactions* CE29:47–53.

Lavoisier, A. 1789. *Traité élémentaire de chimie, présenté dans un ordre nouveau et d'après les découvertes modernes* [Elementary treatise on chemistry]. 2 vols. Translated by Robert Kerr. Paris: Chez Cuchet.

Lee, D.R. 1977. "A Device for Measuring Seepage Flux in Lakes and Estuaries." *Limnology and Oceanography* 22 (1): 140–47.

Lee, D.R., and J.A. Cherry. 1979. "A Field Exercise on Groundwater Flow Using Seepage Meters and Mini-Piezometers." *Journal of Geological Education* 27 (1): 6–10. doi:10.5408/0022-1368-27.1.6.

Lee, J., and W.H. Lim. 1995. "Green-Ampt Equation and the Soil Water Storage Depth in a Watershed Model." In *Proceedings of the Second International Symposium on Urban Stormwater Management*, 157–61. July 11–13, Melbourne, Australia.

Lensky, I., and V. Levizzani. 2008. "Estimation of Precipitation from Space-based Platforms." In *Precipitation: Advances in Measurement, Estimation and Prediction*, edited by S. Michaelides, 195–217. Berlin: Springer-Verlag.

Leonard, G.A. 1962. "Engineering Properties of Soils." In *Foundation Engineering*, edited by G.A. Leonards, 66–240. New York: McGraw-Hill.

Levitt, J. 1956. "The Physical Nature of Transpirational Pull." *Plant Physiology* 31 (3): 248–51.

Likens, G.E. 2013. *Biogeochemistry of a Forested Ecosystem*. New York: Springer-Verlag.

Limerinos, J.T. 1970. "Determination of the Manning Coefficient from Measured Bed Roughness in Natural Channels." US Geological Survey Water-Supply Paper 1898-B, 41 pp.

Linsley, R.K. 1986. "Flood Estimates: How Good Are They?" *Water Resources Research* 22 (9, supp.): 159S–164S.

Linsley, R.K., M.A. Johler, J.H.L. Paulhus. 1975. *Hydrology for Engineers*. New York: McGraw-Hill.

Lotfi, A. 2012. *Lake Uromiyeh: A Concise Baseline Report*. Edited by M. Moser. Tehran: Conservation of Iranian Wetlands Project, Impel Review Initiative Department of Environment, United Nations Development Programme.

Lyne, V., and M. Hollick. 1979. "Stochastic Time-Variable Rainfall-Runoff Modelling." Institute of Engineers, Australian National Conference Publication 79/10, pp. 89–93.

Mahani, S.E., and R. Khanbilvardi. 2009. "Generating Multi-Sensor Precipitation Estimates over Radar Gap Areas." *World Scientific and Engineering Academy and Society, Transactions on Systems* 8 (1): 96–106. http://www.wseas.us/e-library/transactions/systems/2009/28-768.pdf.

Maidment, D.R, ed. 1993. *Handbook of Hydrology*. New York: McGraw-Hill.

Mair, A., and A. Fares. 2011. "Comparison of Rainfall Interpolation Methods in a Mountainous Region of a Tropical Island." *Journal of Hydrologic Engineering* 16 (4): 371–83. doi:10.1061/(ASCE)HE.1943-5584.0000330.

Majidi, A., M. Moradi, H. Vagharfard, and A. Purjenaie. 2012. "Evaluation of Synthetic Unit Hydrograph (SCS) and Rational Methods in Peak Flow Estimation." *International Journal of Hydraulic Engineering* 1 (5): 43–47.

Manning, R. 1891. "On the Flow of Water in Open Channels and Pipes." *Transactions of the Institute of Civil Engineers of Ireland* 20:161–207.

Martonne, E. de, and L. Aufrère. 1928. "L'extension des regions privees d'ecoulement vers l'ocean." *Union Geographique Internationale*, publication no. 3, 3–8.

Massey, B.C., E. Reeves, and W.A. Lear. 1982. "Flood of May 24–25, 1981, in the Austin, Texas, Metropolitan Area." US Geological Survey Hydrologic Investigations Atlas HA-656, 2 sheets. https://pubs.usgs.gov/ha/656/plate-1.pdf and https://pubs.usgs.gov/ha/656/plate-2.pdf.

Matheron, G. 1962. *Traité de géostatistique appliquée. Tome II. Le krigeage* [Treaty of applied geostatistics, vol. 2: kriging]. Paris: Editions Technip.

———. 1963. "Principles of Geostatistics." *Economic Geology* 58 (8): 1246–66. doi:10.2113/gsecongeo.58.8.1246

Maybeck, M. 1995. "Global Distribution of Lakes." In *Physics and Chemistry of Lakes*, 2nd ed., edited by A. Lerman, D.M. Imboden, and J. Gat, 1–35. Berlin: Springer-Verlag.

McBride, M.S., and H.O. Pfannkuch. 1975. "The Distribution of Seepage within Lakebeds." *Journal of Research of the US Geological Survey* 3 (5): 505–12.

McCuen, R.H., and M.W. Snyder. 1986. *Hydrologic Modeling: Statistical Methods and Applications*. Englewood Cliffs, NJ: Prentice Hall.

McKerchar, A.I., and G.H. Macky. 2001. "Comparison of a Regional Method for Estimating Design Floods with Two Rainfall-Based Methods. *Journal of Hydrology New Zealand* 40 (2): 129–38.

McKinney, D.C., X. Cai, M.W. Rosegrant, C. Ringler, and C.A. Scott. 1990. "Modeling Water Resources Management at the Basin Level: Review and Future Directions." SWIM Paper 6. Colombo, Sri Lanka: International Water Management Institute.

McMahon, T.A., B. Finlayson, A.T. Haines, and R. Srikanthan. 1992. *Global Runoff: Continental Comparisons of Annual Flows and Peak Discharge*. Cremlingen-Destedt, Germany: Catena Verlag.

Mein, R.G., and A.G. Goyen. 1988. "Urban Runoff." *Civil Engineering Transactions* 30 (4): 225–38.

Mein, R.G., and C.L. Larson. 1973. "Modeling Infiltration during Steady Rain." *Water Resources Research* 9 (2): 384–94. [Keynote paper at 1988 Hydrology and Water Resources Symposium, Canberra, Australia].

Merkley, G.P. 2004. "Current Metering." Lectures 3 and 4 in *BIE6300: Irrigation & Conveyance Control Systems*, Biological and Irrigation Engineering 5300/6300, Utah State University. Accessed April 10, 2018. https://digitalcommons.usu.edu/ocw_bie/2.

Merritt, M.L., and L.F. Konikow. 2000. "Documentation of a Computer

Program to Simulate Lake-Aquifer Interaction Using the MODFLOW Ground-Water Flow Model and the MOC3D Solute-Transport Model." US Geological Survey Water-Resources Investigations Report 00-4167, 146 pp.

Messager, M.L., B. Lehner, G. Grill, I. Nedeva, and O. Schmitt. 2016. "Estimating the Volume and Age of Water Stored in Global Lakes Using a Geo-Statistical Approach. *Nature Communications* 7:13603. https://www.nature.com/articles/ncomms13603.pdf.

Meyboom, P. 1961. "Estimating Ground-Water Recharge from Stream Hydrographs." *Journal of Geophysical Research* 66 (4): 1203–14. doi:10.1029/JZ066i004p01203.

Meyers, R.L., and J. Ewel. 1990. *Ecosystems of Florida*. Orlando: University of Central Florida Press.

Micklin, P. 2007. "The Aral Sea Disaster." *Annual Review of Earth and Planetary Sciences* 35:47–72. https://www.annualreviews.org/doi/10.1146/annurev.earth.35.031306.140120.

Micklin, P., and N.V. Aladin. 2008. "Reclaiming the Aral Sea." *Scientific American* 298 (4): 64–71.

Mills, W.B., D.B. Porcella, M.J. Ungs, S.A. Gherini, K.V. Summers, L. Mok, G.L. Rupp, G.L. Bowie, and D.A. Haith. 1985. *Water Quality Assessment: A Screening Procedure for Toxic and Conventional Pollutants in Surface and Ground Water: Part 1*. EPA/600/6-85/002a. Athens, GA: US Environmental Protection Agency, Environmental Research Laboratory.

Mockus, V. 1957. *Use of Storm Watershed Characteristics in Synthetic Hydrograph Analysis and Application*. Sacramento, CA: American Geophysical Union, Pacific SW Region.

Montgomery, D.C., and E.A. Peck. 1982. *Introduction to Linear Regression Analysis*. New York: John Wiley & Sons.

Morgali, J.R., and R.K. Linsley. 1965. "Computer Analysis of Overland Flow." *Journal of the Hydraulics Division* 91 (HY3): 81–101.

Morisawa, M.E. 1962. "Quantitative Geomorphology of Some Watersheds in the Appalachian Plateau." *Geological Society of America Bulletin* 73 (9): 1025–46.

Mousavi, S.F., M. Nekoei-Meher, and M. Mahdavi. 1998. "Study and Test of Fitting Natural and Synthetic Unit Hydrographs in Zayandehrud-dam Watershed (Pelasjan Sub-Basin). *Journal of Crop Production and Processing* 2 (2): 93–107. https://jcpp.iut.ac.ir/article-1-260-en.html.

Myers, D.E. 2006. "Reflections on Geostatistics and Stochastic Modeling." In *Stochastic Modeling and Geostatistics: Principles, Methods, and Case Studies*, vol. 2, edited by T.C. Coburn, J.M. Yarus, and R.L. Chamber, 11–22. Tulsa, OK: American Association of Petroleum Geologists. doi:10.1306/CA51063.

Naiman, R.J., ed. 1992. *Watershed Management: Balancing Sustainability and Environmental Change*. Berlin: Springer-Verlag.

National Research Council. 1990. *Ground Water Models: Scientific and Regulatory Applications*. Washington, DC: National Academies Press. doi:10.17226/1219.

Neff, B.P., and J.R. Nicholas. 2005. "Uncertainty in the Great Lakes Water Balance." US Geological Survey Scientific Investigations Report 2004-5100, 42 pp. doi:10.3133/sir20045100.

Neumann, J. 1959. "Maximum Depth and Average Depth of Lakes." *Journal of Fisheries Research Board of Canada* 16 (6): 923–27. doi:10.1139/f59-065.

Nielsen, H.A., and H. Madsen. 2000. *Predicting the Heat Consumption in District Heating Systems Using Meteorological Forecasts*. Copenhagen: Informatics and Mathematical Modelling, Technical University of Denmark.

———. 2006. "Modelling the Heat Consumption in District Heating Systems Using the Grey-Box Approach." *Energy and Buildings* 38 (1): 63–71. doi:10.1016/j.enbuild.2005.05.002.

Nobel, P.S. 1991. *Physiochemical and Environmental Plant Physiology*. San Diego: Academic Press.

Northern Virginia Regional Commission (NVRC). 2001. *The Effect of Urbanization on the Natural Drainage Network in the Four Mile Run Watershed*. Annandale: Published for Virginia Department of Environmental Quality and Virginia Department of Conservation and Recreation.

O'Keefe, T.C., S.R. Elliott, R.J. Naiman. n.d. "Introduction to Watershed Ecology." US Environmental Protection Agency. The Watershed Academy Online Training Program. Accessed January 3, 2017. https://cfpub.epa.gov/watertrain/pdf/modules/WatershedEcology.pdf.

Olson, F.C.W. 1960. "A System of Morphometry." *International Hydrographic Review* 37:147–55.

Pauling, L. 1959. *Hydrogen Bonding*. Edited by D. Hadzi. London: Pergamon Press.

Penman, H.L. 1948. "Natural Evaporation from Open Water, Bare Soil and Grass." *Proceedings from the Royal Society, Series A* 193:120–45.

Pettyjohn, W.A., and R. Henning. 1979. *Preliminary Estimate of Ground-Water Recharge Rates, Related Streamflow and Water Quality in Ohio*. Project Completion Report No. 552. Columbus: Ohio State University Water Resources Center.

Pfannkuch, H.O., and T.C. Winter. 1984. "Effect of Anisotropy and Groundwater System Geometry on Seepage through Lakebeds: 1. Analog and Dimensional Analysis." *Journal of Hydrology* 75 (1–4): 213–37.

Philip, J.R. 1957. "The Theory of Infiltration: 1. The Infiltration Equation and Its Solution." *Soil Science* 83:345–57.

Pilgrim, D.H. 1986. "Bridging the Gap between Flood Research and Design Practice." *Water Resources Research* 22 (9 supplementwet): 165S–176S.

———, ed. 1987. *Australian Rainfall and Runoff: A Guide to Flood Estima-*

tion. Vol. 1, rev. ed. Barton: Institution of Engineers, Australia (Reprinted edition 1998). http://arr.ga.gov.au/arr-guideline.

Pilgrim, D.H., I. Cordery, and R. French. 1969. "Temporal Patterns of Design Rainfall for Sydney." *Civil Engineering Transactions / Institution of Engineers, Australia* CE11:9–14.

Pinder, G.F., and W.G. Gray. 1977. *Finite Element Simulation in Surface and Subsurface Hydrology*. New York: Academic Press.

Planck, M. 1914. *The Theory of Heat Radiation*. 2nd ed. Translated by M. Masius. Philadelphia: P. Blakiston's Son.

Possel, M. 2017. "Waves, Motion, and Frequency: The Doppler Effect." Einstein Online. Accessed April 10, 2018. http://www.einstein-online.info/spotlights/doppler/index.html.

Rawls, W.J., L.R. Ahuja, A. Brakensiek, and A. Shirmohammadi. 1993. "Infiltration and Soil Water Movement." In Maidment, *Handbook of Hydrology*, 5.1–5.51. New York: McGraw-Hill.

Reckhow, K.H., J.T. Clements, and R.C. Dodd. 1990. "Statistical Evaluation of Mechanistic Water Quality Models." *Journal of Environmental Engineering* 116 (2): 250–68.

Richards, L.A. 1931. "Capillary Conduction of Liquids through Porous Mediums. *Physics* 1 (5): 318–33.

Riggs, H.C. 1976. "A Simplified Slope-Area Method for Estimating Flood Discharges in Natural Channels." *Journal of Research of the US Geological Survey* 4 (3): 285–91.

———. 1985. *Streamflow Characteristics*. Amsterdam: Elsevier.

Rorabaugh, M.I. 1964. "Estimating Changes in Bank Storage and Groundwater Contribution to Streamflow." International Association of Scientific Hydrology Publication 63:432–41.

Rosenberry, D.O. 1990. "Inexpensive Groundwater Monitoring Methods for Determining Hydrologic Budgets of Lakes and Wetlands." In *Proceedings of a National Conference on Enhancing the States' Lake and Wetland Management Programs*, sponsored by US Environmental Protection Agency, North American Lake Management Society, and Northeastern Illinois Planning Commission, 123–31. Madison: University of Wisconsin Press.

Rosenberry, D.O., and J.W. LaBaugh, eds. 2008. *Field Techniques for Estimating Water Fluxes between Surface Water and Ground Water*. US Geological Survey Techniques and Methods 4-D2, 128 pp.

Rosenberry, D.O., J.W. LaBaugh, and R.J. Hunt. 2008. "Use of Monitoring Wells, Portable Piezometers, and Seepage Meters to Quantify Flow between Surface Water and Ground Water." In Rosenberry and LaBaugh, *Field Techniques for Estimating Water Fluxes*, 43–70.

Rosenberry, D.O., R.G. Striegel, and D.C. Hudson. 2000. "Plants as Indicators of Focused Ground Water Discharge to a Northern Minnesota Lake." *Groundwater* 38 (2): 296–303.

Rosgen, D.L. 1996. *Applied River Morphology*. 2nd ed. Pagosa Springs,

CO: Wildland Hydrology. In US Environmental Protection Agency, The Watershed Academy Online Training Program, "Fundamentals of Rosgen Stream Classification System."

———. 2011. "Appendix A: Stream Classification & Valley Types." In *Trail Creek Watershed Assessment & Conceptual Restoration Plan*. Fort Collins, CO: Wildland Hydrology. https://www.fs.usda.gov/Internet/FSE_DOCUMENTS/stelprdb5361892.pdf.

Rumer, R.R., Jr. 1969. "Resistance to Flow through Porous Media." In *Flow through Porous Media*, edited by R.J.M. DeWiest, 91–98. New York: Academic Press.

Rupert, F. 2000. *The Disappearing Waters of Lake Jackson*. Florida Geological Survey poster.

Rutledge, A.T. 1998. "Computer Programs for Describing the Recession of Ground-Water Discharge and for Estimating Mean Ground-Water Recharge and Discharge from Streamflow Data—Update." US Geological Survey Water-Resources Investigations Report 98-4148, 43 pp.

Rutledge, A.T., and C.C. Daniel. 1994. "Testing an Automated Method to Estimate Ground-Water Recharge from Streamflow Records." *Groundwater* 32 (2): 180–89.

Salisbury, F.B., and C.W. Ross. 1992. *Plant Physiology*. 4th ed. Belmont, CA: Wadsworth Publishing.

Saravanapavan, T., V. Anbumozhi, and E. Yamaji. 2004. "Using Percent Imperviousness as a Planning Tool in Watershed Management: Case Study of the Shawsheen in USA." *Journal of Rural Planning Association* 23 (23): S73-S78. doi:10.2750/arp.23.23-suppl_73.

Schincariol, R.A., and J.D. McNeil. 2002. "Errors with Small Volume Elastic Seepage Meter Bags." *Groundwater* 40 (6): 649–51.

Schueler, T. 1987. *Controlling Urban Runoff: A Practical Manual for Planning and Designing Urban Best Management Practices*. Washington, DC: Metropolitan Washington Council of Governments.

———. 2003. *Impacts of Impervious Cover on Aquatic Systems*. Watershed Protection Research Monograph No. 1. Ellicott City, MD: Center for Watershed Protection.

Schumm, S.A. 1977. *The Fluvial System*. New York: John Wiley & Sons.

Scofield, R.A., and R.J. Kuligowski. 2003. "Status and Outlook of Operational Satellite Precipitation Algorithms for Extreme-Precipitation Events." *Weather and Forecasting* 18 (6): 1037–51. doi:10.1175/1520-0434(2003)018<1037:SAOOOS>2.0.CO;2.

Seed, A.W., and G.L. Austin. 1990. "Sampling Errors for Rain Gauge Derived Mean-Areal Daily and Monthly Rainfall. *Journal of Hydrology* 118 (1–4): 163–73.

Shaw, E.M. 1988. *Hydrology in Practice*. 2nd ed. London: Van Nostrand Reinhold International.

Shaw, R.D., and E.E. Prepas. 1989. "Anomalous, Short-Term Influx of

Water into Seepage Meters." *Limnology and Oceanography* 34 (7): 1343–51.

Sherman, L.K. 1932. "Streamflow from Rainfall by the Unit-Graph Method." *Engineering News Record* 108:501–5.

———. 1940. "The Hydraulics of Surface Runoff." *Civil Engineering* 10:165–66.

———. 1942. "The Unit Hydrograph Method." Chapter X1E in *Hydrology*, edited by O.E. Meinzer, 514–25. New York: Dover.

Shih, S.F. 1980. "Water Budget Computation for a Shallow Lake: Lake Okeechobee, Florida." *Water Resources Bulletin* 16 (4): 724–28.

Shinn, E.A., C.D. Reich, and T.D. Hickey. 2002. "Seepage Meters and Bernoulli's Revenge." *Estuaries* 25 (1): 126–32.

Sichel, H.S. 1952. "New Methods in the Statistical Evaluation of Mine Sampling Data." *London Institute of Mining and Metallurgy Transactions* 61 (6): 261–88.

Siegel, D.I., and T.C. Winter. 1980. "Hydrologic Setting of Williams Lake, Minnesota." US Geological Survey Open-File Report 80-403, 62 pp.

Simpson, J., C. Kummerow, W.K. Tao, and R.F. Adler. 1996. "On the Tropical Rainfall Measuring Mission (TRMM)." *Meteorology and Atmospheric Physics* 60:19–36.

Sinclair, W.C., and J.W. Stewart. 1985. "Sinkhole Type, Development, and Distribution in Florida." Florida Bureau of Geology Map Series 110.

Singh, V.P., and P.K. Chowdhury. 1986. "Comparing Methods of Estimating Mean Areal Rainfall." *Water Resources Bulletin* 22 (2): 275–82.

Singh, K.P., and J.B. Stall. 1971. "Derivation of Baseflow Recession Curves and Parameters." *Water Resources Research* 7 (2): 292–303.

Siriwardene, N.R., B.P.M. Cheung, and B.J.C. Perera. 2003. "Estimation of Soil Infiltration Rates of Urban Catchments." *Proceedings of the 28th International Hydrology and Water Resources Symposium, Institution of Engineers*, Australia, Wollongong.

Skukla, M.K., R. Lal, and P. Unkefer. 2003. "Experimental Validation of Infiltration Models for Different Land Use and Soil Management Systems." *Journal of Soil Science* 168 (3): 178–91.

Smith, R.B. 1979. "The Influence of Mountains on the Atmosphere." *Advances in Geophysics* 21:87–230.

Smithers, J.C. 2012. "Methods for Design Flood Estimation in South Africa." *Water SA* 38 (4): 633–46.

Snyder, F.F. 1938. "Synthetic Unit-Graphs." *Eos, Transactions, American Geophysical Union* 19 (1): 447–54.

Soil Conservation Service. 1972. "Hydrology." Section 4 of *National Engineering Handbook*, edited by H.F. Moody. Washington, DC: US Department of Agriculture.

Spinello, A.G., and D.L. Simmons. 1992. "Baseflow of 10 South Shore Streams, Long Island, New York 1976–85 and the Effects of Urbanization on Baseflow and Flow Duration." US Geological Survey Water-Resources Investigations Report 90-4205.

Stephens, G.L., and P.J. Webster. 1981. "Clouds and Climate: Sensitivity of Simple Systems." *Journal of the Atmospheric Sciences* 38:235–47.

Stephenson, D.A. 1971. "Groundwater Flow System Analysis in Lake Environments, with Management and Planning Implications." *Water Resources Bulletin* 7 (5): 1038–47.

Strahler, A.N. 1952. "Hypsometric (Area-Altitude) Analysis of Erosional Topology." *Geological Society of America Bulletin* 63 (11): 1117–42.

———. 1957. "Quantitative Analysis of Watershed Geomorphology." *Transactions of the American Geophysical Union* 8 (6): 913–20.

Sverdrup, H.U., M.W. Johnson, and R.H. Fleming. 1942. *The Oceans: Their Physics, Chemistry and General Biology*. New York: Prentice Hall.

Szilagyi, J., and M.B. Parlange. 1998. "Baseflow Separation Based on Analytical Solutions of the Boussinesq Equation." *Journal of Hydrology* 204 (1–4): 251–60.

Taiz, L., and E. Zeiger. 2002. *Plant Physiology*. 3rd ed. Sunderland, MA: Sinauer Association.

Takhar, H.S., and A.J. Rudge. 1970. "Evaporation Studies in Standard Catchments." *Journal of Hydrology* 11 (4): 329–62.

Taube, C.M. 2000. "Instructions for Winter Lake Mapping." Chapter 11 in *Manual of Fisheries Survey Methods II: With Periodic Updates*, edited by J.C. Schneider. Fisheries Special Report 25. Ann Arbor: Michigan Department of Natural Resources.

Teegavarapu, Ramesh S.V. 2012. *Floods in a Changing Climate: Extreme Precipitation*. Cambridge: Cambridge University Press.

Thienemann, A. 1925. Die Binnengewässer Mitteleuropas. Binnengewässer 1: 1–255.

Thiessen, A.H. 1911. "Precipitation for Large Areas." *Monthly Weather Review* 39:1082–94.

Tholin, A.L., and Keifer, C.J. 1959. "The Hydrology of Urban Runoff." *Journal of the Sanitary Engineering Division* 85 (SA2): 47–106.

Timms, B.V. 1992. *Lake Geomorphology*. Glen Osmond, South Australia: Gleneagles.

Tobler, W. 1970. "A Computer Movie Simulating Urban Growth in the Detroit Region." *Economic Geography* 46 (2): 234–40.

Toth, J. 1963. "A Theoretical Analysis of Groundwater Flow in Small Drainage Basin." *Journal of Geophysical Research* 68 (16): 4795–812.

Trescott, P.C., G.F. Pinder, and S.P. Larson. 1976. "Finite-Difference Model for Aquifer Simulation in Two Dimensions with Results of Numerical Experiments." In *Techniques of Water-Resources Investigations of the United States Geological Survey*, book 7, chapter C1, 116 pp.

Trimble, S.W. 1997. "Contribution of Stream Channel Erosion to Sediment Yield from an Urbanizing Watershed." *Science* 278 (5342): 1442–44.

Triola, M.F. 2001. *Elementary Statistics*. 8th ed. Boston: Addison-Wesley Longman.

Turner, E. 2006. "Comparison of Infiltration Equations and Their Field Validation by Rainfall Simulation." Master's thesis, Department of Biological Resources Engineering, University of Maryland, College Park.

US Bureau of Reclamation. 2001. *Water Measurement Manual: A Water Resources Technical Publication*. Chapters 10 and 11. Washington, DC: US Department of the Interior. Accessed April 10, 2018 https://www.usbr.gov/tsc/techreferences/mands/wmm/index.htm.

US Department of Agriculture (USDA), Natural Resources Conservation Service. 1986. *Urban Hydrology for Small Watersheds*. Technical Release 55. Washington DC: USDA. https://www.nrcs.usda.gov/Internet/FSE_DOCUMENTS/stelprdb1044171.pdf. Updated in 2015 and available for download at https://www.nrcs.usda.gov/wps/portal/nrcs/detailfull/national/water/?&cid=stelprdb1042901.

———. 2004. "Estimation of Direct Runoff from Storm Rainfall." Chapter 10 in *National Engineering Handbook: Part 630 Hydrology*, edited by A.T. Hielmfelt and H.F. Moody. Washington, DC: USDA. https://www.wcc.nrcs.usda.gov/ftpref/wntsc/H&H/NEHhydrology/ch10.pdf.

———. 2005. *National Soil Survey Handbook*. Title 430-VI. Washington, DC: USDA. http://soils.usda.gov/technical/handbook.

———. 2007. "Hydrographs" Chapter 16 in *National Engineering Handbook: Part 630 Hydrology*, edited by W.H. Merkel and H.F. Moody. Washington, DC: USDA. https://www.wcc.nrcs.usda.gov/ftpref/wntsc/H&H/NEHhydrology/ch16.pdf.

———. 2015. "Storm Rainfall Depth and Distribution." Chapter 4 in *National Engineering Handbook: Part 630 Hydrology*, edited by W.H. Merkel, H.F. Moody, D.E. Woodward, and C.C. Hoeft. Washington, DC: USDA. https://www.wcc.nrcs.usda.gov/ftpref/wntsc/H&H/NEHhydrology/ch4_Sept2015draft.pdf.

US Department of Energy. 1986. *Environmental Assessment: Reference Repository Location, Hanford Site, Washington*. Vol. 2. From Nuclear Waste Policy Act, Section 112. DOE/RW-0070. Washington, DC: US Department of Energy.

US Environmental Protection Agency (EPA). 1998. *Guidelines for Ecological Risk Assessment*. EPA/630/R-95/002F. Washington, DC: EPA.

———. 2017. "What Is Nonpoint Source?" Archived website content. Accessed July 16, 2020. https://19january2017snapshot.epa.gov/nps/what-nonpoint-source_.html.

———. n.d. The Watershed Academy Online Training Program. Accessed January 3, 2019. https://www.epa.gov/watershedacademy/online-training-watershed-management#themes

———. n.d. "Watershed Modeling." The Watershed Academy Online Training Program. Accessed January 3, 2019. https://cfpub.epa.gov/watertrain/pdf/modules/WshedModTools.pdf.

US Geological Survey (USGS). 2002. "USGS 02336300 Peachtree Creek at Atlanta, GA." National Water Information System: Web Interface. Accessed July 27, 2015. https://waterdata.usgs.gov/ga/nwis/dv/?site_no=02336300.

———. 2009. "Largest Earthquake in Montana." Historic Earthquakes. Accessed January 3, 2018. https://earthquake.usgs.gov/earthquakes/states/events/1959_08_18.php.

———. 2017. "USGS 07329610 Lake Fuqua near Duncan, OK." National Water Information System: Web Interface. Accessed July 27, 2018. https://waterdata.usgs.gov/nwis/uv?site_no=07329610.

———. n.d. "How Streamflow Is Measured." Water Science School. Accessed July 27, 2016. http://water.usgs.gov/edu/measureflow.html.

———. 2018. *National Field Manual for the Collection of Water-Quality Data*. Handbooks for Water-Resources Investigations, book 9. Reston, VA: US Geological Survey. http://pubs.water.usgs.gov/twri9A.

Van der Heijde, P., Y. Bachmat, J. Bredehoeft, B. Andrews, D. Holtz, and S. Sebastian. 1980. *Groundwater Management: The Use of Numerical Models*. Water Resources Monograph No. 5. Washington, DC: American Geophysical Union.

Van Dyke, J. 1994. Personal communication. Florida Northwest Water Management District, Aquatic Plant Management Office.

Verworn, A., and U. Haberlandt. 2011. "Spatial Interpolation of Hourly Rainfall: Effect of Additional Information, Variogram Inference and Storm Properties." *Hydrology and Earth System Sciences* 15:569–84. https://www.hydrol-earth-syst-sci.net/15/569/2011.

Wagner, J. R. 1984. *Hydrogeologic Assessment of the October 1982 Draining of Lake Jackson, Leon County, Florida*. Water Resources Special Report 81-1. Havana: Northwest Florida Water Management District.

Wang, H.F., and M.P. Anderson. 1982. *Introduction to Groundwater Modeling Finite Difference and Finite Element Methods*, San Francisco: W. H. Freeman.

Ward, J.V. 1989. "The Four-Dimensional Nature of Lotic Ecosystems." *Journal of the North American Benthological Society* 8 (1): 2–8.

Ward, R.C. 1971. "Measuring Evapotranspiration: A Review." *Journal of Hydrology* 13:1–21.

Wei, T.C., and J.L. McGuinness. 1973. "Reciprocal Distance Squared Method: A Computer Technique for Estimating Areal Precipitation." ARS-NC-8. Washington, DC: Agricultural Research Service.

Weiler, M., J.J. McDonnell, I. Tromp-van Meerveld, and T. Uchida. 2005. "Subsurface Stormflow." In *Encyclopedia of Hydrological Sciences*, edited by M.G. Anderson, 10:112. New York: John Wiley & Sons. doi:10.1002/0470848944.hsa119.

Welch, P.S. 1948. *Limnological Methods*. Philadelphia: Blackstone.

Westlake, D.F. 1965. "Some Problems in the Measurement of Radia-

tion under Water: A Review." *Photochemistry and Photobiology* 4 (5): 849–68.
Wetzel, R.G. 1975. *Limnology*. Philadelphia: Saunders College Publishing.
———. 2001. *Limnology: Lake and River Ecosystems*. 3rd ed. New York: Academic Press.
Wetzel, R.G., and G.E. Likens. 2000. *Limnological Analyses*. 3rd ed. New York: Springer.
Williams, W.D. 1996. "What Future for Saline Lakes?" *Environment: Science and Policy for Sustainable Development* 38 (9): 12–39.
Winter, T.C. 1976. "Numerical Simulation Analysis of the Interaction of Lakes and Ground Water." US Geological Survey Professional Paper 1001, 45 pp.
———. 1978. "Numerical Simulation of Steady State Three-Dimensional Groundwater Flow Near Lakes." *Water Resources Research* 14 (2): 245–54.
———. 1981. "Uncertainties in Estimating the Water Balance of Lakes." *Water Resources Bulletin* 17 (1): 82–115.
———. 1995. "Hydrological Processes and the Water Budget of Lakes." In *Physics and Chemistry of Lakes*, 2nd ed., edited by A. Lerman, D.M. Imboden, and J. Gat, 37–60. Berlin: Springer-Verlag.
———. 2001. "The Concept of Hydrologic Landscapes." *Journal of the American Water Resources Association* 37 (2): 335–49.
Winter, T.C., J.W. LaBaugh, and D.O. Rosenberry. 1988. "The Design and Use of a Hydraulic Potentiomanometer for Direct Measurement of Differences in Hydraulic Head Between Groundwater and Surface Water." *Limnology and Oceanography* 33 (5): 1209–14.
Woessner, W.W., and K.E. Sullivan. 1984. "Results of Seepage Meter and Mini-Piezometer Study, Lake Mead, Nevada." *Groundwater* 22 (5): 561–68.
WOW. 2004. "Water on the Web: Monitoring Minnesota Lakes on the Internet and Training Water Science Technicians for the Future. A National On-line Curriculum Using Advanced Technologies and Real-Time Data." University of Minnesota-Duluth. Accessed July 2016. http://www.waterontheweb.org.
Wright-McLaughlin Engineers. 1969. *Urban Storm Drainage Criteria Manual*. 2 vols. Denver, CO: Denver Regional Council of Governments.
Wurtsbaugh, W., C. Miller, S. Null, P. Wilcock, M. Hahneneberger, and F. Howe. 2016. "Impacts of Water Development on Great Salt Lake and the Wasatch Front." *Watershed Sciences Faculty Publications*. Paper 875. https://digitalcommons.usu.edu/wats_facpub/875.
Xu, C.Y. 2003. "Approximate Infiltration Models." Section 5.3 in *Hydrologic Models*. Uppsala, Sweden: Uppsala University Department of Earth, Air and Water Sciences.

SUBJECT INDEX

Page numbers in boldface indicate definitions. Page numbers in italics indicate figures. Page numbers with *t* indicates tables.